Introduction to CORRECTIONS

Third Edition

RICHARD W. SNARR
Eastern Kentucky University

Brown & Benchmark
PUBLISHERS
Madison Dubuque, IA Guilford, CT Chicago Toronto London
Caracas Mexico City Buenos Aires Madrid Bogota Sydney

Book Team

Publisher *Bevan O'Callaghan*
Acquisitions Editor *Michael Alread*
Managing Editor *Sue Pulvermacher-Alt*
Production Editor *Donna Nemmers*
Proofreading Coordinator *Carrie Barker*
Art Editor *Rachel Imsland*
Photo Editor *Leslie Dague*
Permissions Coordinator *Mavis Oeth*
Production Manager *Beth Kundert*
Production/Costing Manager *Sherry Padden*
Visuals/Design Freelance Specialist *Mary L. Christianson*

Basal Text *10/12 Times Roman*
Display Type *Bembo*
Typesetting System *Aviion*
Paper Stock *50# Solutions*

Brown & Benchmark PUBLISHERS

President and Chief Executive Officer *Thomas E. Doran*
Vice President of Production and Business Development *Vickie Putman*
Vice President of Sales and Marketing *Bob McLaughlin*
Director of Marketing *John Finn*

A Times Mirror Company

Cover design by Kay Fulton Design

Cover image by © Robert E. Daemmrich/Tony Stone Images

Copyedited by Cynthia Greene; proofread by Barb Callahan

Library of Congress Catalog Card Number: 95-75331

ISBN 0-697-25117-9

Printed in the United States of America by Times Mirror Higher Education Group, Inc., 2460 Kerper Boulevard, Dubuque, IA 52001

10 9 8 7 6 5 4 3 2

Contents

Preface xv

Unit 1

Placement and Goals of Corrections 1

1 Corrections and the Criminal Justice System 3

Learning Objectives 3
Key Terms 3
Introduction 4
Features of Criminal Justice 4
Law in the United States 7
Civil and Criminal Law 7
Felonies and Misdemeanors 8
Substantive and Procedural Law 9
Activities Constituting Crime 9
Measurement of Crime 11
Recidivism 11
Caseflow into Corrections 12
Funnel Effect 15
Criminal Justice Information 16
Summary 17
Discussion Topics 17
Additional Readings 18
Notes 18

2 Setting for Corrections 19

Learning Objectives 19
Key Terms 19
Introduction 20
Differences in Government Structure 20
United States Government Structure 21
The Constitution 21
Bill of Rights 25
Fourteenth Amendment 25
Fragmentation of Corrections 26
Corrections at the Federal Level 27
Corrections at the State and Local Level 29
Correctional Standards and Accreditation 31
Correctional Accreditation 32
Corrections and Social Science 33
The Scientific Method 33
Careers in Corrections 34
Demand 35
Types of Jobs 35
Professional Ethics 37
Summary 40
Discussion Topics 40
Additional Readings 40
Notes 41

3 History and Competing Aims of Corrections 42

Learning Objectives 42
Key Terms 42
Background 43
What Is Corrections? 44
Deviance and Social Control 45
The Response to Deviation 46
The Continuum of Corrections 46
Philosophies of Corrections 46
Classical View 47
Positivist View 47
Comparison of the Classical and Positivist Views 48

Rationales and Models of Corrections 49
Retribution and a Justice Model 50
Incapacitation and a Custodial Model 51
Reintegration and a Least Restrictive Alternative Model 52
Resocialization and a Rehabilitation Model 53
Punishment 54
Methods of Punishment 54
Banishment 55
Capital Punishment 56
Corporal Punishment 56
Economic Punishment 57
Incarceration 57
Deterrence 57
Simple Deterrence 58
General Deterrence 58
Special Deterrence 59
Marginal Deterrence 59
The Deterrent Effect of Punishment 59
Summary 61
Discussion Topics 62
Additional Readings 62
Notes 62

4 Sentencing 64

Learning Objectives 64
Key Terms 64
Sentencing 65
Nonincarceration Sentences 66
Incarceration Sentences 68
Disparity in Sentencing 68
Changes in Sentencing 70
Federal Sentencing Guidelines 71
State Sentencing Laws 73
Summary 75
Discussion Topics 75
Additional Readings 75
Notes 75

Unit 2

Incarceration 77

5 Jails 79

Learning Objectives 79
Key Terms 79
Overview 80
Historical Development of Jails 80
Jails in England 80
American Colonial Jails 82
Early Changes 82
Jails in the United States 84
Definition 84
Facilities 85
Jail Populations 85
Legal Status 86
Annual Activity 87
Socioeconomic Characteristics of Inmates 87
Needs of Residents 90
Social Service Needs 90
Special Needs 90
Female Jail Residents 92
Jail Operations 92
Intake 92
Classification 93
Orientation 93
Court Appearances 94
Release 94
Management of Jails 94
Essential Services 94
Security 95
Controlling Contraband and Fires 95
Programs 95
Staffing and Funding 96
Inmate Rights 97
The Bail System 98
Problems and Change 99
Alternatives to Bail: Pretrial Release Options 100
New Facilities 103

Future of Jails 104
Regional Jails 104
Jail Standards and Court Intervention 106
Continuing Role of Jails 106
Summary 106
Discussion Topics 107
Additional Readings 107
Notes 107

6 Prison Facilities and Populations 109

Learning Objectives 109
Key Terms 109
Introduction 110
Development of Prisons in the United States 111
Early Prison Changes 112
Resolving a Debate 115
Building Reformatories 115
Prison Facilities 117
Custody Levels 117
Design 119
Prison Costs 120
Prison Population 121
Classification of the Prison Population 124
Security Level Designations 124
Special Populations 125
Persistent Felons 131
Summary 132
Discussion Topics 133
Additional Readings 133
Notes 133

7 Inmate Life and Prisoner Rights 135

Learning Objectives 135
Key Terms 135
Introduction 136
Total Institutions 136

Inmate Subculture: Traditional Views 137
Deprivation Model 137
Importation Model 138
Inmate Subculture: Contemporary Views 138
Daily Life and Problems in Prison 139
Making It in Prison 139
Violence and Victimization in Prison 140
Sexual Deprivation 141
Significant Others 143
Prison Discipline 143
Disciplinary Processes 144
Appeal and Review 149
Rewards versus Punishments 149
Prisoner Rights 150
Sources of Rights 151
Categories of Rights 152
Access to Court 152
No Cruel and Unusual Punishment 153
Civil Rights 154
Protection When Adverse Consequences Possible 155
Outcomes of Court Intervention 156
Summary 159
Discussion Topics 160
Additional Readings 160
Notes 160

8 Prison Programs and Management 163

Learning Objectives 163
Key Terms 163
Overview 164
Prison Industry Programs 164
Sheltered versus Open Markets 166
Opposition to Prison Industries 167
Model Programs 168
UNICOR 168
Future of Prison Industries 169
Rehabilitation Programs 172
Right to Treatment 172
Education Programs 173

Mental Health Programs 174
Psychological Service 174
Counseling 175
Eclectic Approach 176
Prison Chaplains 176
Leisure Time Programs 176
Self-Help Groups 177
Management of Prisons 177
Managers 178
Organizations 178
Setting for Prison Management 179
Management Functions 181
Managing People in a Prison Community 181
Correctional Officers 182
Women as Correctional Officers 183
Styles of Prison Management 184
Autocratic Style 185
Participative Style 186
Autocratic versus Participative Style 188
Summary 190
Discussion Topics 191
Additional Readings 192
Notes 192

9 Women in Prison 194

Learning Objectives 194
Key Terms 194
Introduction 195
Historical Review 195
Reform Efforts 196
Population Profile 197
Prison Life for Women 200
Inmate Subculture: Women's Prison 202
Women Prisoners with Children 206
Retaining Mother/Child Ties 209
Prison Programs for Incarcerated Women 210
Some Specific Inequalities 210
Court Challenges to Work Program Inequalities 211

Alternatives for More Equitable Treatment 212
Summary 214
Discussion Topics 215
Additional Readings 215
Notes 216

Unit 3 Alternatives to Incarceration 217

10 Community-Based Corrections 219

Learning Objectives 219
Key Terms 219
Introduction 220
Reintegration 220
Reasons for the Development of Community-Based Corrections 222
Dissatisfaction with Institutions 222
Humanitarianism 222
Cost Effectiveness 223
More Adequate Justice Administration 223
Intermediate Sanctions 224
Boot Camps 225
Offender Involvement 226
Strategies for Development 228
Georgia's Alternatives to Incarceration 234
Basic Probation 235
Community Service 236
Intensive Probation Supervision (IPS) 236
Diversion Centers 237
Detention Centers 237
Shock Incarceration 237
Electronic Monitoring 238
Initial Problems 239
Assessing Community-Based Corrections 240
Summary 240
Discussion Topics 241
Additional Readings 241
Notes 241

11 Probation 243

Learning Objectives 243
Key Terms 243
Overview 244
Time Line 244
Definition 244
Probation Population 244
Antecedents of Present-Day Probation 245
Benefit of Clergy 245
Judicial Reprieve 246
Recognizance 248
Bail 249
Filing of Cases 249
John Augustus 249
Legal Basis 250
Rationale 251
The Presentence Investigation 251
Preparation of PSI Reports 252
Disclosing Information Contained in PSI Reports 253
Granting Probation 253
Conditions of Probation 257
Revocation of Probation 260
Service Delivery Strategies 261
Casework Approach 261
Brokerage Approach 261
Community Resource Management Team (CRMT) 262
Probation/Parole Officers 262
Differing Roles 263
Outcomes of Probation 264
Split Sentences 266
Summary 267
Discussion Topics 268
Additional Readings 268
Notes 268

12 Parole 270

Learning Objectives 270
Key Terms 270
Overview 271
Administrative Organization 271
Interstate Compact 272
Method of Release From Prison 272
Historical Perspective 273
English Experience 275
Alexander Maconochie 275
Sir Walter Crofton 276
Experience in the United States 276
Zebulon R. Brockway 277
Discretionary Granting of Parole 277
Release Eligibility 277
Good Time 278
Parole Plan 280
Preliminary Hearing 281
Parole Boards and Hearing 281
Court Impact on Granting Parole 283
Return to the Community 284
Employment and Parole 285
Conditions of Parole 285
Final Release 287
Parole Officer Supervision 287
Interviewing 287
Investigating 288
Enforcing 289
Helping 289
Revocation of Parole 289
Morrissey v. Brewer 290
Gagnon v. Scarpelli 291
Parole Outcomes 291
Changes in Parole 294
Determinate Sentencing 294
Parole Guidelines: Risk Assessment 295
Maintaining Parole 295
Special Forms of Early Release 296

Executive Clemency 296
Pardon 297
Amnesty 297
Reprieve 297
Commutation 297
Summary 298
Discussion Topics 298
Additional Readings 299
Notes 299

Unit 4
Juvenile Offenders 301

13 Juvenile Law and Offenders 303

Learning Objectives 303
Key Terms 303
Background 304
Overview of Juvenile Justice Systems 304
Juvenile Law and Jurisdiction 305
Age 306
Behavior 306
Waiver of Jurisdiction 307
Juvenile Legal Terms 308
Juvenile Court 308
Caseflow 309
Due Process Rights 311
Impact of *in re Gault* 312
Juvenile Offenders 313
Self-Reports 314
Summary 318
Discussion Topics 318
Additional Readings 318
Notes 318

14 Treatment for Juvenile Offenders 320

Learning Objectives 320
Key Terms 320
Competing Treatment Philosophies 321
Rehabilitation Philosophy 321
Due Process Philosophy 321

Nonintervention Philosophy 321
Crime Control Philosophy 322
Blended Philosophies 322
Background 322
Example of Deinstitutionalization 323
Dispositions 325
Probation 326
Facilities for Juveniles 326
Programs 329
Nonresidential Programs 330
Residential Programs 331
Treatment Methods 332
Individual Treatment Methods 333
Group Treatment Methods 333
Assessment of Outcomes 334
Aftercare 334
Summary 335
Discussion Topics 335
Additional Readings 336
Notes 336

Unit 5

Selected Issues 337

15 Selected National Issues Involving Corrections 339

Learning Objectives 339
Key Terms 339
Introduction 340
Rate of Imprisonment 340
Race and Imprisonment 341
Privatization of Prisons and Jails 342
The Death Sentence 345
Concluding Remarks 348
Summary 351
Discussion Topics 351
Additional Readings 351
Notes 352

Glossary 353
Index of Court Cases Cited 360
Index 362

Preface

The third edition of this successful textbook is designed to be comprehensive, yet affordable. Included throughout this edition are a variety of additional and updated materials and information presented in tables and figures. New topics and expanded coverage include

• Learning objectives	All chapters
• Updated tables and figures	All chapters
• Corrections and social science	Chapter 2
• Careers in corrections	Chapter 2
• Professional ethics	Chapter 2
• Expanded correctional law	Chapter 7
• Additional management theory	Chapter 8
• Expanded chapter on women	Chapter 9
• Expanded community-based corrections	Chapter 10
• Intermediate sanctions	Chapter 10
• Boot camps	Chapter 10

The third edition is ''chunked'' into five logical units providing an overall direction and purpose. Unit 1 is devoted to the placement and goals of corrections; Unit 2 examines incarceration; Unit 3 covers alternatives to incarceration; Unit 4 deals with juvenile offenders; and Unit 5 concludes the text by identifying several controversial issues involving corrections in the United States. Materials presented in this logical manner enhance learning and retention, benefiting both student and teacher.

This textbook is directly targeted toward a clear pedagogical or teaching function. To begin, considerable thought and preparation have been given to organizing materials from beginning to end for the most effective teaching presentation. Beginning each chapter are *learning objectives* that help to clarify and organize ideas. Students and faculty have indicated this is a very effective learning aid. Key terms appear in *bold type* within each chapter. A *summary* is also provided for each chapter. The inclusion of thoughtful and timely *discussion topics* provides opportunity for various group or individual learning activities. *Additional readings* for each chapter are carefully selected to provide relevant supplementary information.

Appropriate and up-to-date *tables, charts, figures, graphs, maps* and *photographs* are included as teaching aids. Most of these materials have been tested in the classroom by the author teaching introductory students. Each of these illustrations, including photographs, is included for the express purpose of helping to communicate and teach ideas expressed in narrative form. This third edition contains a comprehensive *glossary* appropriately focused on corrections in order to be more useful for introductory corrections students and their teachers.

This textbook is written at a level to be read and understood by students who have little or no academic background in corrections. It has been used successfully in traditional classrooms throughout the United States and more recently in distance learning classes conducted via satellite television. After completing this text, students have a foundation from which to pursue additional study.

Instructor's Manual

A well organized and easy to use Instructor's Manual accompanies this textbook. It was written by the textbook author whose career includes many years in classrooms teaching introduction to corrections.

A set of learning objectives, important terms, and a summary are provided for each chapter. An array of multiple-choice and essay questions are compatible with chapter contents and stated objectives. Questions have been thoughtfully designed to enhance learning and provide comprehensive coverage; they are written in a form most appropriate for the nature of the material covered. Utilization of this manual will help reduce time needed by instructors to organize classroom presentations and to assemble and grade examinations.

Acknowledgments

The preparation and publication of this third edition involved a number of people and the commitment of their time and resources. I have been fortunate in the support and encouragement I received, and sincerely appreciate these efforts.

First, my thanks to Sue Alt and all of the staff at Brown & Benchmark Publishers who assisted in so many ways. Their input in working on all the details of the manuscript was first rate, and their cooperation was most helpful in generating this final product.

I would also like to acknowledge the role of Eastern Kentucky University and the Department of Correctional Services. My colleagues and the administration were very supportive of my efforts, and I express my gratitude to them.

In addition, I would like to thank the following colleagues for their valuable reviews of the manuscript:

Tom Cook Wayne State College
Henry E. DeLuca Westfield State College
Thomas W. Lenahan Herkimer Community College
Mary L. Parker University of Arkansas–Little Rock
Victor C. Ortloff Marshall University
Thomas C. Calhoun Ohio University–Athens

Their comments and suggestions were very useful in developing a final manuscript.

Although running the risk of an inadvertent omission, I would like to identify additional people who helped this effort succeed. They are Valerie Barnett, Amy Coleman, John Haynes, LaDonna Koebel, James Manion, Jan Mays, Kevin Minor, Jennifer Moher, Jennifer Moore, Truett Ricks, Jane Snarr, and my introductory students.

Unit 1

Placement and Goals of Corrections

These introductory materials discuss the location of corrections as one of the components within the criminal justice system and its operation within a democratic form of government. Selected topics provide historical background and differing viewpoints, culminating with a discussion of the all-important decision-making process of sentencing.

Chapter 1 Corrections and the Criminal Justice System
Chapter 2 Setting for Corrections
Chapter 3 History and Competing Aims of Corrections
Chapter 4 Sentencing

Chapter 1

Corrections and the Criminal Justice System

Learning Objectives

After studying this chapter students should be able to:

1. Provide a definition of corrections.
2. Provide a definition of criminal justice.
3. Explain features of the criminal justice system.
4. Understand the flow of cases into corrections.
5. Identify basic criminal justice procedures.
6. Distinguish categories of law.
7. Understand two commonly used measures of crime.
8. Describe what is meant by the "funnel effect."
9. Understand the meaning and importance of recidivism.

Key Terms

Civil law
Corrections
Courts
Crime
Criminal justice procedures:
  Investigation
  Arrest and booking
  Initial appearance
  Preliminary hearing
  Information
  Indictment
  Arraignment
  Plea negotiations
  Trial
  Sentencing
  Appeal
Criminal justice system
Criminal law
Discretion
Exclusionary rule
Felony
Funnel effect
"Hidden" crime
Index crimes
Law
Misdemeanor
National Crime Victim Survey (NCVS)
National Criminal Justice Reference Service (NCJRS)
Police
Procedural law
Recidivism
Substantive law
Uniform Crime Reports (UCR)

Introduction

Corrections is one segment of the **criminal justice system** by which American society seeks to protect the public, punish offenders, change offenders' behavior, and in some cases, compensate victims. Since a wide range of social and political objectives are expressed in criminal justice and correctional processes, various degrees of supervision are in operation. Supervision ranges from a minimal amount within the community to maximum security incarceration.

All societies place limits on human behavior and establish both formal and informal means to deal with those who exceed the limits. Formal means find expression in written laws and institutional structures that identify, charge, convict, and sentence violators; this crime control apparatus is referred to as the criminal justice system.

"The response to crime is a complex process that involves citizens as well as agencies, levels, and branches of government."[1] The criminal justice system is designed to control crime and contribute toward the goal of a safe and orderly society. There are really many criminal justice systems, and the response to crime usually involves local officials. The total system seeks to properly identify law violators, establish guilt, issue an appropriate sanction, and change offender behavior. These responsibilities are carried out by **police, courts,** and corrections components within a framework of democratic principles designed to protect all individuals' civil rights. Officials are legally permitted to exercise wide discretion in determining a person's entry into, movement through, and exit from the system. The criminal justice system thus displays five prominent features.

Features of Criminal Justice

First, the system operates according to certain democratic fundamentals specified by the United States Constitution, the Bill of Rights, and our system of laws. Governmental power is subdivided at federal, state, and local levels; separate enforcement, court, and correctional functions exist at all levels. In addition, numerous due process rights have been established throughout the system to protect the civil rights of all persons, important features examined in more detail in chapter 2.

Second, criminal justice is primarily a function of state and local government.[2] Although reference is commonly made to the criminal justice system, it is important to understand the fragmentation that does exist. There are thousands of local criminal justice systems throughout the nation. Although only a few crimes are under exclusive federal jurisdiction, federal enforcement, courts and corrections comprise an important segment of this system. Responsibility to respond to most crimes, however, rests with state and local governments. Police protection is mainly a function of cities, towns, and counties.

Table 1.1

Percent of Criminal Justice Employment by Level of Government

	Local	State	Federal
Police	77%	15%	8%
Judicial (courts only)	59	31	10
Prosecution and legal services	53	49	21
Public defense	49	47	4
Corrections	34	62	4
Total	59%	34%	8%

Source: Adapted from "Justice Expenditure and Employment, 1990" in BJS Bulletin, September 1992, p. 6.

Corrections is primarily a function of state governments. As shown in table 1.1, nearly two-thirds of criminal justice personnel are employed at the local level.

Third, the justice system is only one portion of a larger economic and political system and in large measure functions according to power decisions made by other entities. Legislators play key roles in this respect; they reflect public sentiment, enact legislation, and authorize funding from limited resources. During recent years legislators have greatly increased their power and influence over criminal justice and corrections. Politicians are usually most eager to display that they are "tough on crime." Legislators have prompted new policies and passed new laws that have had major impacts on corrections. For example, additional crimes have been defined, especially in the drug area; some mandatory prison terms have been enacted and other prison terms lengthened; new capital punishment laws have been enacted; prison construction has been funded; and parole has been modified or eliminated. Ultimately, correctional policies and operations result largely from decisions made by legislators in the political arena.

The role of the judicial branch includes interpreting law, which impacts on the procedures by which criminal justice personnel perform their jobs and are held accountable. For example, beginning in the 1960s, courts dropped their hands-off doctrine toward intervening in correctional operations; their intervention has brought considerable change in such areas as prison discipline, revocation of parole, and visitation rights. More recently, the rise of privately funded and managed correctional facilities contracting with the public sector indicates an emerging and strengthening of this linkage between corrections and the rest of society.

Fourth, the system is composed of three major components—police, courts, and corrections. The role of police includes enforcing laws through investigation and arrest; the courts interpret law, consider charges, determine whether a person is guilty, and issue sentences; the responsibilities of corrections include carrying out sentences, protecting the public, and attempting to change offender behavior. Critics maintain that throughout the system, victims' needs and rights are not satisfactorily met.

Table 1.2

Who Exercises Discretion?

Officials	*Discretionary Power*
Legislators	Primary power brokers who define policy; enact laws and determine type of sentence; authorize funding.
Police	Enforce specific laws; investigate specific crimes; search people, vicinities, buildings; arrest or detain people.
Prosecutors	File charges or petitions for adjudication; seek indictments; drop cases; reduce charges or plea bargain.
Judges or magistrates	Set bail or conditions for release; accept pleas; dismiss charges; impose sentence; revoke probation.
Correctional officials	Assign to type of correctional facility; award privileges; punish for disciplinary infractions.
Paroling authority	Determine date and conditions of parole; revoke parole.

Source: Bureau of Justice Statistics, Report to the Nation on Crime and Justice *2d ed., (Washington, D.C.: United States Department of Justice, March, 1988), pp. 2–3.*

Fifth, the system is marked by a high degree of discretionary decision-making. In many instances officials are permitted wide latitude regarding a person's entry into, movement through, and exit from the system. For example, police exercise **discretion** in deciding whether to arrest; prosecutors decide which charges to file; judges decide whether to incarcerate offenders; parole board members decide whether to release prisoners before they have served full terms. A comparison of decisions often indicates significant disparity between cases, which generates considerable criticism and ongoing controversy. The importance of understanding this discretionary nature of the system was effectively described by Albert J. Reiss, Jr., when he stated that:

> Criminal matters necessarily involve decisions. Where alternative courses of action exist, choice may be authorized and limited by legitimate authority, and by laws or administrative rules, but they may also arise as unauthorized options and rules within operating organizations. Where an agent is free to choose among alternatives in making a decision, we speak of his exercising choice. When that choice is not open to review . . . we speak of the choice as discretionary. Discretion exists, then, whenever an organization and its agents make choices that are not generally open to reexamination by others.[3]

Such is the nature of the criminal justice system. Considerable discretion is allowed, review of decisions is infrequent, and considerable disparity can result. As shown in table 1.2, numerous individuals and agencies are involved directly or indirectly with discretionary decisions. These include police, prosecutors, legislators, judges, juries (in some states), parole boards, and correctional officials. A decision by the police whether to arrest and, if an arrest is made, what charges to file has an impact, including a possible sentence. The prosecutor may play a major role in the process. This occurs as the prosecutor determines specific charges to place against an offender. Whatever charge is decided upon often dictates a different sentence if convicted. Additionally, prosecutors often have a direct influence on the movement of cases as a result of the plea bargaining process. In this process, a reduced sentence is offered a defendant in exchange for a guilty plea.

Law in the United States

The United States is a country in which there has always been substantial reliance on the **law.** In a very real sense the birth of our nation resulted from a legal struggle. Colonists rebelled against numerous English laws—rules and regulations—which eventually resulted in the formation of a nation independent from England. This new nation was founded with a reliance on strong legal ideas and principles set forth in the Constitution. As problems of governing a new nation were encountered, people and their leaders turned to the development of additional laws to manage these problems. As our nation has become increasingly complex, reliance on laws and legal solutions to our problems has expanded.

Law may be defined as rules of conduct prescribed and enforced by governments that control relations among citizens and between citizens and government. Law thus provides an important means of social control. It is through law that offenders are sentenced for violations, the rights of citizens are defined, justice is obtained, and a more peaceful and orderly society is possible.

Civil and Criminal Law

There are many divisions and categories of law. One major division is between civil and criminal law. **Civil law** is designed to regulate matters of a private nature involving issues such as contracts and domestic relations. A major purpose of **criminal law** is to protect the public and prevent or deter criminal behavior. The government brings suit and imposes penalties on those found guilty of violation. For example, since armed robbery is a violation of criminal law, the state acts as prosecutor and imposes sentences on those found guilty. In civil law cases, individuals rather than government bring suit in which redress is sought for some harmful action. Breach of contract could be grounds for a civil suit. A contractor being sued for failure to complete a building is a civil action. In many instances the behavior that violates criminal law also provides a basis for civil action. A person committing arson is liable under the criminal law and could also be liable for civil law damages as well. Major features of each are outlined below.

Civil Law

- It provides a means for settling private disputes.
- The individual citizen (rather than government) brings suit; this is called a civil suit.

Many criminal justice activities take place in and around local courthouses across the nation.

- It is concerned with payment for damages.
- The government provides a forum for settlement of disputes but is not a prosecutor.

Criminal Law

- It defines conduct considered to be a public threat.
- It is concerned with providing punishment and sentences such as fines and imprisonment.
- The government acts as prosecutor rather than the individual citizen bringing suit.
- It provides legal safeguards for the accused.

Felonies and Misdemeanors

Criminal law is subdivided into categories of felonies and misdemeanors. This distinction is somewhat arbitrary and may vary from state to state. A **felony** is designated as a crime of a more serious nature and carries a potentially more severe punishment than a **misdemeanor.** Major differences are highlighted below.

Felony

- A major criminal offense
- Examples—murder, rape, and arson

- Possible imprisonment for more than one year, usually in a state penitentiary
- Possible capital punishment in some instances
- May include fines

Misdemeanor

- A minor criminal offense
- Examples—petty theft, disorderly conduct, and traffic violations
- Possible imprisonment for less than one year, usually in a local jail
- No capital punishment
- May include fines

Substantive and Procedural Law

Another distinction is between substantive and procedural law. This distinction applies to both civil and criminal law. **Substantive law** is concerned with the content of law. This portion of law creates and defines what is illegal and prescribes penalties for violation. Laws defining crimes of murder or robbery provide an example of substantive law.

On the other hand, **procedural law** prescribes procedures by which substantive law may be enforced. This enforcement cannot be arbitrary and must be according to the rules set forth by procedural law. This affects the entire administration of justice from the time a suspect encounters the system until he or she is released. One example of procedural law is the **exclusionary rule.** This rule prohibits using evidence at a trial that was obtained in violation of the constitutional rights of the defendant. For instance, if it were determined that evidence was obtained by an illegal search of a defendant's house, this evidence would be excluded from trial. Another item of procedural importance is the opportunity for appeal. A conviction may be appealed to a higher court. This permits an examination to determine if proceedings to obtain the conviction conformed to legal requirements. Some areas where procedural law has major implications for corrections include sentencing, granting or revoking probation and parole, and conditions of imprisonment.

Activities Constituting Crime

We have just noted that the substantive portion of criminal law defines various behaviors constituting **crime.** A listing and definition of serious or **index crimes,** accompanied by summary data, are presented in table 1.3. Behaviors resulting in personal injury, such as rape, homicide, and assault, are referred to as crimes against persons. Crimes such as burglary and motor vehicle theft, which do not involve the use or threat of force against a person, are commonly designated as property offenses.

Table 1.3

Characteristics of the Most Common Serious Crimes

Crime	Characteristics
Homicide: Causing the death of another person without legal justification or excuse.	• Homicide is the least frequent violent crime. • 93% of the victims are slain in single-victim situations. • At least 55% of murders are carried out by relatives or acquaintances of the victims. • 24% of all murders occur or are suspected to occur as the result of some felonious activity.
Rape: Unlawful heterosexual intercourse by force or without legal or factual consent.	• Most rapes involve a lone offender and a lone victim. • About 36% of rapes are committed in victims' homes. • 58% of rapes occur at night between 6:00 P.M. and 6:00 A.M.
Robbery: Unlawful taking or attempted taking of property that is in the immediate possession of another, by force or threat of force.	• Robbery is a violent crime that typically involves more than one offender (in about half of all cases). • Slightly fewer than half of all robberies involve the use of a weapon. • Fewer than 2% of the robberies reported to the police are bank robberies.
Assault: Unlawful, intentional inflicting or attempted inflicting of injury upon another person. Aggravated assault is the unlawful threat or attempt to inflict bodily injury or death by means of a deadly or dangerous weapon, with or without actual infliction of injury. Simple assault is the unlawful intentional inflicting of less than serious bodily injury without a deadly or dangerous weapon or an attempt or a threat to inflict bodily injury without a deadly or dangerous weapon.	• Simple assault occurs more frequently than aggravated assault. • Assault is the most common type of violent crime.
Burglary: Unlawful entry of any fixed structure, vehicle, or vessel used for regular residence, industry, or business, with or without force, with the intent to commit a felony or larceny.	• 42% of all household burglaries occur without forced entry. • In the burglary of several million U.S. households, the offenders enter through an unlocked window or door or use a key (for example, a key "hidden" under a doormat). • About 34% of no-force household burglaries occur between 6:00 A.M. and 6:00 P.M. • Residential property is targeted in 67% of reported burglaries; nonresidential property accounts for the remaining 33%. • Three-quarters of nonresidential burglaries for which the time of occurrence is known take place at night.
Larceny: Unlawful taking or attempted taking of property other than a motor vehicle from the possession of another, by stealth, without force and without deceit, with intent to permanently deprive the owner of the property.	• Pocket picking and purse snatching most frequently occur inside nonresidential buildings or in street locations. • Unlike most other crimes, pocket picking and purse snatching affect the elderly. • Most personal larcenies with contact occur during the daytime, but most household larcenies occur at night.
Motor vehicle theft: Unlawful taking or attempted taking of a self-propelled road vehicle owned by another with the intent of permanently or temporarily depriving the owner of it.	• Motor vehicle theft is relatively well reported to the police because reporting is required for insurance claims, and vehicles are more likely than other stolen property to be recovered. • About three-fifths of all motor vehicle thefts occur at night.
Arson: Intentional damaging or destruction or attempted damaging or destruction by means of fire or explosion of the property without the consent of the owner, or of one's own property or that of another by fire or explosives with or without the intent to defraud.	• Single-family residences are the most frequent targets of arson. • More than 17% of all structures where arson occurs are not in use.

Source: Bureau of Justice Statistics, Report to the Nation on Crime and Justice, *2d ed. (Washington, D.C.: United States Department of Justice, March, 1988), pp. 2–3.*

Measurement of Crime

Corrections exists because society seeks to intervene when persons violate laws. Increases or decreases in the number of persons sentenced can affect the number of people corrections has to supervise. Corrections has a keen interest in whether those who have undergone a period of correctional supervision in prisons and/or community-based programs commit crimes after their release (recidivism). For these reasons correctional personnel have an interest in the occurrence of crime and how it is measured. We now turn attention to some methods by which crime is measured and what the resulting figures indicate. As we shall see, these measurements are subject to some problems and limitations.

Two widely used measures of the extent of crime in the United States are the **Uniform Crime Reports (UCR)** and the **National Crime Victim Survey (NCVS).** Each approaches measurement differently and therefore yields a "more comprehensive assessment of crime in the United States than could be obtained from either statistical series alone."[4]

These two measures do not include all possible criminal events and they use commonly understood rather than legal definitions. Each measure has limitations because each is based on different sources and each serves different purposes. For example, many victims do not report to the police crimes that have been committed against them. Often this is designated as unreported or **"hidden" crime** because it is not counted in the Uniform Crime Report.

Recidivism

Recidivism is a term with many definitions that reflect individual differences of opinion regarding crime, corrections, and the criminal justice system. Simply stated, recidivism is a general label assigned to statistical data that reports rates of repeated criminal activity. The rate of recidivism is a measure of how frequently former criminals participate in future criminal activity. A simple example would be to consider one hundred convicted criminals, fifty of whom commit new criminal acts and fifty of whom do not commit new criminal acts. In this case, the recidivism rate for the group is 50 percent.

When reading about various recidivism rates, great differences are found. There are many explanations for these differences, but certainly one key factor may be that different measures of recidivism are utilized. For example, suppose we want to measure the recidivism rate for a group of one hundred offenders guilty of robbery by a court in one county. What events justify placement of the case in a recidivism category? Would the criteria include arrest for any new crime, or arrest only for robbery? How long will we watch the offender's behavior (six weeks, one year, or the rest of their lives)? What if the person commits a new crime in another county or state?

The point is that a simple percentage rate of recidivism, such as 30 percent, tells us very little unless we know how the researcher arrived at that figure. There are three factors that should be considered when trying to determine what is meant by a particular measure of recidivism:

1. The nature of the events to be counted
2. Categories of the behaviors and degrees of seriousness to be included
3. Duration of the follow-up period[5]

Because recidivism is frequently considered the ultimate criteria of success in corrections, it is important to know exactly what is meant by the various rates reported. If corrections is to be measured by the rate of recidivism of its clients, then it behooves those who work in corrections to fully understand the meaning of recidivism.

Caseflow into Corrections

The intake of persons into corrections is dependent upon their being processed by the justice system. This involves a series of steps in a rather long journey. Figure 1.1 shows the route of cases as they are processed in the criminal justice system. Keep in mind this figure represents a general view of the movement of cases as they are processed from investigation through sentencing and possible appeal. For a variety of reasons and at any point in the process a case may "exit" the system. Also remember that our system of justice assumes a person is innocent until proven guilty (or guilt is freely admitted), and that a defendant has many legal safeguards. Specific processes in a given jurisdiction might vary from those shown.

A. Investigation Criminal activity may or may not involve an investigation. Depending on circumstances, an investigation often takes much time and expenditure of considerable resources. In many jurisdictions this function is carried out by specialized officers and detectives. Investigative personnel proceed through a series of steps in the hope of linking evidence with a suspect. Increasingly, sophisticated scientific crime labs assist this process.

B. Arrest and Booking Some criminal activity and encounters with police result in an arrest being made, while in other cases no arrest is made. When official arrests are made, suspects are taken into custody. At the police station or jail, a record of the arrest is made, a process known as booking the arrest. Temporary release on bail or recognizance may be available at this point.

Corrections may be heavily involved at this stage, charged with administering thousands of jails across the nation. Expanded coverage of jails is provided in chapter 5.

C. Initial Appearance This is the first appearance of arrested suspects before a court official, such as a magistrate or justice of peace. During this procedure a formal notice of charges is given and suspects are advised of their rights. Bail, if any, is set.

Figure 1.1 Sequence of events in the criminal justice system

Source: From "The Response to Crime" by M. W. Zawitz, T. R. Mina, C. M. Kuykondall, L. A. Greenfield, and J. L. White in Report to the Nation on Crime and Justice: The Data, *(pp. 42–43) by the Bureau of Justice Statistics, 1983. Washington, D.C.: U.S. Department of Justice.*

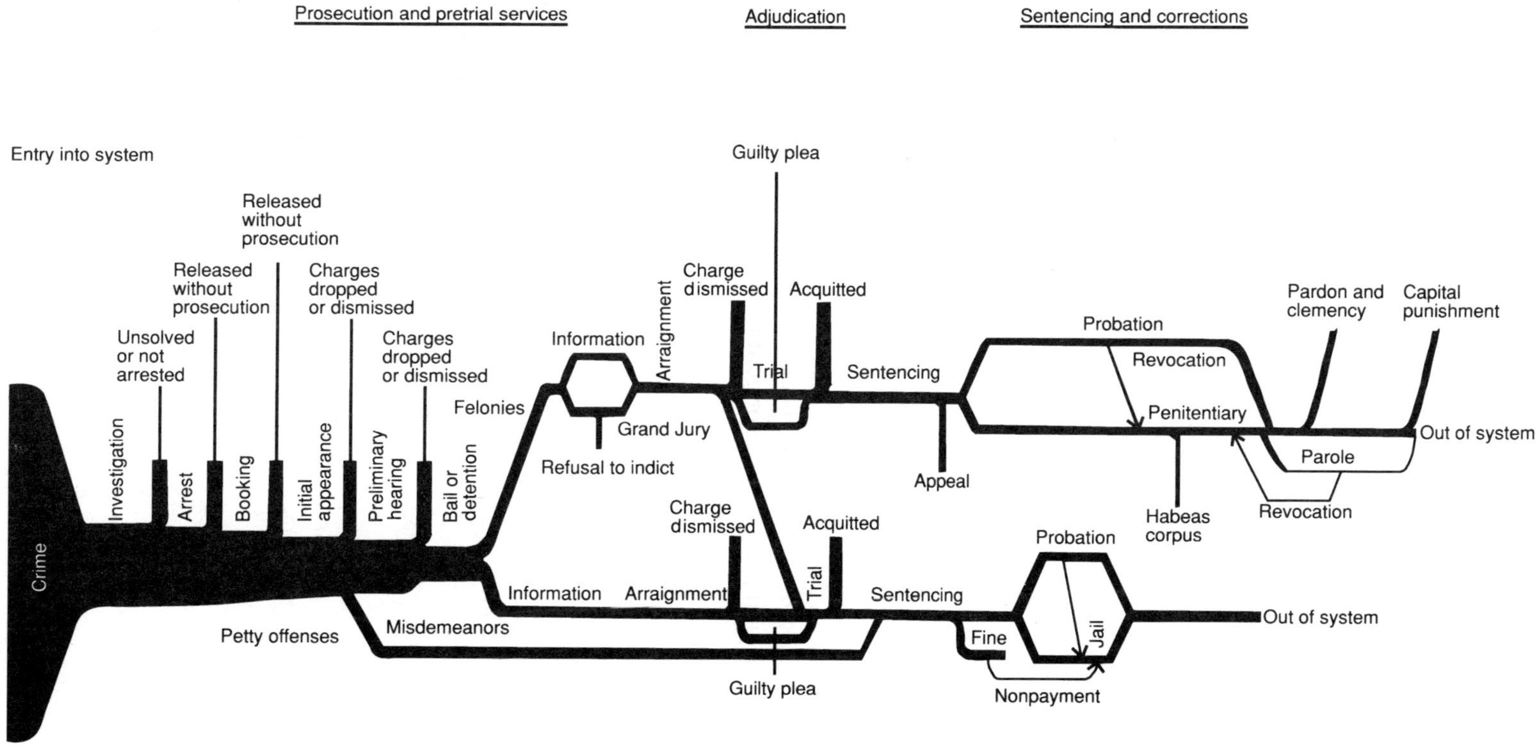

D. Preliminary Hearing A preliminary hearing is held before a legal officer, such as a magistrate. A major purpose of this hearing is to determine whether evidence against the defendant is sufficient to process the case toward a trial or if charges should be dismissed. (In some jurisdictions preliminary hearings are not held for misdemeanor cases.)

E. Information or Indictment Secured In preparation for cases to be brought before the court, charges must be prepared and filed. This is done through one of two methods. One is by returning what is commonly called a complaint. This is a written document charging the accused with a crime. It is prepared by a prosecutor based on evidence from police or private citizens.

Another method is by grand jury indictment. In this instance persons serving on a grand jury (usually twelve to twenty-three in number) hear evidence presented by a prosecutor against a suspect accused of a crime. The grand jury decides whether the evidence warrants a suspect's standing trial for the accused crime. If so, they vote and return what is called a ''true bill'' which ''indicts'' the accused to stand trial. If they vote not to return an indictment, the person goes free or might be brought before another grand jury.

Many jurisdictions provide only for an information procedure and do not have a grand jury procedure. Where grand juries do operate, it is usually in felony cases and rarely misdemeanor cases. All felony defendants at the federal level have a right to a grand jury, but may waive that right, and in that event they are charged by complaint as opposed to indictment. States may use a grand jury procedure but are not required to do so.

F. Arraignment During the arraignment segment, a suspect appears in court and hears the formal charges contained in either a bill of information or grand jury indictment. At this point a defendant pleads guilty or not guilty. If the accused pleads guilty, the case moves to sentencing. If a not guilty plea is entered, the case moves toward trial. It is at this point that **plea negotiations** or bargaining often occurs. Such negotiations involve the prosecutor and the defense. Usually a defendant agrees to plead guilty in exchange for a reduction in sentence or charge. If a case is to move on to trial, some jurisdictions have an evidentiary hearing to discuss what evidence can be used at trial.

G. Trial A trial provides an opportunity for the prosecution and defense to present their arguments and for the judge and jury to examine evidence and testimony. In a trial the state must prove the defendant guilty ''beyond a reasonable doubt.'' The Sixth Amendment to the United States Constitution provides the right to trial by jury in all criminal prosecutions.

If found not guilty, the defendant goes free and is said to be acquitted. If found guilty, the accused has been convicted as charged. The case now moves to sentencing.

H. Sentencing Sentencing is a very important stage in the criminal justice process. It is especially crucial for offenders since sentencing decisions will

Figure 1.2 Persons under correctional supervision

Source: Adapted from Bureau of Justice Statistics, press release September 11, 1994.

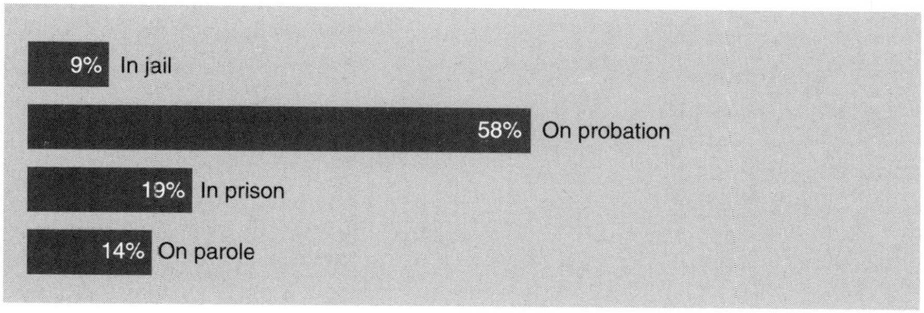

obviously have a major impact on their lives. As a general rule, judges play a major part in determining what sentence to impose. In most jurisdictions, corrections also plays a major role in this process as probation officers prepare presentence reports which include a sentencing recommendation. The role of a probation officer is explained more fully in chapter 11. Depending on the crime a person is guilty of committing, possible sentences fall into the following general categories:

- A fine, restitution or forfeiture of property
- Community-based supervision
- Incarceration
- Capital punishment
- Combination of the above

As a result of the jailing of suspects and the sentencing process, cases flow into corrections for some type of supervision. A breakdown of the relative proportion of adults at both the state and federal level under some form of correctional supervision is shown by figure 1.2. Nationwide, nearly 60 percent of the total group are on probation. The remainder are in jail, in prison, or on parole.

I. Appeal The appeal process allows a losing defendant to carry a case from a lower court to a higher court. An appeal is designed as a check for injustice done as a result of legal procedural or substantial errors that may have occurred in lower court proceedings. There is a guaranteed right to one level of appeal although some cases are appealed and others are not. Sometimes the lower court decision is affirmed. In other instances, an appeals court may reverse a lower court decision.

Funnel Effect

Figure 1.3 indicates that at each stage in the criminal justice process the volume of cases remaining in the system decreases. This has been identified by such terms as "**funnel effect,**" "screening effect," or "sieve effect." What this

Figure 1.3 The "funnel effect"—typical outcome of 100 felony arrests

Source: Bureau of Justice Statistics, The Prosecution of Felony Arrests, 1988 *(Washington, D.C.: USGPO, February, 1992), p. 3.*

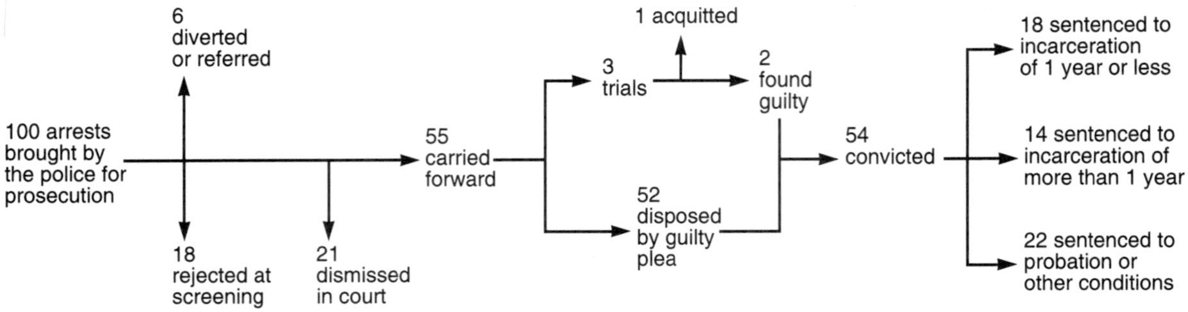

means is that only a proportion of those arrested reach the point of being sentenced for correctional supervision. This occurs because charges are sometimes dropped, cases are dismissed, and some defendants are acquitted. The funnel effect reflects some basic principles of the justice system. One of these principles is safeguarding the rights of citizens. The purpose of these rights is to protect innocent persons, even though this sometimes makes prosecution more difficult. Some jurisdictions operate with a philosophy of applying the least restrictive alternative. In such instances suspects may be diverted out of the system. There is also the pressure of time and numbers. Great resources would be required to process all defendants through a trial. Good or bad, this is a major reason many cases are handled by plea negotiations.

While it is difficult to estimate the number of suspects who "exit" at successive stages, figure 1.3 indicates likely outcomes per one hundred felony arrests. While various analyses can be done with these data, it is of interest to corrections to note that from the one hundred arrests eighteen cases were incarcerated for one year or less; fourteen cases were incarcerated for more than one year; and twenty-two cases were sentenced to some form of probation.

Criminal Justice Information

Up-to-date corrections and criminal justice data is available from the Bureau of Justice statistics/**National Criminal Justice Reference Service (NCJRS).** Information such as prison populations and the number of probationers, for example, are constantly changing. Regular bimonthly and yearly reports issued by this agency enable students and scholars to remain current. Examples of this data are presented judiciously in this textbook because much information is quickly dated. We feel that students are best served by utilizing the latest Bureau of Justice statistics information in conjunction with this text.

The NCJRS operates as a centralized national clearinghouse of criminal justice information. It maintains a computerized data base of many thousand criminal justice documents, operates a public reading room, and offers complete information and referral services. Associated clearinghouses included as components of NCJRS are the Juvenile Justice Clearinghouse for the Office of Juvenile Justice and Delinquency Prevention, the Justice Statistics Clearinghouse for the Bureau of Justice Statistics, and the National Victims Resource Center for the Office for Victims of Crime.

Registered users receive bimonthly reports as a free service. For information on becoming a registered user, write: User Services, Bureau of Justice Statistics/NCJRS, Box 6000, Rockville, MD 20850 or call 1–800–732–3277.

Summary

Corrections is one segment of the criminal justice system by which American society seeks to protect the public, punish offenders, change offenders' behavior, and in some cases compensate victims. Features of the criminal justice system include operation according to democratic fundamentals within the context of the other social institutions. Police, courts, and corrections comprise the three major components of the system with a high degree of discretion permitted for decision-making officials. The criminal justice system is sensitive to political influences and legislators in particular wield major power and influence over policies and operations. Current data reflecting activity with the criminal justice system is available from the National Institute of Justice.

Law provides a means of social control by prescribing written rules of conduct and is categorized in numerous ways. Criminal law defines crimes, sets penalties, and seeks to protect citizens from harmful acts. Civil law regulates private matters. Major criminal offenses are classified as felonies whereas lesser criminal offenses are misdemeanors. Laws establishing how enforcement can legally be achieved is procedural law. Law that explains and defines what is illegal is substantive law.

The National Crime Victim Survey (NCVS) reports on victims of crime throughout the United States. This procedure provides a measure of criminal activity whether or not it is known to the police. The Uniform Crime Report (UCR) measures the amount of crime known to law enforcement agencies. This measure is influenced by whether crimes are reported to the police. Society becomes involved in various ways in response to criminal activity. An important part of this response is encompassed by corrections.

The criminal justice system can be seen as a series of steps. Only a proportion of those who enter by an arrest remain in the system for possible correctional supervision. This ''sieve effect'' results from such factors as acquittals, procedures to safeguard citizens' rights, and lack of evidence to prosecute.

Discussion Topics

1. Does the criminal justice system permit too much discretion? Why or why not?
2. To what extent should political decisions shape correctional policy?
3. To what extent should criminal justice respond to crime as a unified system?

4. In some instances the power and discretion permitted criminal justice officials leads to corruption and favoritism. How might this risk be reduced?
5. Some have the opinion that our nation relies too heavily on the written law to settle our every dispute. Do you agree or disagree? Why?
6. Design a measure of recidivism and explain the reason for your decisions regarding length of follow-up period and type of crime included.

Additional Readings

Cox, Steven M., and Wade, John E. *The Criminal Justice Network: An Introduction.* 3d ed. Madison, WI: Brown & Benchmark Publishers, 1996.

Johnson, Herbert A. *History of Criminal Justice.* Cincinnati, OH: Anderson Publishing Company, 1988.

Rush, George E. *The Dictionary of Criminal Justice.* 4th ed. Guilford, CT: The Dushkin Publishing Group, Inc., 1994.

Schmalleger, Frank. *Criminal Justice Today.* 2d ed. Englewood Cliffs, NJ: Prentice–Hall, 1993.

Notes

1. Bureau of Justice Statistics, *Report to the Nation on Crime and Justice,* 2d ed. (Washington, D.C.: United States Department of Justice, 1988), 56.
2. Ibid., 59.
3. Albert J. Reiss, Jr., ''Discretionary Justice,'' in *Handbook of Criminology,* ed. Daniel Glaser (Chicago, IL: Rand McNally College Publishing Company, 1974), 679.
4. Bureau of Justice Statistics, *Highlights From 20 Years of Surveying Crime Victims* (Washington, D.C.: United States Department of Justice, October, 1993), 4.
5. National Advisory Commission on Criminal Justice Standards and Goals, *Corrections* (Washington, D.C.: Government Printing Office, 1973), 512.

Chapter 2

Setting for Corrections

Learning Objectives

After studying this chapter students should be able to:

1. Understand how federalism fragments corrections.
2. Accurately review United States government structure.
3. Explain the relationship between corrections and the scientific method.
4. Explain the origin and purpose of correctional standards and accreditation and identify suggested improvements.
5. Understand the types of jobs available to and demand for correctional workers.
6. Cite examples and describe the content of several Codes of Ethics found in corrections.

Key Terms

Bill of Rights
Branches of government
Correctional accreditation
Correctional standards
Democratic government
Dual system of courts
Due process
Executive branch
Federal Bureau of Prisons
Federalism
Fourteenth Amendment
Fragmentation of Corrections
Hypotheses
Judicial branch
Jurisdiction
Legislative branch
Levels of government
Organizational categories
Penitentiary
Scientific method
Totalitarian governments
United States Constitution

Introduction

Correctional activities in the United States take place within a justice system and within the setting of **democratic government.** Materials in chapter 1 described the placement and relationship of corrections with other activities and processes in the overall administration of justice. This chapter examines corrections while considering the nature of governmental structure and policies in the United States. The Constitution distributes governmental power among the **levels of government** (federal, state, and local) and the **branches of government** (legislative, executive, and judicial). This presence of a democratically organized government has major implications concerning the organization and operation of corrections—including the processing of suspects as well as those found guilty of law violation. In this sense, corrections in the United States operates as an open system within a democratic form of government. Corrections thus comprises one subsystem in an overall system of democratic governmental structure. This chapter focuses on correctional activities as they take place within this type of government.

Differences in Government Structure

It has been estimated by Garcitoral[1] that about forty-three nations of the world have democratic government, forty-eight are partially democratic, and forty-four are totalitarian. This accounts in large measure for the differing forms of governments seen throughout the world. In **totalitarian governments,** power tends to be more centralized and controlled by a smaller number of persons. Less concern is generally given to individual freedoms or rights. There are more limited safeguards to protect persons in such matters as arrest, conviction, and incarceration. In a democracy such as the United States, power is separated and resides in the hands of citizens through elected officials. In legal proceedings, an accused person is assumed innocent and must be proven guilty. Seeking equality of rights and **due process** is a major consideration that extends to many areas, including operations in the area of corrections. Therefore, it is useful to be mindful that the setting for corrections in the United States is within this framework of democratic government. Although issues and conflicts abound, this accounts for the reason many correctional procedures follow democratic philosophies that would not be emphasized under totalitarian governments. Thus, under our government we concern ourselves, for example, with such matters as right to treatment, due process regarding parole revocation, the question of cruel and unusual punishment, rights while imprisoned, and rights after being released from incarceration. A fundamental and formal expression of democratic values is found in the United States Constitution and the Bill of Rights. In the next section, these documents are briefly reviewed.

In the United States, the accused is assumed innocent until proven guilty. Due process guarantees that a person cannot be deprived of life, liberty, or property without legal procedures that are fair and reasonable.

United States Government Structure

The Constitution

The **United States Constitution,** which is the supreme law of the land, went into effect March 4, 1789. This written document sets forth the fundamental rules by which our government is organized. The Constitution specifies the form of government at the same time it limits and separates the powers of those who govern.

A basic political idea written into the Constitution was that no one person or group of persons should possess all the power to govern. For this reason the power of the federal government was divided. The first three articles of the Constitution formed three branches of government. Article I granted legislative power to a Congress of the United States consisting of two houses, the Senate and the House of Representatives. Article II established an executive branch, vesting power in the offices of president and vice-president. A third branch, the judicial, was granted power according to Article III. This article provided for a United States Supreme Court and other inferior courts as necessary. The Constitution was also written in such a manner that power was divided among different levels of government. This division of power was made especially clear by the Tenth Amendment to the Constitution. This amendment stated that

Table 2.1

Division of Governmental Power			
Branches of Government	***Levels of Government***		
	Federal	***State***	***Local***
Legislative	U.S. Congress	State Legislature	City Council
Executive	President	Governor	Mayor
Judicial	Federal Courts	State Courts	Local Courts

powers not granted by the Constitution to the federal government, nor denied to state governments, are reserved to the states or the people. These divisions of governmental power according to levels and branches of government are depicted in table 2.1. This chart also depicts examples of respective officials or units within each of these divisions.

Branches of Government

The legislative branch has the power and responsibility to enact or pass laws. Many of these laws affect corrections. The **legislative branch** sets forth laws defining illegal behavior and sentences for violating these laws. A very direct effect is that legislatures must approve the amount of money available for correctional operations. The implications of this are great. The amount of money available determines the number of employees that may be hired, what programs can be implemented, and what facilities can be built. Legislators have great power and influence and correctional policies and operations are largely determined within the political environment of the legislative branch.

It is the responsibility of the **executive branch** to implement laws that have been passed. Corrections (except granting probation which is considered to be a judicial function) is considered to be a function of the executive branch of government. The executive branch is headed by a chief executive who is expected to play a leadership role. At the federal level this is the president; at the state level the governor; and at the local level the mayor. Before a law passed by the legislature can be placed into effect, it must be approved by the executive branch. Therefore, the prevailing view within the executive branch toward corrections is important because this can influence the type and level of support for correctional operations. The judicial branch has responsibility for interpreting laws as controversies arise.

The **judicial branch** is comprised of a system of courts and judges. Controversies over the application and meaning of laws arise in many areas including corrections. Beginning in the late 1960s, there was an increase in lawsuits involving various aspects of corrections. Many of the legal decisions made by the judiciary have had major implications for corrections management and for persons under correctional supervision. In addition, granting probation is considered to be a responsibility of the judicial branch of government.

Levels of Government

The Constitution was also written in such a manner that power was divided among different levels of government. Dividing governmental power at several levels rather than centralizing power at only one level is a system of government known as **federalism.** In a large country like the United States, federalism represents a compromise between extreme concentration of power and extreme scattering of power among independent states. In the United States, governmental system power is divided into three levels: federal, state, and local.

Although many powers were granted to the federal government by the Constitution, many others were retained at the state and local levels of government. Because laws are enacted, enforced, and interpreted at each of these three levels, law violators are dealt with at the appropriate level. At each of these levels an important relationship exists between corrections and the courts. To understand this relationship it is first necessary to examine the structure and organization of the court system.

Consistent with the idea of federalism, the Federal Judiciary Act of 1789 created separate state and federal judicial systems. There are federal courts at one level and state and local courts at another level. Thus, we have a **dual system of courts** for handling criminal matters, a simplified diagram of which is shown in figure 2.1.

There are two broad categories of courts: trial and appellate. Trial courts are called courts of original jurisdiction since they hear cases for the first time and issue a decision. Courts having appellate jurisdiction review decisions made by lower courts, checking mainly for errors in procedures.

Courts at the federal level are arranged into three tiers. Courts with original jurisdiction are known as district courts. These courts hear cases arising from violation of federal law.

The courts of appeal review cases on appeal from the district courts. The Supreme Court of the United States serves as court of last resort for both the federal and state system and is the ultimate interpreter of the Constitution and federal laws. The Supreme Court has the authority to review cases from the lower federal courts and from courts of final decision at the state level, thereby linking the dual system of courts at the top tier. (It has original jurisdiction in matters where a state is a party and cases involving ministers and ambassadors.)

Figure 2.1 also provides a sketch of prominent features found in the structure of most courts within the states. The specific organization, duties, and even the name of courts vary considerably from one state to another, but states have trial courts at the local level. In most states they are divided into major and minor trial courts. Minor trial courts have limited jurisdiction; that is, they are authorized to hear only those cases on certain subjects, such as small claims, traffic violations, and probate issues. Minor misdemeanor cases are also generally heard in these courts. Although limited in jurisdiction, these courts

Figure 2.1 Dual system of courts in the United States

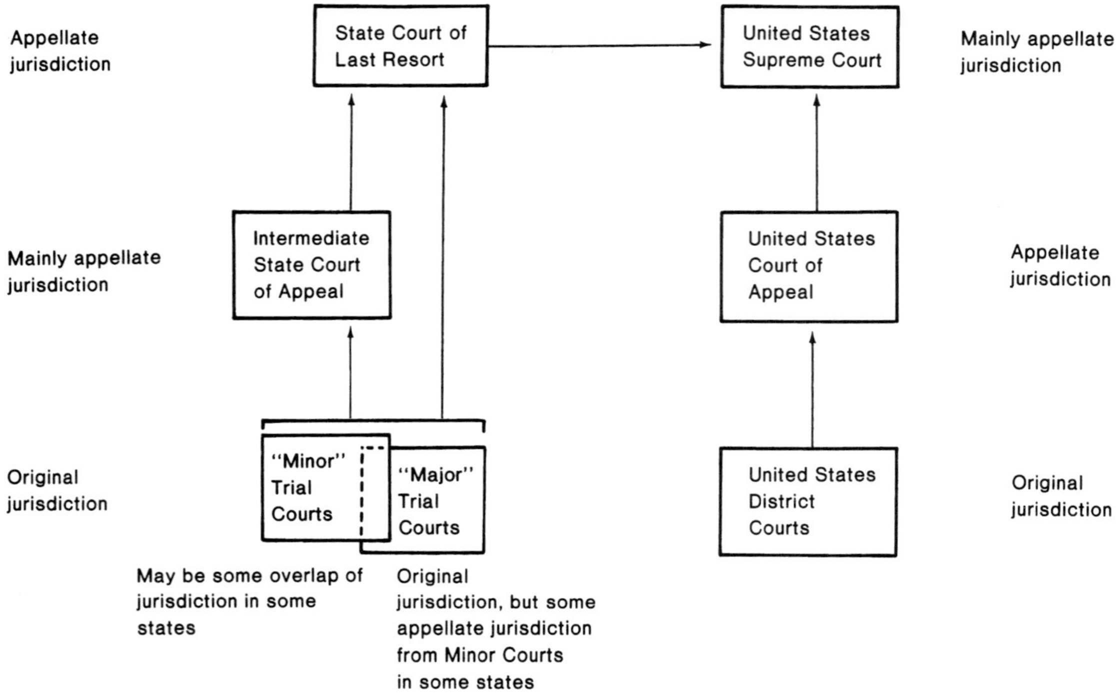

number in the thousands nationwide and actually hear a majority of all criminal cases. All states have major trial courts possessing general jurisdiction in which there is no special limitation regarding the subject matter of cases they may hear. Serious criminal offenses are usually heard in these courts. The state court of last resort is usually named the state supreme court but in some instances it is known as the court of appeals or supreme court of appeals.

Impact of Federalism on Corrections

Dividing governmental power at several levels (known as federalism) means there are federal crimes and state crimes. Federal crimes violate federal law; state crimes violate state law. The authority to hear and decide a case is known as having **jurisdiction.** In many cases it is clear as to whether the state or federal level has jurisdiction. However, in other cases it may be unclear, and in some instances the same act may violate both state and federal law. Such would be the case if a stolen automobile (violation of state law) were driven across state lines (violation of federal law). In this instance the defendant could be tried in both a federal and state court. Usually, however, the defendant is prosecuted in one or the other. The essential point for corrections is that defendants found guilty by a state court and sentenced for correctional supervision will be supervised by some facet of the correctional system at the state or local level.

Similarly, a defendant found guilty by a federal court and sentenced for correctional supervision will be the administrative responsibility of the federal correctional system. Since there are three levels of government and corresponding jurisdictions, the nature and organization of corrections follow a similar pattern. It is logical, then, that we think in terms of corrections at the federal, state, and local level.

Bill of Rights

The first ten amendments, or additions, to the Constitution of the United States are popularly called the **Bill of Rights.** These additions went into effect very soon after the Constitution was ratified. They became effective December 15, 1791. These amendments grant specific rights to the people and prohibit Congress from passing any laws violating these civil liberties. An abbreviated outline of the Bill of Rights listing those guaranteed rights most likely to be involved in correctional processes appears below:

- First Amendment
  - Freedom of religion
  - Freedom of speech
  - Freedom of press
  - Freedom of peaceable assembly
- Fourth Amendment
  - No unreasonable searches and seizures
- Fifth Amendment
  - Freedom from self-incrimination
  - Not being tried twice for the same offense (also known as no double-jeopardy)
  - No deprivation of life, liberty, or property without due process of law
- Sixth Amendment
  - Right to speedy and public trial
  - Right to confront witnesses
  - Right to assistance of counsel
- Eighth Amendment
  - No cruel and unusual punishment
  - No excessive bail
  - No excessive fines

Fourteenth Amendment

An amendment having far-reaching consequences, including corrections, is the **Fourteenth Amendment** which became effective July 28, 1868. This amendment reads as follows:

> All persons born or naturalized in the United States, and subject to the jurisdiction thereof, are citizens of the United States and of the states where they reside. No state shall make or enforce any law which shall abridge the

> privileges or immunities of citizens of the United States; nor shall any state deprive any person of life, liberty, or property, without due process of law; nor deny to any person within this jurisdiction the equal protection of the laws.[2]

A main purpose of this amendment was to grant citizenship to former slaves. However, the phrases "all persons" and "any persons" are sufficiently broad to mean any *citizen* is entitled to due process of law as originally stated in the Fifth Amendment. Furthermore, no state can deny any citizen within its jurisdiction the equal protection of the law. The judicial branch, through the courts, has ruled on a very large number of cases involving this amendment, some of which have involved the correctional area.

Fragmentation of Corrections

Based on materials presented to this point, we can see a **fragmentation of corrections** according to a number of factors that make administrative coordination and linkage to other systems difficult. Major sources of fragmentation are summarized in the following outline.

1. By jurisdiction
   a. Federal
   b. State
   c. Local
2. By criminal justice function
   a. Police
   b. Courts
   c. Corrections
3. By location
   a. Institutional
   b. Noninstitutional (Community-based)
4. By age
   a. Adult
   b. Juvenile
5. By other factors
   a. Size of operation
   b. Sex of offender
   c. Type of offense
   d. Special programs

Furthermore, it has been emphasized that:

> The response to crime is founded in the intergovernmental structure of the United States.
>
> Under our form of government, each State and the Federal Government has its own criminal justice system. All systems must respect the rights of

individuals set forth in court interpretation of the U.S. Constitution and defined in case law.

State constitutions and laws define the criminal justice system within each State and delegate the authority and responsibility for criminal justice to various jurisdictions, officials, and institutions. State laws also define criminal behavior and groups of children or acts under jurisdictions of the juvenile courts.

Municipalities and counties further define their criminal justice systems through local ordinances that proscribe additional illegal behavior and establish the local agencies responsible for criminal justice processing that were not established by the State.

Congress also has established a criminal justice system at the Federal level to respond to Federal crimes such as bank robbery, kidnaping, and transporting stolen goods across State lines. The response to crime is mainly a State and local function.

Very few crimes are under exclusive Federal jurisdiction. The responsibility to respond to most crime rests with the State and local governments. Corrections is primarily a function of State governments.[3]

Corrections at the Federal Level

Congress established the **Federal Bureau of Prisons** on May 14, 1930, in the Department of Justice. This department is headed by the attorney general, an official appointed by the president of the United States. The Bureau of Prisons is managed by a director. Paul W. Tappan[4] has recalled the situation regarding federal corrections during the late 1800s and early 1900s. Because there were no federal prisons, most of the prisoners sentenced under federal law were actually held in state prisons and county jails. A few were held in United States marshals' jails for the federal territories. Pressures mounted for the federal government to establish prisons at the federal level. As a result, the military prison at Fort Leavenworth, Kansas was temporarily used for civil prisoners beginning in 1894. A federal penitentiary was later built at Leavenworth, opening in 1905. Other authorized federal institutions during the next twenty-five years included McNeil Island at Puget Sound, a women's reformatory at Alderson, West Virginia, and a male reformatory at Chillicothe, Ohio. Alcatraz, a military base in San Francisco Bay, was converted into a Federal Prison for the most hardened criminals in 1934. It was closed in 1963 and has since become a popular tourist attraction.

At the time the Bureau of Prisons was established by Congress in 1930, there were about 12,000 offenders housed in seven federal prisons. Each institution operated under its own policies with regulations set by individual wardens. The congressional act establishing the Bureau of Prisons sought to develop a more unified system and to provide individualized custody and treatment. Although not free from problems and episodes of controversy, the federal system is considered by many to be a leader and innovator in the field of corrections. As shown by figure 2.2, the federal Bureau of Prisons has a network of facilities ranging from penitentiaries to prison camps.

Past and present directors of the Federal Bureau of Prisons, left to right: J. Michael Quinlan, Director 1987–1992; Kathleen M. Hawk, Director 1992–present; Norman A. Carlson, Director 1970–1987.

Courtesy of the Federal Bureau of Prisons

Figure 2.2 U.S. Department of Justice Federal Bureau of Prisons

Source: Federal Bureau of Prisons, "Population Report," September 8, 1994.

Corrections at the State and Local Level

During colonial times, corrections was a concern of local governments. People were not imprisoned. Instead, they were subjected to fines, physical pain (corporal punishment), or death (capital punishment). However, following the Revolutionary War in 1783, more enlightened social ideas brought change. Imprisonment and solitude were developed as a punishment or penance in place of corporal and capital punishment. Offenders were considered to be placed in a state of penitence wherein they were to regret their wrongdoing and become a contrite and penitent person. Such practices led to the term **penitentiary**—an idea from which two problems developed.

The first problem was that buildings had to be located or constructed in which to house offenders. Secondly, it was difficult for each local government to provide long-term imprisonment for serious offenders. Therefore, during the late 1700s and early 1800s state legislatures began to authorize the construction and development of prisons. In 1773, Connecticut began using an abandoned copper mine, known as Newgate, as a state prison. Another example, as described by McKelvey,[5] was the approval given by the Pennsylvania State Legislature in 1790 to build a cell house in the yard of the Walnut Street Jail. This cell house was for the solitary confinement of males convicted of felony violations. Approval was also given for the Walnut Street Jail to receive convicts from other counties, which in effect established a prison at the state level.

Other states soon followed the action taken in Pennsylvania. State legislators began to plan and authorize money for facilities to imprison convicted offenders. The manner in which these institutions were managed and organized within the structure of state government was very different. Commenting on these early management arrangements, Carter, McGee, and Nelson stated:

> While these institutions were authorized by state legislatures and supported from state funds, the administrative arrangements for their relationships to the states' elective officials assumed many forms. To attempt to trace the development of these administrative arrangements for each of the states would involve one in an almost unbelievable thicket of political expediencies.[6]

Over the years, these various management forms were organized and recognized numerous times. Nevertheless, the Council of State Governments has identified five basic **organizational categories** that have emerged.[7] These patterns illustrate the settings existing among correctional agencies at the state level. These patterns exist according to which governmental unit has responsibility for providing correctional services. In some cases one state agency has responsibility for all operations, including juvenile services, probation/parole, and jails. In other instances, there is fragmentation among three or more state agencies.

A census of state correctional facilities included a vast amount of information on state prisons and inmates.[8] Table 2.2 focuses on the location and increasing number of state facilities, the actual number of inmates and the number of inmates per 100,000 population. Some states have relatively few

Table 2.2

Number of State Correctional Facilities and Inmates by Region and State

Jurisdiction and Region	Number of Facilities		Number of Inmates		Percent Change 1984–90	Inmates per 100,000 Population
	1984	1990	1984	1990		
Total	903	1,207	395,309	658,828	66.7%	262.6
Northeast Region	134	182	68,000	117,865	73.3%	231.2
Connecticut	13	20	5,055	9,577	89.5	295.0
Maine	6	7	1,074	1,503	39.9	121.3
Massachusetts	17	20	4,764	8,282	73.8	139.4
New Hampshire	3	5	551	1,441	161.5	127.5
New Jersey	13	25	10,103	16,721	65.5	215.5
New York	45	62	32,138	56,251	75.0	312.4
Pennsylvania	24	30	12,606	20,822	65.2	172.2
Rhode Island	7	7	1,203	2,460	104.5	245.0
Vermont	6	6	506	808	59.7	140.3
Midwest Region	193	255	80,298	141,726	76.5%	234.4
Illinois	34	39	16,228	26,712	64.6	228.1
Indiana	18	23	9,277	12,618	36.0	224.1
Iowa	16	24	2,824	4,522	60.1	158.6
Kansas	9	16	3,825	5,500	43.8	217.3
Michigan	45	67	13,800	31,812	130.5	341.6
Minnesota	8	9	2,260	3,239	43.3	73.6
Missouri	13	17	8,173	14,600	78.6	281.7
Nebraska	8	7	1,765	2,390	35.4	147.5
North Dakota	2	2	436	557	27.8	85.2
Ohio	16	22	16,433	31,808	93.6	290.3
South Dakota	2	2	796	1,247	56.7	174.0
Wisconsin	22	27	4,481	6,721	50.0	137.0
South Region	435	534	175,444	253,453	44.5%	293.1
Alabama	23	28	8,774	12,433	41.7	300.6
Arkansas	11	13	4,005	6,455	61.2	267.1
Delaware	7	8	1,898	3,449	81.7	502.8
District of Columbia	8	11	2,737	7,290	166.4	1,225.2
Florida	80	100	26,382	42,306	60.4	325.1
Georgia	29	32	13,138	18,540	41.1	283.6
Kentucky	10	15	4,536	6,900	52.1	185.0
Louisiana	11	20	10,409	13,939	33.9	319.8
Maryland	18	21	12,742	17,057	33.9	358.0
Mississippi	11	24	4,695	6,852	45.9	261.1
North Carolina	87	91	16,603	18,346	10.5	275.6
Oklahoma	21	23	7,493	10,449	39.5	324.8
South Carolina	27	32	8,434	15,091	78.9	423.8
Tennessee	15	18	7,745	8,333	7.6	167.1
Texas	27	43	34,761	49,815	43.3	290.3
Virginia	44	48	9,484	14,649	54.5	236.8
West Virginia	6	7	1,608	1,549	−3.7	84.2
West Region	141	236	71,567	145,784	103.7%	275.4
Alaska	12	13	1,590	2,414	51.8	456.3
Arizona	15	20	7,239	13,903	92.1	382.9
California	38	100	39,274	92,604	135.8	310.5
Colorado	11	14	2,636	5,593	112.2	167.7
Hawaii	7	10	1,709	2,569	50.3	227.5
Idaho	4	7	1,127	1,767	56.8	172.2
Montana	3	5	836	1,273	52.3	157.7
Nevada	10	17	3,129	5,620	79.6	481.2
New Mexico	6	11	1,725	3,118	80.8	201.6
Oregon	7	11	3,421	5,994	75.2	208.5
Utah	8	8	1,678	2,799	66.8	162.4
Washington	17	16	6,518	7,036	7.9	144.4
Wyoming	3	4	685	1,094	59.7	231.8

Source: Adapted from Bureau of Justice Statistics, Census of State and Federal Facilities, 1990 *(Washington, D.C.: Government Printing Office, May, 1992), p. 2.*

inmates, such as Vermont and North Dakota; others, such as California, New York, and Florida have high numbers of inmates. Similarly, the number of inmates per 100,000 population shows much diversity. Minnesota, West Virginia, and North Dakota have the lowest rates of incarceration while District of Columbia, Delaware, Nevada, Alaska, and South Carolina have among the highest rates of incarceration.

Some additional highlights from the census included:

- 55 percent of all State inmates were housed in large institutions with an average daily population of 1,000 or more;
- All facilities housing 2,500 or more inmates were State-operated;
- State facilities were operating 22 percent above design capacity;
- State facilities provided an average of 56 square feet of living space per inmate;
- More than one in four State facilities were under court or consent decree;
- 245,750 people were employed in state correctional facilities.[9]

Substantial local level correctional activity involves jails, a topic covered in chapter 5. Chapters 11 and 12 present the topics probation and parole which also have major local level correctional activity.

Correctional Standards and Accreditation

Standards are guidelines established as a basis of comparison in measuring or judging the adequacy of programs, facilities, or activities. These guidelines can be very formal and supported by the force of law, or can be rather informal with only moderate social pressure to comply. Standards are important for many reasons, such as maintaining and supporting quality, encouraging improvement, and providing both a means to recognize excellence and to identify and possibly eliminate inferior programs.

For a profession, standards provide a mechanism for self-regulation. Many powerful and well-recognized professional organizations, such as the American Medical Association and the American Bar Association, have highly developed sets of standards their members must follow. Members who fail to meet the standards of the profession can be sanctioned and ultimately lose their licenses to practice. Other professional groups, such as corrections, have recognized both the need for and advantages of self-regulating standards.

Correctional standards have been developed by many groups including the United Nations, the American Medical Association, The United States Department of Justice, and many state and local governmental agencies. The most comprehensive and widely recognized set of standards that apply to correctional institutions and probation/parole divisions have been developed by the Commission on Accreditation for Corrections (CAC). In 1967, the Ford Foundation provided a grant to the American Correctional Association (ACA) to develop a self-evaluation procedure of corrections.[10] Then, in 1974, the ACA, with a grant from the Law Enforcement Assistance Administration, established the

Commission of Accreditation for Corrections which was charged with the responsibility of developing comprehensive national standards for corrections.[11] Correctional professionals from throughout the nation worked in committees to develop the various standards for corrections. The first sets of standards were published in 1977; later revised editions were published in 1981.

The standards have been divided into four categories: mandatory, essential, important, and desirable. These descriptive categories let the correctional agency personnel know what emphasis should be placed on each area of the standards. The standards are recognized as minimum guidelines an agency should attempt to meet. It is the hope of most professionals that eventually all correctional institutions will not only meet but exceed these guidelines.

Correctional Accreditation

Correctional accreditation is a voluntary program to measure compliance with the ACA standards. A correctional institution or agency wishing to receive accreditation first completes a self-evaluation to determine their level of compliance. After making internal adjustments to reach as many standards as possible, the institution submits a report to the CAC for review. The next step in the process involves a CAC audit team (made up of selected correctional professionals from other jurisdictions) visiting the institution to conduct an official evaluation. Following the submission of the audit team's report, the correctional institution submits a plan of action for achieving compliance in those areas where facilities or programs were found to be substandard. These materials are reviewed by a panel of commission members who make the final determination on accreditation. Following the awarding of accreditation, yearly reports must be submitted by the correctional institution and periodic reevaluations are conducted.

Many agencies and institutions have received accreditation. Generally the program has received a favorable reception, although some within the profession have criticized the process. Some groups such as the National Prison Project expressed concerns that the CAC standards fell short of common constitutional guarantees.[12] Other criticisms were directed at the CAC because it does not require compliance with all of the standards in order to gain accreditation. It has been noted that some correctional institutions have received accreditation while under court order to improve conditions.[13]

There are older, clearly outdated facilities that have been accredited. A conflict has developed between those who view the standards as goals that correctional agencies and staffs should work toward and those who see in the standards very minimum level conditions needed for prisons to operate. The program of accreditation has prompted some correctional systems to seriously reevaluate the operation of their prisons and to make significant changes to comply. In other instances, correctional institutions have relied upon Band-Aid efforts to receive accreditation in the hope that such recognition would relieve pressure from federal and state courts to make fundamental changes in the operation of their institution.

As experience is gained and the CAC accreditation process matures, suggestions for improvement and change have been forthcoming. It has been suggested that the objectives of accreditation could be better met by including the following:

- Ensuring that ACA standards meet constitutional requirements,
- Being alert to possible unconstitutional conditions if a facility undergoing the accreditation process is under a court order,
- Conducting private interviews with inmates,
- Upholding the intent of standards,
- Establishing a multicategory system of accreditation.[14]

Despite criticisms raised, the development of standards for correctional institutions and agencies must be recognized as one of the major developments in the profession. Differences exist within the profession regarding how the standards may most benefit corrections. Some view receiving accreditation as a sort of immunity against future litigation over substandard conditions; others view the standards as a mechanism to spur change and eliminate woefully inadequate facilities. Achieving accreditation requires considerable time and money. Correctional standards and accreditation will likely remain both important and controversial throughout the years ahead.

Corrections and Social Science

The social sciences are those areas of study focusing on human behavior and the interaction of people and their various social groupings. Traditional social sciences include anthropology, economics, political science, psychology, and sociology. As a component of the criminal justice system, corrections has frequently been referred to as an applied science. The term ''applied'' refers to a close association between study and practical applications by working professionals in corrections. The study of corrections has many direct practical applications for professionals in the field, but it also shares many of the elements common to traditional social science areas.

The traditional social science disciplines frequently employ the scientific method in their studies, and the application of this method as a tool in the study of corrections has been a key concern of many scholars.

The Scientific Method

The **scientific method** provided us with a means to test our notions of the world. Many behaviors and ideas are justified on the basis of common sense. How often do you hear the phrase, common sense would tell you . . . ? This notion of common sense refers to one way of knowing—intuition. Often, common sense is very sound and application of the scientific method to the problem would merely confirm one's earlier notion. Frequently, however, common sense is nonsense, and, when subjected to an empirical test, fails to be supported by the evidence collected. Basing decisions and policies in the field of corrections on common sense without any empirical testing of the theories behind the action is not a scientifically sound approach.

Countless examples of the application of the scientific method to the research and study of corrections can be cited. Cesare Lombroso utilized the scientific method in his attempts to identify the "born criminal." Lombroso's notions of criminal behavior did not prove to be accurate; he did, however, apply the scientific method in an attempt to explain his beliefs.

The scientific method is only one way of knowing (others include spiritualism, dogmatism, intuition, revelation, and authoritarianism).[15] What differentiates this method from other ways of knowing is the process that one goes through to arrive at a conclusion. The scientific method is an objective, logical, and a systematic method of analysis devised to permit the accumulation of reliable knowledge.[15] When faced with a problem for which an explanation is sought, the scientific approach calls for the development of a **hypotheses.** A hypotheses is a speculative notion or idea about the relationship between two or more variables and serves as a guide for research. The next step would be the collection of data related to the problem.

After data have been collected and analyzed, conclusions are drawn regarding the problem and possible explanations. Conclusions lead to confirmation, modification, or rejection of ideas contained in hypotheses. This information may or may not be directly used in forming correctional policies.

Careers in Corrections

Effective corrections presents a challenge for America and all of its citizens regardless of race, sex, or national origin. Offenders in the United States represent both sexes and all races. Correctional employees represent both sexes and many ethnic groups, although a current need exists to increase employment for women and minorities. Regardless of your sex or race, a career in corrections can represent an individual challenge for you. Dedicated and enthusiastic people are needed for an effective correctional system; a wide variety of jobs and opportunities are available. A career in corrections offers the opportunity to transfer feelings of concern for others into positive action.

Most authorities suggest that a necessary early step in career selection is to honestly determine what kind of person you are. One way to accomplish this is through self-assessment. This involves determining your interests, values, what makes you happy, your dislikes, abilities, and limitations. As a result, you should increase your self-awareness and thereby be in a better position to make a more informed and satisfying career choice. Numerous publications and aids are available to help in this process.[16]

Demand

Figures indicate that nearly 350,000 persons are employed by correctional systems at the adult and juvenile level.[17] An elaboration of employees by sex, race, and level of government of these employees is shown in table 2.3. These correctional employees are involved in the supervision and treatment of convicted adults and adjudicated juveniles. It is evident that substantial numbers of women and minorities are included among these employees.

Demand for correctional workers has been increasing and will continue to increase for the foreseeable future. This is due to a number of reasons discussed throughout this textbook, including such matters as public and political attitudes, higher rates of incarceration, longer prison terms, and drug problems.

Types of Jobs

Work in corrections is varied and numerous skills and academic disciplines are involved in one way or another at the adult and juvenile level in either an institutional environment, such as prison, jail or juvenile reform facility, or a community-based setting which includes probation, parole, or some related form of supervision. Corrections does not draw people from a single discipline, but utilizes many types of knowledge and skills.

It has been suggested that correctional work can be divided into four main categories:

1. custodial
2. treatment
3. administrative
4. support[18]

These categories cover a broad range of occupations and professions.

Most persons enter corrections as a Correctional Officer (CO), performing a custodial role in an institutional setting, such as a prison or jail. This role, described in more detail in chapter 8, is quite complex and involves monitoring inmates, enforcing rules, and maintaining order. For many, this becomes a stepping stone to subsequent work in a treatment role, as an administrator, or in the probation and parole field. These operations and the nature of working in these areas is presented in more detail in chapters 8, 11, and 12.

A wide array of occupations and job skills are required to support correctional operations within institutions and in community-based corrections. A few diverse examples include, medical and legal personnel, computer operators, food service workers, electrical workers, industrial technicians, printers, agricultural workers, transportation workers, and safety officials.

Corrections is not known for high entry-level pay. Typically, entry pay for a full-time CO is in the range of $15,000 to $20,000 per year. With experience and movement into administrative roles or probation and parole work pay may be in the $20,000 to $40,000 range.

Table 2.3

Personnel in Adult and Juvenile Corrections by Sex, Race, and Level of Government

Adult Corrections

	State Level		Federal Level	
	Number	**Percent**	**Number**	**Percent**
White				
Male	163,760	53	12,713	53
Female	62,682	20	4,363	18
Black				
Male	36,445	12	2,840	12
Female	21,566	7	1,492	6
Hispanic				
Male	11,695	4	1,715	7
Female	4,365	1	442	2
All Other				
Male	4,722	2	390	2
Female	2,011	.7	139	.6
Total	307,246	100	24,094	100

Juvenile Corrections

	State Level[1]	
	Number	**Percent**
White		
Male	16,320	44
Female	8,289	22
Black		
Male	5,983	16
Female	3,901	10
Hispanic		
Male	1,507	4
Female	738	2
All Other		
Male	428	1
Female	269	.7
Total	37,435	100

0 = *Rounding error*

[1]There is no federal level juvenile corrections systems

Source: Adapted from American Correctional Association, *Directory Juvenile and Adult Correctional Departments, Institutions, Agencies and Paroling Authorities,* (Laurel, MD, American Correctional Association 1994), xlii–xliii.

Administrative jobs may range from $30,000 to $70,000, or above in a few instances. Corrections work is generally considered stressful and some choose not to continue in this area. Those who do make a career in corrections often indicate a long-term interest and desire to perform this type of work and do receive satisfaction from their job.

Correctional work is not for everyone and as with any occupational choice you should learn as much as possible about yourself and the job you are considering before making a commitment. There are many sources of information about careers in corrections. These include, but are not limited to:

- Colleges and Universities
- American Correctional Association
- Federal Bureau of Prisons
- State Employment Office
- State Department of Corrections
- Local Jails
- American Probation and Parole Association

Professional Ethics

Corrections has become a more open system and court intervention has demanded more accountability. The need for education and training has increased. Parallel efforts have been made to establish a professional status for correctional workers. This is a gradual and continuing process. There are many components of professionalism including specialized education, a sense of calling to the field, and self-regulation. As occupational groups seek to be recognized as having professional status, a Code of Ethics tends to emerge. The purpose of such a code is to develop moral guidelines for ideal and appropriate behavior. Those contemplating a career in corrections or those already working in the field might look to such codes for guidance.

Examples of three Codes of Ethics are shown in tables 2.4, 2.5, and 2.6. Shown in table 2.4 is the American Correctional Association Code of Ethics, which was adopted in 1975, and represented a first in the field of corrections. A second example, shown in table 2.5 was written specifically by the American Probation and Parole Association for probation and parole officers. A third example in table 2.6 was written for workers in a large detention facility at the local level. As can be seen in the table, common themes include respect for the dignity of all people, the decision not to exploit a job for personal gain, and a desire to serve the public.

Table 2.4

Code of Ethics—American Correctional Association

Preamble

The American Correctional Association expects of its members unfailing honesty, respect for the dignity and individuality of human beings and a commitment to professional and compassionate service. To this end, we subscribe to the following principles:

Members will respect and protect the civil and legal rights of all individuals.

Members will treat every professional situation with concern for the person's welfare and with no intent of personal gain.

Relationships with colleagues will be such that they promote mutual respect within the profession and improve the quality of service.

Public criticisms of colleagues or their agencies will be made only when warranted, verifiable and constructive in purpose.

Members will respect the importance of all disciplines within the criminal justice system and work to improve cooperation with each segment.

Subject to the individual's right to privacy, members will honor the public's right to know and will share information with the public to the extent permitted by law.

Members will respect and protect the right of the public to be safeguarded from criminal activity.

Members will not use their positions to secure personal privileges or advantages.

Members will not, while acting in an official capacity, allow personal interest to impair objectivity in the performance of duty.

No member will enter into any activity or agreement, formal or informal, which presents a conflict of interest or is inconsistent with the conscientious performance of his or her duties.

No member will accept any gift, service or favor that is or appears to be improper or implies an obligation inconsistent with the free and objective exercise of his or her professional duties.

In any public statement, members will clearly distinguish between personal views and those statements or positions made on behalf of an agency or the Association.

Each member will report to the appropriate authority any corrupt or unethical behavior where there is sufficient cause to initiate a review.

Members will not discriminate against any individual because of race, gender, creed, national origin, religious affiliation, age or any other type of prohibited discrimination.

Members will preserve the integrity of private information; they will neither seek data on individuals beyond that needed to perform their responsibilities, nor reveal nonpublic data unless expressly authorized to do so.

Any member who is responsible for agency personnel actions will make all appointments, promotions, or dismissals in accordance with established civil service rules, applicable contract agreements and individual merit, and not in furtherance of partisan interests.

Adopted August 1975 at the 105th Congress of Correction
Revised August 1990 at the 120th Congress of Correction

Reprinted by permission of the American Correctional Association, Laurel, MD.

Table 2.5

Code of Ethics—American Probation and Parole Association

Code of Ethics

1. I will render professional service to the justice system and the community at large in effecting social adjustment of the offender.
2. I will uphold the law with dignity; displaying an awareness of my responsibility to offenders with recognizing the right of the public to be safeguarded from criminal activity.
3. I will strive to be objective in the performance of my duties, recognizing the inalienable right of all persons, appreciating the inherent worth of the individual, and respecting those confidences which can be reposed in me.
4. I will conduct my personal life with decorum, neither accepting nor granting favors in connection with my office.
5. I will cooperate with my co-workers and related agencies and will continually strive to improve my professional competence through the seeking and sharing of knowledge and understanding.
6. I will distinguish clearly, in public, between my statements and actions as an individual and as a representative of my profession.
7. I will encourage policy, procedures and personnel practices which will enable others to conduct themselves in accordance with the values, goals and objectives of the American Probation and Parole Association.
8. I recognize my office as a symbol of public faith and I accept it as a public trust to be held as long as I am true to the ethics of the American Probation and Parole Association.
9. I will constantly strive to achieve these objectives and ideals, dedicating myself to my chosen profession.

Source: Reprinted by permission of the American Probation and Parole Association, Lexington, KY.

Table 2.6

Code of Ethics—Orange County Corrections Division

The Orange County Corrections Division expects of its employees unfailing honesty, respect for the dignity and individuality of human beings, and a commitment to professional and compassionate service. To this end we subscribe to the following principles.

As a member of the Corrections Division, I will regard myself as a member of an important and honorable profession.
I will perform my duty with efficiency at all times.
I will be exemplary in my conduct, direct in my conversation, honest in my dealings and obedient to the laws of the city, state and country.
I will not, in the performance of my duty, work for personal advantage or profit.
I will, at all times, recognize that I am a public servant.
I will give the most efficient and impartial service of which I am capable at all times.
I will be courteous in my contacts at all times.
I will regard my fellow employees with the same standards as I hold myself.
I will guard my fellow employees' honor and life as I guard my own.
I will accept responsibility for my actions.
I will do only those things that will reflect honor on my fellow employees, the Division, and myself.
I will maintain the integrity of private information. I will neither seek personal data beyond that needed to perform my responsibilities, nor reveal personal information of employees, inmates, or clients, to anyone not having proper professional use for such.

Source: Reprinted by permission of the Orange County Florida Corrections Division, Orlando, FL.

Summary

Corrections in the United States is fragmented and reflects the division of power as provided by the Constitution. Correctional services are provided at the federal level as well as at the state and local levels. Corrections is considered to be a function of the executive branch of government.

Corrections at all levels must be sensitive to due process protection contained in the Constitution and the Bill of Rights. The recent decline of the hands-off doctrine serves to emphasize this point. Correctional standards and accreditation are designed to improve conditions and to develop greater management accountability. Application of the scientific method may provide a more reliable basis for making decisions.

The organization of corrections at the state level lacks uniformity. In some states all services for adults and juveniles are provided by a single department. In other states the responsibility for providing correctional service rests with two, three, or even more departments. Corrections at the federal level is the responsibility of the Federal Bureau of Prisons which operates a number of institutions and treatment centers throughout the United States.

Choice of a career begins with a self-assessment and a career in corrections presents challenges for a variety of individuals. Numerous jobs, with a wide range of pay, are available at the adult and juvenile level, either in an institutional setting or in a community environment. Many, but not all persons enter the field of corrections as a correctional officer (CO) performing mainly a custodial role. Codes of Ethics have emerged as correctional workers seek professional status.

Discussion Topics

1. How can corrections be sensitive to due process protection contained in the Constitution and the Bill of Rights?
2. Would it be desirable to have less fragmentation of corrections?
3. To what extent should federal money help pay for corrections at the state and local level?
4. Is achieving accreditation for correctional operation worth the time and money involved? Why or why not?
5. Divide into groups of two to three people and list advantages and disadvantages of choosing a career in corrections.
6. What is the likelihood of Codes of Ethics being an effective guide to employee behavior?

Additional Readings

Pollock, Joycelyn M. *Ethics in Crime and Justice: Dilemmas and Decisions.* 2d ed. Belmont, CA: Wadsworth Publishing Company, 1994.

Roberts, John W., Ed., *Escaping Prison Myths: The History of Federal Corrections.* Lanham, MD: University Press of America, Inc., 1994.

Shover, Neal and Werner J. Einstadter. *Analyzing American Corrections.* Belmont, CA: Wadsworth Publishing Company, 1988.

Williamson, Harold E. *The Corrections Profession.* Newbury Park, CA: Sage Publications, 1990.

Notes

1. Alicio Garcitoral, ''Can Our Democracy Survive?'' *Modern Maturity,* (December–January 1979–1980): 49–50.
2. United States Constitution, Fourteenth Amendment.
3. Bureau of Justice Statistics. *Report to the Nation on Crime and Justice,* 2d ed. (Washington, D.C.: United States Department of Justice, 1988), 59.
4. Paul W. Tappan, *Crime, Justice and Corrections* (New York: McGraw-Hill Book Company, Inc., 1960), 619–620.
5. Blake McKelvey, *American Prisons* (Chicago: University of Chicago Press, 1936), 7–8.
6. Robert M. Carter, Richard A. McGee, and E. Kim Nelson, *Corrections in America* (Philadelphia, PA: J.B. Lippincott Company, 1975), 234.
7. The Council of State Governments, *Reorganization of State Corrections: A Decade of Experience* (Lexington, KY, 1977), 12–15.
8. Bureau of Justice Statistics, *Census of State and Federal Correctional Facilities, 1990* (Washington, D.C.: Government Printing Office, May, 1992).
9. Ibid., 2–16.
10. Stephen Gettinger, ''Accreditation on Trial,'' *Corrections Magazine* 8, no. 1 (Feb. 1982): 11.
11. Commission on Accreditation for Corrections, *Accreditation: Blueprint for Corrections* (Rockville, MD: Commission on Accreditation for Corrections,1978), 1.
12. Gettinger, Stephen, ''Accreditation on Trial,'' 12.
13. Ibid, 11–16.
14. Lynn S. Branham, ''Accreditation: Making a Good Process Better,'' *Federal Probation* (June 1993): 11.
15. Carlo L. Lastrucci, *The Scientific Approach* (Cambridge, MA: Schenkman Publishing Company, 1967), 29.
16. Richard Nelson Rolles, *What Color is Your Parachute?* (Berkeley, CA: Ten Speed Press, 1994).
17. American Correctional Association, *Directory Juvenile and Adult Correctional Departments, Institutions, Agencies and Paroling Authorities* (Laurel, MD: American Correctional Association, 1994), xlii–xliii.
18. Harold E. Williamson, *The Corrections Profession* (Newbury Park, CA: Sage Publications, 1990), 94.

Chapter 3

History and Competing Aims of Corrections

Learning Objectives

After studying this chapter students should be able to:

1. Recall a definition of corrections.
2. Differentiate between classical and positivistic views.
3. Understand how changing attitudes in colonial America lead to the development of prisons.
4. Compare and contrast four models of corrections.
5. Explain and evaluate the concept of deterrence.

Key Terms

Banishment
Capital punishment
Classical view
Corporal punishment
Corrections
Custodial model
Determinism
Deterrence
Deviance
Economic punishment
Free will
Incapacitation
Incarceration
Just deserts
Justice model
Least restrictive alternative model
Lex talionis
Mala in se
Mala prohibita
Norms of behavior
Positivist view
Punishment
Rehabilitation model
Reintegration
Resocialization
Retribution
Social control
Social learning
Utilitarian doctrine

Background

Criminals in colonial America were subject to both corporal punishment and capital punishment for their offenses. They were not imprisoned for a given period of time as is so familiar to us today. Corporal punishment was often harsh, practiced by using such methods as the stocks, branding, and whippings. The punishments were often administered within public view, designed to bring shame and ridicule upon the offender. Capital punishment also played a role. In addition to murder, laws provided execution for such offenses as adultery, witchcraft, and blasphemy.

These punishments were closely linked with religious viewpoints and thoughts of the colonists. These were God-fearing people, and there was widespread belief in the concept of sin. Crime was considered to be a sin, and the perpetrator (sinner), needed to be punished in accordance with God's law. These "corrections", in the form of punishment, were primarily seen as a necessary retribution for the transgressor. If the community failed to punish, they feared that everyone might be subject to God's wrath. Because Calvin's doctrine of predestination was widely accepted, punishment was viewed as having a limited deterrent effect.

> ". . . the colonists did not anticipate widespread success in deterring people from wrongdoing. Their Christian sense of crime as sin, their belief that men were born to corruption, lowered their expectations and made deviant behavior a predictable and inevitable component of society. The causes of crime were not difficult for them to understand and the theory of its origin stirred little controversy.[1]

However, during the latter 1700s and early 1800s these views were challenged, and the response to crime and criminals began to change. New attitudes and new thinking were inspired by a number of events, including the American Revolution and circulation of Beccaria's essay *On Crime and Punishment.*[2]

The origins of crime were increasingly considered to be from within society, rather than from a predestined criminal sinner. Crime was seen as rooted in environment and something that could be prevented and controlled. Harsh and retributive punishments began to be seen as counterproductive, and Americans began to generate what was at that time a new approach to corrections—the development of prisons for extended periods of incarceration for offenders.

In 1773, the state of Connecticut began using an abandoned copper mine to incarcerate serious criminals for an extended period of time. In 1790, a portion of the Walnut Street Jail in Philadelphia was converted to a prison. Several other states constructed prisons during the early 1800s. During the 1820s Pennsylvania and New York developed state prisons that were viewed as competing models. In Pennsylvania the system was based on complete isolation (separate system), whereas at Auburn prison in New York congregate work with enforced silence was emphasized. By the mid 1800s this move to build prisons had spread to the midwest and western states.

The combined effects of isolation, obedience to authority, an orderly routine, and hard work found in these prisons represented an American approach to the problem of crime, an approach that has continued in a substantial way through the present day.

It is probably difficult for us to imagine a time when prisons did not exist. We would probably also be shocked by a public whipping or a stockade. We thus see different approaches to crime and criminals. Differing assumptions, values and beliefs influence society's response.

We now turn our attention to examining the nature and aims of corrections and to some of the assumptions and viewpoints that underlie various correctional approaches and programs.

What Is Corrections?

Corrections is concerned with and operates as society's primary formal dispenser of punishment. Corrections, however, is more than simply a nice term for punishment. Corrections, as the root of the word implies, focuses on correcting a problem or series of problems in society. It has come to stand for a broad category of activities ranging from incarceration of offenders, to assisting ex-offenders in securing employment and education in the community, to providing assistance for the victims of crimes. The working definition of corrections used in this text is

> *the systematic and organized efforts directed by a society that attempt to punish offenders, protect the public from offenders, change offender behavior, and in some cases may compensate victims.*

This is by no means the only available definition of corrections, as many authors have sought to redefine the concept. The four key elements in this definition are punish, protect, change, and compensate. These activities reflect the full range of activities in the field of corrections.

Corrections is a broad field and includes divergent activities, goals, and directions. Within the field of corrections and law enforcement, in general, there are a great many philosophies, approaches, and techniques. The student of corrections should recognize and learn to accept that there may not be an absolute right way to achieve a particular goal. In fact, it may not be uncommon to find some correctional workers within the same system taking nearly opposite approaches to the same problem. Some point to the different approaches as a weakness of corrections because there is often little consistency between and among systems. Others point to the differences as strengths and cite the need for multiple approaches to the complex problems facing corrections. Whatever your views of the differences, it is important that you recognize them and begin to identify where you fit into the scheme of corrections. Your views should be based on facts and solid evidence.

Deviance and Social Control

Any act designed to diminish **deviance** could be categorized as a form of **social control.** Two key issues emerge from this opening sentence: What is deviance? What are the various forms of social control? The intent of laws, both public and private, is the same: the control of deviance from established **norms of behavior.** To deviate means to depart from the normal or acceptable standard. The courts have often referred to the reasonable person as a measure of acceptability. Put another way, what would the normal (reasonable) person have done in a particular situation? Would a reasonable person have been justified in using deadly force to protect his home against unwanted intrusions by strangers?

It is important to note that not all deviation from the norm is unacceptable. Societies and individuals within societies are willing to tolerate varying degrees of deviance. All nations have established laws that state that the taking of human life is illegal, but there are acceptable deviations from these laws. The taking of human life as an act of self-defense of one's own life is generally accepted as reasonable, as well as military actions taken during times of war.

Not only is there great latitude in what constitutes deviance, but there is similar variance in what is normal. Norms, or the normal acceptable behaviors of individuals, are not constant. Norms represent expectations regarding what behavior is considered socially acceptable. They provide guidelines for behavior appropriate and applicable for particular social situations. The norms and standards of one social group may be vastly different from those established by another. Not only do norms vary between groups, but they can and do change over time. A norm today may be considered unacceptable in five years. We think nothing of changes in fashion and rapid advances in technology (in fact, we have come to expect change). These rapid advances in technology have encouraged the rethinking of some of our standards and even long-held beliefs related to basic values. Normal is a relative term that refers to the prevailing standard, the dominant assumption. Change is constant, and students who study forms of social control are repeatedly reminded of the differences in what is normal and acceptable behavior among societies and throughout history.

Social control takes many forms and can range from the scornful stare of a disapproving parent at a child who is misbehaving to the execution of a convicted killer. The varying mechanisms of social control can be identified as either informal (private) or formal (public). The private forms of social control are all those everyday actions that humans use to express approval or disapproval for the behavior of those around them. These informal controls form the basis of the process of **social learning.**[3] That process which allows individuals to acquire knowledge, form beliefs, and adopt values consistent with those around them is social learning.

The process of social learning and the informal control associated with it form the foundation for formal (public) controls. When expectations reach a crucial limit of importance to a society, they merit being codified as laws. The laws are formal control mechanisms that require or prohibit certain conduct. The conduct prohibited by the law may be considered wrong in itself (**mala in se**), such as murder, robbery, or incest. Other prohibited conduct may only be considered wrong because of the law (**mala prohibita**), such as status offenses for juveniles, traffic ordinances, and alcoholic beverage regulations. The difference between these two divisions of the law relates to the moral culpability of the offender and the threat that his behavior poses to the society in which he lives.

The Response to Deviation

When an individual or group of individuals deviates from the established patterns of behavior of a society, a reaction generally follows. This response may often not be formal, even if there exist codified laws that apply. However, when an act is viewed as threatening to the fabric of a society, there will be some formal response in most cases.

The Continuum of Corrections

The rationales for and operation of corrections in the United States can be viewed on a continuum. There have existed throughout the nation's history periods of major reform in correctional practices (the reformatory and rehabilitation movements), and there have been similar periods of harsh and, in fact, brutal administration of corrections. The notion of a continuum of correctional thought allows one to place the historical development of corrections in perspective. Anchored firmly at either end of this continuum are two fundamental philosophies or world views relating to human behavior.

Philosophies of Corrections

Philosophies are conceptual frameworks designed to account for particular behaviors, actions, or events. They provide us with a general view of a method of operation. Approaches range from very abstract notions of the world based upon unproven beliefs to very concrete notions rooted in time-honored ideas. Philosophies help to define basic directions, purposes, and goals. By understanding a particular philosophical approach, an observer gains insight when studying behavior and events.

Frequently, philosophies are founded upon basic world views. This is the case in the field of corrections. There are basically two world views that underlie all philosophies of corrections—the classical and positivist views. These two views represent opposite ends of the continuum of correctional thought in terms of the nature and causes of human behavior. Few people hold strictly to either view but rather favor one over the other and tend to act and think accordingly.

Classical View

A **classical view** of the world assumes that citizens are very rational and logical. This view is premised upon the writings of Rousseau (1712–1778), the author of *The Social Contract*. Rousseau wrote of an implicit contract that exists between the citizen and the state. This so-called contract involves the exchange of some individual freedoms for the benefits of security and tranquility.[4]

Cesare Beccaria (Cesare Bonesana, Marquis of Beccaris, 1738–1794) in his work *An Essay on Crimes and Punishments* (1767), applied the Rousseauian idea of the social contract to punishment and thus created what has come to be known as the classical view. This view maintained that "laws are conditions by which free and independent men unite to form society."[5]

"Let the pain of prospective punishment so exceed the pleasure of committing crime that the individual will weigh the two in balance and choose to remain law-abiding."[6] Beccaria, as the preceding passage indicates, viewed citizens as having **free will** to choose the pursuit of pleasure and the avoidance of pain. He called for swift and certain punishment and was not concerned with the intent of the offender who committed a crime: "equal crimes—equal penalties." Beccaria's view was that there be extreme equality regardless of age, sex, or sanity. Modifiers of this extreme equality were dubbed neoclassicalist; they excluded children and the insane from the equal treatment because these groups were unable to comprehend the notion of pleasure and pain.

The concept of free will is central to the classical view. The citizen must function in a world where choices can be made and acted upon. The laws of this nation are likewise premised upon the notion of free will. With noteworthy exceptions, such as infancy and insanity provisions, the law recognizes the citizen as having free will to choose between conformity and nonconformity to the established social order. This notion of the citizen as a rational individual provides the basis for the classical view of deterrence as the aim of punishment.

"The **utilitarian doctrine** is based on the principle that actions are right insofar as they contribute to maximizing the happiness of people, wrong insofar as they decrease that happiness."[7] The doctrine maintains that there should be the greatest happiness shared by the greatest number. This view of the world developed by Jeremy Bentham and others, coupled with the notion of free will and the classical view, provide a rationale for the use of strict punishments in response to crime. This view maintains that social cohesion and the good of the many (conformist) takes precedent over the pains inflicted upon the few (nonconformists). This view of the world has been modified, rejected, and readopted many times since it was first developed. The major opposing world view to this has been termed the positivist view.

Positivist View

There developed increasing opposition to the classical view during the late 1800s. Many scholars began to question the basic notions upon which the classicalists based their assumptions regarding punishment. Chief among these new thinkers was Cesare Lombroso (1835–1909). Lombroso developed the notion of vestigial atavistic traits, which he maintained were inherited and thus led to the concept of the "born criminal." He maintained that there were recognizable human traits, such as an asymmetrical cranium, long lower jaw, a flattened nose, and other characteristics that were signs of potential criminality. He maintained that such individuals had a predisposition to crime and that such persons needed exceptionally favorable circumstances of life to avoid criminal behavior.[8]

Although Lombroso's early work has been proven to be flawed by inadequate research methodology, which led to incorrect conclusions, he nonetheless helped to develop a new school of thought regarding criminal behavior. Through the writings of scholars such as Enrico Ferri, the **positivist view** of criminology was expanded. This view has built on Lombroso's early notion of the effect of environmental influences on criminal behavior. The positivist view adopted the concept of determinism.

Determinism maintains that human behavior is the product of a multitude of environmental and cultural influences. The view regards crime as a consequence, not of a single but of many factors. Among the influences considered were population density, economic status, and the legal definition of crime. The multiple-factor causation theory brought the positivist view into direct conflict with the less complex free-will notion of the rejection of pain and the seeking of pleasure concept.

What evolved from the positivist view was a new approach to dealing with crime. This new approach maintained that the punishment should fit the criminal rather than be assigned categorically to a particular type of crime. This belief in the need for individualized punishment led to a call for greater discretion in dealing with the criminal. Discretion is the latitude of free choice within certain legal bounds. The push for greater discretion in dealing with criminals ran counter to the classical notion of certainty and absolute equality of punishment.

The proponents of the positivist view influenced many changes in the way criminals were punished including the adoption of indeterminate sentencing procedures and the acceptability of a medical model of treatment for offenders. The approach taken to crime during much of the twentieth century has been premised upon the positivist view of human behavior.

Comparison of the Classical and Positivist Views

> A hungry donkey stands between two haystacks, equally fragrant, equal in size, and at equal distances away. The poor animal, having no will to decide (if the Positivist view is right), or having no motive to influence its decision (if the Classical view is right), eventually reaches a condition of absolute indifference and, being unable to choose one of the haystacks, dies of hunger.[9]

The preceding analogy may place the two opposing views of human behavior in perspective for the reader. Although the views are clearly on opposite ends of the continuum of correctional thought, both have contributed significantly to the field.

Determinism maintains that human behavior is the product of environmental, economic, and cultural influences.

© Mark Antman/The Image Works

The classical view has encouraged fair and orderly legal procedures (due process of law), has mitigated the severity of punishment, and has advocated limits on the arbitrary use of judicial authority.

The positivist view helped to introduce the scientific method into the study of crime and developed the experimental method of research in the field. The multiple-factor theory helped to demonstrate the effect of biological, environmental, and social origins of criminal behavior. The positivist view supported the expansion of programs and constructive activities as a component of punishment.

The classical view is crime oriented, while the positivists place emphasis upon the criminal. Although many subsequent theories and approaches to dealing with crime have been developed, the basic emphasis remains either on dealing with the crime or the criminal; seldom has society been willing or capable of addressing both issues simultaneously.

Rationales and Models of Corrections

When considering the rationale or justification for the use of punishment by a society against individuals who fail to adhere to its rules, one basic question arises: Why must a wrongful act be punished? This may seem such an obvious question that it does not deserve further discussion, but the rationale behind a society's use of punishment can indicate a great deal about the character of its response to deviance. Justifications for punishments can be either proactive or reactive in nature. The proactive justifications focus attention on the present or future. Reactive justifications place emphasis upon the past.

Figure 3.1 A comparison of correctional rationales and models

Rationale		Model
1. Retribution	→	Justice model (JM)
2. Incapacitation	→	Custodial model (CM)
3. Reintegration	→	Least restrictive alternative model (LRAM)
4. Resocialization	→	Rehabilitation model (RM)

When particular elements of a rationale are combined into an operational framework, a correctional model emerges. "A model is a conceptual representation of how something works or should work."[10] Correctional models form idealized views of the manner in which corrections could operate. Models differ greatly in their scope and goals. Examples of correctional programs based upon each of these models as shown in figure 3.1 will be discussed.

Retribution and a Justice Model

The rationale of **retribution** has been applied when the singular justification for punishment is the fact that a crime has been committed.[11] At the foundation of the notion of retribution is the concept of revenge. Revenge or vengeance appears to be a basic and understandable justification for punishment. During various periods of history, revenge in the form of blood feuds and other types of personal reprisals has been carried out by individuals in response to criminal behavior. The Latin phrase **lex talionis** (the law of retaliation) is frequently cited in reference to the concept of revenge. An individual's feeling of anger, grief, or pain has led many to contemplate and seek revenge.

Retribution clearly provides a reactive rationale for punishment. The rationale of retribution implies the payment of debts to society. Retribution differs from revenge in that the punishment is prescribed by law and there exists a formal mechanism (judicial process) to determine who is guilty of criminal offenses and the level of punishment that the individual should receive.

Retribution represents the state's fulfillment of the social contract that removes the right of individual response to criminal acts. Retribution supports a modifying influence in making the punishment fit the crime. The state does not typically respond to a crime in the "eye for an eye" fashion but rather establishes levels of punishment for various offenses.

A concept of **just deserts** is closely related to retribution. This rationale maintains that the punishment should be administered because the criminal deserves it. The focus of the concept of just deserts is also the social contract, which obligates citizens to certain behavior patterns, deviation from which results in punishment.

A **justice model** as outlined by David Fogel in his book *We Are the Living Proof*[12] has been recognized as comparable to other models such as the equity model, punishment model, and neoclassicism. The justice model's basic principles are that the threat of punishment is needed to implement the law. The model supports an active system of rewards and punishments for criminals to channel their behavior in acceptable directions. The model recognizes that there are many divergent values in society but that the law represents the dominant views of the ruling social-political power group. It is these dominant views of the world that criminal law enforces as the prescribed morality. The law is based upon a belief in the free will of people. Corrections represents a deprivation of individual liberty. Fogel stresses the notion that justice and fairness are equitable. He views the central problem in corrections as the unrestricted use of discretion. The justice model would limit the use of discretion by correctional personnel.

A justice model views corrections as more than just control but at the same time reaffirms that society should and must punish criminals for illegal acts. The punishment must be administered fairly and justly. This approach supports limited discretion, which in most cases does not permit predictions of potential behavior. Restraint under the justice model is consistent among similar crimes and types of criminals. Corrections under this model is less concerned with changing criminal behavior than with doing less harm to incarcerated offenders.

Incapacitation and a Custodial Model

Incapacitation as a rationale for punishment provides a transition between reactive and proactive approaches to corrections. Incapacitation, or a society's attempt to render a criminal incapable of further illegal acts, may take on a variety of forms. The most common form of incapacitation in the United States is incarceration (detention in a jail or prison). By physically removing a criminal from society for short or long periods of time, it is possible to deny that individual the opportunity to directly threaten the society outside the correctional institution.

The rationale of incapacitation provides support for the use of a variety of harsh punishments, such as the death sentence, castration of rapists, and the removal of a thief's hand. It can also be argued that far less punitive responses such as probation, substance abuse treatment programs, and even work and education activities could be construed as justifiable under an incapacitation rationale, if in fact these measures made the criminal incapable of further illegal acts.

The focus of incapacitation is on the present. The criminal is prevented from participating in criminal activity through a restriction of individual freedoms. The denial of the opportunity to engage in criminal activity is the key to the rationale of incapacitation.

A **custodial model** of corrections is the least complex of the four models discussed. The key terms associated with this model are control and restraint. Supporters of this model tend to believe that the primary goal of corrections should be the control and restraint of convicted criminals.

All of the models view the protection of society as an essential goal of corrections, but in the custodial model, this becomes the dominant influence.

The custodial model directs corrections toward the control of criminal behavior through inhibiting means such as incarceration. This model measures success in how well corrections is able to perform basic custodial and supervisory tasks. Associated with this model is the notion of predictive restraint. One question facing all of the corrections is, who should be corrected and for how long? Predictive restraint refers to the process of determining which offenders will likely return to crime and then extending the period of correctional control to protect society from such individuals. The custodial model does not assume that the offender's behavior will necessarily be changed by the correctional process but rather that for a specified time the individual will be unable to participate in activities which pose a direct threat to society.

Reintegration and a Least Restrictive Alternative Model

A **reintegration** rationale for punishment is proactive in its response to crime. The emphasis under this rationale is not on the criminal act but rather on the offender and the community. Reintegration is directed toward retrofitting the individual to the community and enhancing the community's acceptance of the offender. Punishments designed in the reintegration framework emphasize helping the individual to fit into the community. This **least restrictive alternative model** advocates minimal intervention.

This rationale places the responsibility for change on both the individual and the community. The correctional system becomes a facilitator that identifies where and why the individual does not fit into the community. Once the areas of conflict are identified, the task becomes to seek change in individual behavior or community acceptance of the offender. The rationale of reintegration views traditional correctional practices such as incarceration as short-term periods of intervention which allow for the identification of areas of conflict between the offender and the community. Reintegration places the responsibility for the criminal behavior and the changing of that behavior upon both the individual and the community. It is the shared responsibility for both the deviance and the change of behavior that clearly distinguishes reintegration from the other rationales for punishment.

This model advocates a minimum amount of intervention by society into the life of convicted criminals. This approach views punishment as counterproductive and therefore only to be used as absolutely necessary.

To operationalize this approach calls for corrections to incarcerate or intervene in individuals' lives only when no other alternatives exist. This model views punishment as a necessary evil that must be used in some cases because of the extreme nature of some individuals' criminal behavior. But the guiding principle of this model is that individuals should be free from unwanted restraints. No more restraint should be utilized in the correctional process than is absolutely necessary to accomplish its task.

This model moves away from the use of the traditional correctional practice of incarceration and seeks whenever possible to address the problem of crime and criminals in the community. This model's concern for protection of the public is greatly modified by its emphasis upon the individual's freedom from punitive restraints. The model seeks to change deviant behavior but emphasizes that this can most effectively be accomplished not in isolation from but rather in concert with the community.

Resocialization and a Rehabilitation Model

The rationale for punishment identified as **resocialization** is directed at the future and at the needs of the offender. This rationale views punishment as an intervention into the offender's life that can act as a catalyst for change. Resocialization supports punishments designed to change the offender's need or desire to commit crime.

This rationale assumes that criminal behavior is learned and that through specific interventions directed at the offender's needs, a relearning process can take place. The offender who is resocialized can then successfully reenter society without posing a threat to himself (or herself) or others. Resocialization is a reactive response to criminal behavior in that the individual must first exhibit criminal behavior to be identified. Once the offender is identified, resocialization operates via a variety of correctional and court programs such as counseling, education, and job training. The means to prevent future deviant behavior is sought through a process that removes the offender's motivation toward crime.

A **rehabilitation model** of corrections is best described by reviewing the five basic steps in the medical approach to a problem: examine, diagnosis, prescribe, treat, and reexamine. When applied to the correctional setting, this model has correctional personnel examining offenders for maladies or flaws, followed by a diagnosis of problems discovered. Next the correctional system prescribes a treatment for offenders and administers that treatment. Finally the correctional system reexamines offenders and determines if they can be released or need additional diagnosis and treatment.

Under the rehabilitation model, the offender has lost the stigma of being "bad" and has been identified as "sick." The sickness may be physical, emotional, or social. The focus of attention is upon the sickness (the cause of the offender's behavior) not the society or social institutions that may have failed to help him nor the individual's own choice to commit a criminal act. Part of the inherent definition of the rehabilitation model is a presumption of nonculpability for the disease. It is not the criminal's fault he or she is sick. The cause of the deviant behavior and the criminal act itself are important only in so much as they contribute information and direction to the treatment process.

The focus of corrections under the rehabilitation model is on the restoration of offenders to a socially acceptable condition identified as "well." Corrections is assigned a strong change-agent role under this model. In fact, corrections' primary tools to protect society become the programs and activities designed to rehabilitate the offenders.

Elements from all four of these models may be found in corrections. The models are not always mutually exclusive but have some overlap; they are not "pure types." In some instances different models may form a blended approach.

Categorizing rationales and models in this manner serves as an analytical tool. Any program, legislative proposal, agency policy, or court decision may be examined and evaluated in terms of its philosophical thrust and the direction in which it leads. It is likely that over time elements of each of these four models will continue to provide a mix of policies and approaches for corrections. Changing political opinions are a major force in shaping a dominant philosophy.

Punishment

"The law threatens certain pains if you do certain things. . . . If you persist in doing them, it has to inflict the pains in order that its threat may continue to be believed,"[13] said Oliver Wendell Holmes. The pains that Justice Holmes referred to are punishments. **Punishment** is "the infliction by the state of consequences normally considered unpleasant, on a person in response to his having been convicted of a crime."[14] A more detailed description of punishment identifies six basic elements:

> First, punishment is a privation (evil, pain, disvalue). Second, it is coercive. Third, it is inflicted in the name of the State; it is "authorized." Fourth, punishment presupposes rules, their violation, and a more or less formal determination of that, expressed in a judgment. Fifth, it is inflicted upon an offender who has committed a harm, and this presupposes a set of values by reference to which both the harm and the punishment are ethically significant. Sixth, the extent or type of punishment is in some defended way related to the commission of the harm, and aggravated or mitigated by reference to the personality of the offender, his motives and temptation.[15]

There is a universal practice of responding to wrongful or criminal acts. The use of punishment by the state is an organized attempt to take the responsibility for retaliation away from the victim or individual citizen. The concept that a crime committed against any member of society is a crime against all members of society is crucial to the modern notion of punishment. In the United States and most nations of the world, private or individual vengeance is not condoned and, in fact, is prohibited by statute.

There are many issues that enter into any consideration of a society's use of punishment. Influential factors in the use of punishment include circumstances and culpability, the interests of those harmed by the offense, the motivation and characteristics of the offender, the notion of consistency and fairness, and the various alternative forms of punishment available. Partly because of the many issues involved, wide degrees of latitude exist in the administration of punishment in the United States.

Methods of Punishment

As the complexity of a society increases, there is a corresponding increase in the variety of methods utilized for the punishment of deviant behavior. Punishment also carries with it a stigmatization. An individual's image in the community and personal sense of self-worth can be negatively affected by criminal prosecution, even if he is not proven guilty. The classic story of Hester Prin and the scarlet letter *A* that adorned her clothing signifying her as an adulteress illustrates the stigmatizing nature that punishments have taken for centuries.

Punishments in many forms have provided societies with a formal means to respond to criminal behavior. The challenge facing the field of corrections is to fairly and humanely administer these various punishments to the convicted offenders. Changing social, economic, and technological aspects of society have provided new forms of punishment and made some traditional approaches acceptable. In very limited technological, agrarian societies of the past, the luxury of incarcerating large numbers of noncontributing citizens did not exist. As societies developed, there were increasing needs for labor in the production of goods, the mining of ores, and such demanding tasks as working as oarsmen on military ships. The needed labor was often viewed as undesirable and dangerous. These tasks frequently were assigned to convicted criminals.

There has long been disagreement regarding whether punishments should fit the crime or the criminal, but what may have been the case over time is that punishments have fit the needs of society. Punishments, like all other aspects of society, have been altered with changes in the times. Six general methods of punishment are presented here: banishment, capital punishment, corporal punishment, economic punishment, incarceration, and other forms of intervention.

Banishment

Banishment, the exclusion of an offender from the social group, has existed since the earliest times. Banishment as a response to deviance has been utilized by nations, churches, schools, tribes, families, and most social groupings. In people's early existence the social group was essential to survival, and exclusion meant near certain death.

Banishment, which has been referred to as a social death, may be the harshest and most painful punishment devised by the human race. Banishment was common in classical Greece and Rome. Frequently citizens of that time were given a choice between death and banishment. It has been reported that Socrates chose to drink a cup of hemlock rather than accept banishment from his country.

The British developed a sophisticated form of banishment known as penal transportation. This practice was utilized from 1597 until the 1850s. During this time thousands of convicted criminals were placed on ships and sent to colonial America and Australia. The criminals were given fixed sentences which ranged from a few years to life. Upon their arrival in the colonies, the criminals were often forced into hard labor for the duration of their sentences.[16]

The British have not been alone in the use of banishment. The French operated penal colonies on Corsica and in French Guinea. It was not until 1953 that the infamous Devil's Island was finally closed. As detailed in such works as the *Gulag Archipelago* by Alexander Solzhenitsyn, the banishment of prisoners to remote penal camps is an ongoing process in the former Soviet Union.

Although not as widespread a practice as in earlier times, banishment should be recognized as a major method in the history of punishment. It should also be noted that the concept of isolating or expelling criminal offenders from society underlies the modern practice of imprisonment. A strong argument can be made for viewing prisons as the penal colonies of the twentieth century.

Although banishment was a common practice for centuries, it is not generally considered a viable alternative today. The "shrinking world" has made it increasingly difficult to use banishment as a form of punishment because we no longer have a frontier to send the offenders to. Discounting, for the time being, the possibilities of future outer space alternatives, it would appear that banishment will be beyond the range of available punishment options for some years to come.

Capital Punishment

Along with banishment, the most common punishment in earlier times was death. **Capital punishment,** the judicially ordered execution of convicted criminals, has been used in varying degrees throughout history. One example of the widespread use of capital punishment occurred in England in the early 1800s. Under English law of that period, over 350 crimes ranging from petty theft to murder were punishable by death.

Although death penalty provisions are included in the statutes of over 70 percent of the jurisdictions in the United States, there has been only limited use of this punishment in recent years. There was, in fact, a ten-year moratorium on executions in the United States from 1967–1977. This moratorium was brought about by changing public attitudes, the repeal of some states' death penalties, and the landmark 1972 United States Supreme Court case *Furman v. Georgia,* which struck down all of the then existing state capital punishment statutes. Subsequent court decisions and the enactment of new state legislation have reinstated capital punishment as a sentencing option in many states. The issue of capital punishment is discussed in chapter 15.

Corporal Punishment

Corporal punishment has tended to be symbolic in nature. Punishments that inflict physical pain and/or mutilation are considered to be corporal in nature. Frequently these punishments were designed to symbolize the individual's crime, that is, scandalmongers had their lips sewn together and thieves had their hands cut off. Public whippings and the use of pillories were designed to both punish the offender and serve notice to the public of the ramifications of crime.

Corporal punishment is not a legally authorized sanction in our criminal justice system today. To assume that abuses of authority do not at times lead to such acts of violence against offenders would be erroneous, but they are neither legal nor socially acceptable by current standards.

Economic Punishment

Economic punishment has been utilized for centuries. It was common practice in both ancient Greece and Rome to confiscate the property of convicted criminals. The extensive use of fines was also common in these ancient governments. For centuries, the practice of restitution (repaying victims of crime for their losses) has been recognized as an appropriate response to crime.

There are highly developed systems of fines in the United States and most other nations today for those convicted of criminal activity. The high cost of professional legal counsel needed to defend oneself in a criminal case poses an unofficial form of economic punishment for some accused criminals.

The loss of income resulting from criminal prosecution can be considered a form of economic punishment. Individuals convicted of crime may well find their job, property, and other assets in jeopardy. Given the importance our society places on money and other material possessions, economic sanctions seem to offer an effective form of punishment, particularly for misdemeanors and other less serious crimes. These economic punishments have limitations, however, when one considers the number of indigent offenders who cannot pay even moderate fines or afford to hire their own counsel. There are also the super rich for whom a sizable fine by normal standards represents only a nuisance payment. There are limitations to the effectiveness of these sanctions for it must be recognized that the threat of economic deprivation only operates on those who have economic resources.

Incarceration

Incarceration of convicted criminals in prisons and jails is considered a relatively recent addition to the range of punishments available to societies. Although the church maintained punishment cells in many European monasteries from 1200 through the 1600s, the widespread use of incarceration as a form of punishment is associated with the United States.[17] During the late eighteenth century, the Quakers led a correctional reform movement in the United States. As a keystone, they used imprisonment under decent conditions as a substitute for the corporal and capital punishments prescribed at that time by the harsh Anglican codes.[18] The incarceration of convicted criminals has remained a major form of punishment in the United States and most of the nations of the world.

Deterrence

The end of punishment ". . . is no other than to prevent the criminal from doing further injury to society and to prevent others from committing the life offense."[19] Beccaria equated deterrence with prevention. His view of free will led him to believe that laws clearly proscribed with known and certain punishment work to deter future crime.

Incarceration is a major form of punishment in the United States and most nations of the world.

© James L. Shaffer

Deterrence, as it is now recognized, is a far more complex concept than outlined by Beccaria. Deterrence could occur in two basic ways: "through reinforcing the anti-criminal values of society and changing criminals into non-criminals."[20] When considering the impact of a particular punishment on future criminal activity, the term deterrent effect is referred to. "The deterrent effect of a particular legal threat is the total number of threatened behaviors it prevents."[21] Punishment as a deterrent can be directed at either criminal or non-criminal citizens and can be measured by the number of crimes it prevents. Having established these basic principles of deterrence, let's examine different types of deterrence.

Simple Deterrence

Simple or direct deterrence is similar to the early notion of Beccaria. In the example of simple deterrence, a citizen refrains from committing an offense because the pleasure that could be obtained would be more than offset by the risk of great unpleasantness communicated through the law. This simple deterrence applies equally to both the previously criminal and non-criminal citizen.

General Deterrence

General deterrence refers to the effect that the threat of punishment has in inducing citizens to refrain from illegal activity. The impact that the threat of law has on all individuals other than those who are being punished is general deterrence. A citizen who refrains from criminal activity for fear of incarceration or other punishment is said to have been deterred by the law.

Special Deterrence

Special or individual deterrence applies specifically to the convicted offender who is being punished. The offender who refrains from future criminal activity because of the personal experience of an earlier punishment is said to have been individually deterred. Any single act of punishment could provide both general and special deterrence effects. The consideration of the preventive capability of punishment does not stop at this point.

Marginal Deterrence

The purported measures to prevent criminal activity by both criminals and non-criminals have been identified. But what of the differences between the types of punishments to be utilized? Marginal deterrence refers to ". . . reducing the rate of threatened behavior below that experienced under lesser penalty."[22] The difference in the prevention of crime that would result from changing the punishment for first-degree armed robbery from three years in prison to ten years in prison would be the marginal deterrent effect. The marginal deterrent effect of greater or lesser penalties could be applied to either special or general deterrence. A related concept is that of partial deterrence, which attempts to account not for the prevention of crime, so much, but for the reduction in the level of severity of the future offense. If, for example, it could be demonstrated that many citizens reduced their maximum driving speed from seventy-five miles per hour to sixty-five miles per hour because of the imposition of the fifty-five miles per hour speed limit, then a partial deterrent effect would have existed. Full deterrence would have been measured by those complying with the fifty-five miles per hour speed limit, while partial deterrence accounts for reductions in the level of criminal activity that fall short of full compliance with the law.

The Deterrent Effect of Punishment

The arguments supporting and questioning the deterrent effect of punishment have gone on for generations and in all likelihood will continue for the foreseeable future. The following is a brief review of some of the salient points in the debate over deterrence.

In Support of Deterrence

When the "get tough" or "law and order" stance becomes the political, social issues of the day, it is possible to hear a great deal of rhetoric in support of stricter law enforcement and greater punitive measures to be taken in response to crime. One basic premise supporting the concept of deterrence is that individuals engage in a rapid and rational consideration of pain and pleasure prior to taking a certain action. This notion of free will decision-making is central to the concept of deterrence, but there are also additional rationales that support this approach to punishment.

The threat of punishment can be viewed as a teaching or socializing tool. The association of negative consequences with certain behaviors can provide an attention-getting mechanism to aid in the formation of socially acceptable behavior and habits. The threat of punishment, it can be argued, helps to build respect for the law through the assignment of penalties for certain behavior. The penalties may help to reduce the temptation to engage in criminal activity and can assist the individual in developing a personal rationale for conformity.

For many individuals, not endorsing the concept of deterrence would be admitting that there was no solution to the problem of crime. Franklin Zimring said, "It is difficult for the official to feel or announce that the problem is either insoluble or that the solution lies outside his sphere of influence."[23] Emile Durkheim argued that unpunished deviance tended to demoralize the conformist.[24] The notion of deterrence provides non-criminal, law-abiding citizens with a vicarious reward for their conforming behavior and a sense of rightness in that socially unacceptable behavior is punished.

Questions about Deterrence

Returning again to the motivation and thought process of the criminal prior to the commission of a crime, it can be questioned whether planning or impulse controls behavior. Has the criminal planned out his actions and their consequences, or does he act largely on impulse without giving full consideration to the implication of his behavior? "Observation has established that deviant subcultures place less emphasis on future events than representative members of the middle class."[25] Are criminals less future oriented than non-criminals? If in fact the typical criminal feels he has little to look forward to in life, how then can a future punishment be a deterrent? What of the differences between types of crimes and deterrence? It can be argued that planning and rational decision-making more frequently occur prior to instrumental crimes than in the case of expressive crimes.[26] Instrumental crimes are those that result in material gain, such as theft, fraud, or forgery. Expressive crimes or crimes of passion are instances in which the act itself is what the potential criminal wants, such as murder, rape, or assault. If, in fact, the deterrent effect of punishment is greatest in the case of instrumental crime, then the present structuring of the legal codes in most jurisdictions seem out of order. Most penal codes provide far greater penalties for expressive, potentially less deterrable crimes, than they do for crimes of an instrumental nature.

Another serious consideration is that of objective probability.[27] Does the potential criminal believe his chances of apprehension are great or small? The effect of deterrence, even if harsh penalties are mandated, may be negated if the criminal believes that his apprehension and a conviction are unlikely. For there to be a deterrent effect, there must be both knowledge of and an appreciation for the punishments prescribed for certain behavior.

It can be argued that many criminals possess an unrealistic world view, or at least one that varies greatly from the view held by a majority of citizens. The criminal may have realistically determined that his likelihood of apprehension for criminal behavior is limited and, because he has little stake in conformity, that there is no reason to fear the threat of legal sanctions. After adding to this world view the aggravating influences of alcohol and other chemical substances, it is possible to raise serious questions regarding the true deterrent effect of punishment.

Measuring Deterrences

The deterrent effect of punishment is difficult to measure accurately. There is a vast array of variables that influence human behavior, and any attempt to assign a causal relationship between a legal sanction and the actions of humans is fraught with problems. It has been noted that new deterrent programs are generally tried during periods of high crime. Times of high crime are generally peak intervals followed by relative periods of calm. Who is to say that harsher penalties enacted during high-crime periods had any effect in reducing violations? But on the other hand, it can be argued that crime would have continued to increase had not the new measure been introduced.

It is doubtful whether legislative penalties for major crimes have much impact as a deterrent. Yet, no one has been able to determine how many conformists would have become deviants if there was not the fear of punishment. The debate over the deterrent effect of punishment will probably never be decided based upon statistical studies because the support for or opposition to deterrence is predicated more on one's world view than on measurement devices. One major explanation for why corrections' policy and thought have continued to swing from one extreme to another on the continuum is that decisions on basic issues, such as the deterrent effect of crime, are made largely on a belief, rather than on a logical and scientific basis.

Summary

Various correctional activities are directed toward protecting society, punishing offenders, and changing offenders' behavior. Different approaches have been utilized in the past as well as during current times. Varying philosophies are premised on certain assumptions embodied in the concepts of free will or determinism.

In colonial America, criminal penalties included corporal and capital punishment rather than a substantial period of incarceration. However, changing attitudes lead to the development of prisons, where offenders would spend extended periods of confinement, an approach relied on heavily during present times.

Varying rationales and models have formed the underlying framework for correctional operations. Specific examples include retribution and a justice model, incapacitation and a custodial model, reintegration and a least restrictive alternative model, and resocialization and a rehabilitation model.

The deterrent effect of punishment remains a complex issue. Numerous supporters as well as detractors have debated this issue for a long time but with no lasting resolution. Much of the discussion has been based on an emotional rather than a scientific basis.

Discussion Topics

1. Are your notions of human behavior closer to the concept of free will or determinism? Explain.
2. Which of the four correctional models most nearly reflect your views of corrections? Why?
3. To what extent do you think punishment acts as a deterrent? Why?
4. Does isolating offenders in prison away from the community do more harm or good for that person in the long run? Explain.
5. Should fines be based on ability to pay? Are fines a deterrent?

Additional Readings

American Friends Service Committee. *Struggle for Justice.* New York: Hill and Wang, 1971.

Beccaria, Cesare. *On Crime and Punishment.* Translated by Henry Paolucci. Indianapolis: Bobbs-Merrill Educational Publishing, 1977.

Durham, Alexis M., III. ''Newgate of Connecticut: Origins and Early Days of an American Prison.'' *Justice Quarterly.* March, 1989.

Rothman, David J. *The Discovery of the Asylum.* Boston: Little, Brown and Company, 1971.

Notes

1. David J. Rothman, *The Discovery of the Asylum* (Little, Brown and Company, Boston, 1971), 17.
2. Alexis M. Durham III, ''Newgate of Connecticut: Origins and Early Days of an American Prison,'' *Justice Quarterly* vol. 6, no. 1 (March 1989): 92.
3. Maxwell Jones, *The Therapeutic Community: A New Treatment Method in Psychiatry* (New York: Basic Books, 1953).
4. David Dressler, *Practice and Theory of Probation and Parole,* 2d ed. (New York: Columbia University Press, 1969), 6.
5. Ibid.
6. Cesare B. Beccaria, *An Essay on Crimes and Punishments* (London: J. Alman, 1767), 165.
7. Sidney L. Barnes, ''A Review of the Philosophical Theoretical and Practical Aspects of Legalized Punishment,'' in *Penology: The Evolution of Corrections in America,* ed. George C. Killinger et al. (St. Paul, MN: West Publishing Co., 1797), 31.
8. Dressler, *Practice and Theory of Probation and Parole,* 8.
9. Stephen Schafer, *Theories in Criminology* (New York: Random House, 1969), 82.
10. Louis B. Carney, *Corrections: Treatment and Philosophy* (Englewood Cliffs, NJ: Prentice–Hall, Inc., 1970), 92.
11. Barnes, ''A Review of the Philosophical Theoretical and Practical Aspects of Legalized Punishment,'' 26.
12. David Fogel, *We are the Living Proof* (Cincinnati, OH: Anderson Publishing Company, 1975), 183–184.
13. Oliver Wendell Holmes, *The Common Law,* ed. Mark De Wolfe Howe (Boston: Little, Brown, 1963), 40.

14. Andrew von Hirsch, *Doing Time* (New York: Hill and Wang, 1976), 3.
15. Jerome Hall, *General Principles of Criminal Law,* 2d ed. (New York, 1960), 310.
16. Paul W. Tappan, *Crime, Justice and Corrections* (New York: McGraw Hill, 1960), 103–8.
17. Torsten Eriksson, *The Reformers* (New York: Elsevier, 1976), 6.
18. Edward Eldefonso, *Issues in Corrections* (Beverly Hills: Glencoe Press, 1974), 78.
19. Beccaria, *An Essay on Crimes and Punishments,* 165.
20. Barnes, "A Review of the Philosophical Theoretical and Practical Aspects of Legalized Punishment," 39.
21. Franklin E. Zimring, *Perspectives on Deterrence* (Rockville, MD: National Institute of Mental Health, 1973), 2.
22. Ibid., 3.
23. Zimring, *Perspectives on Deterrence,* 9.
24. Emile Durkheim, *The Division of Labor in Society* (New York: 1947), 365.
25. Ibid., 35.
26. Ibid., 54.
27. Ibid., 67.

Chapter 4

Sentencing

Learning Objectives

After studying this chapter students should be able to:

1. Provide a definition of sentencing.
2. Understand the importance and impact of a sentencing decision.
3. Identify and describe various types of sentences.
4. Recognize what constitutes disparity in sentencing.
5. Explain the purpose and application of sentencing guidelines.

Key Terms

Aggravating circumstances
Community-based sentences
Community service
Criminal history category
Determinate sentence
Federal sentencing guidelines
Fines
Incarceration
Incarceration sentence
Indeterminate sentence
Nonincarceration sentence
Mandatory sentence
Mitigating circumstances
Offense level
Presentence investigation report
Presumptive sentence
Probation
Restitution
Sentencing
Sentencing disparity
Sentencing guidelines
Shock incarceration

Sentencing

Sentencing is a formal means by which society deals with offenders guilty of a crime. A sentence is an authorized judicial decision that places some degree of penalty on a guilty person. The responsibility for administering this judicial decision is placed with corrections. In this section we examine some ways in which sentencing decisions have an impact on corrections and discuss some issues surrounding sentencing.

Sentencing involves making a choice among many alternatives, such as probation, short-term incarceration, long-term incarceration, a fine, or imposing no penalty. Making this choice also involves making a philosophical choice. Probably more than at any other point, sentencing decisions reflect political and philosophical viewpoints. Table 4.1 indicates the proportion of felony sentences imposed by state courts that resulted in incarceration or probation.

Sentencing decisions are usually made by a judge within applicable provisions of the law (in a few jurisdictions, juries make this decision). Judges customarily base a sentencing decision on information contained in a **presentence investigation report** (PSI). Such reports contain a variety of information, including the seriousness of the crime, any history of criminal activity, any substance abuse, a threat posed to community safety, and any mitigating or aggravating circumstances. **Aggravating circumstances,** such as being on parole when committing a new crime or torturing a victim, act to increase the seriousness of the offense and could result in a more severe penalty. **Mitigating circumstances,** such as cooperating with police or mental illness, act to decrease seriousness and possibly lead to a less severe sentence. Preparing PSI reports comprises an important part of the work for probation departments and probation officers and is discussed with more detail in chapter 11, ''Probation.''

You will recall that several rationales and models of corrections have been discussed. Sometimes different and conflicting viewpoints emerge. Although different, these approaches share a general goal. However, they differ in means and methods by which to accomplish this goal.

Some approaches, such as retribution, are more likely to place greater emphasis on lengthy incarceration. On the other hand, the idea of reintegration de-emphasizes incarceration and considers probation and community-based programs as being more effective. Therefore, the choice of sentence authorized by the court carries a significant message. This message carries an implied viewpoint concerning the purpose society and the courts want corrections to fulfill in penalizing guilty persons. For this reason, the impact of sentencing on corrections is considerable since corrections is responsible for implementing the particular decision made.

Because of differing philosophies and discretionary justice, the purpose of corrections is not always agreed upon. Disagreement may be found between judges and correctional personnel, among judges, or among correctional personnel. For example, some hold the view that corrections should be concerned mainly with retribution, while others argue for a rehabilitative role.

Historically, attitudes toward corrections constantly shift. At times, thinking has been punitive and geared mainly to strict punishment. At other times, the ideal of rehabilitation emerged as a dominant theme. It is safe to say that sentencing will always be affected by prevailing philosophical views and political pressures, which in turn carry implications for corrections.

It is obvious that major changes in sentencing practices bring considerable change in the operation of corrections. A significant decrease in the number of defendants sentenced to probation could cause a substantial increase in prison populations. Similarly, a greater use of fines and community corrections (which some authorities recommend) could decrease prison populations. Such changes would carry consequences in terms of prison space required, personnel needs such as probation officers, and type of budgetary expenses. For example, legislators in recent years have enacted a variety of new laws and sentences in response to problems posed by substance abuse in our society. Coupled with existing laws, these changes more than any other factor have contributed to dramatic increases in the nation's prison population.

Sentences are often differentiated by whether or not incarceration results. Table 4.2 details the average amount of time various types of convicted state offenders spend in prison and on probation.

Nonincarceration Sentences

Legislative bodies have authorized a variety of sentences that the courts may give to offenders. Some sentences result in an institutional commitment (incarceration) whereas others do not. The types of **nonincarceration sentences** include:

- **Fines:** an economic penalty that requires the offender to pay a specific sum of money within the limit set by law. Fines are often imposed in addition to probation, as an alternative to incarceration, and sometimes in addition to incarceration.
- **Community service:** the requirement that the offender provide a specific number of hours of public service work, such as collecting trash in parks or other public facilities.
- **Restitution:** the requirement that the offender provide financial remuneration for the losses incurred by the victim.
- **Probation:** the sentencing of an offender to community supervision by a probation agency, often the result of suspending a sentence of confinement. Such supervision normally requires adherence to specific rules of conduct while the offender is in the community. If violated, a sentencing judge may impose a sentence of confinement. Probation is the most widely used correctional disposition in the United States.
- Other **community-based sentences:** a variety of programs were developed and enhanced beginning in the 1970s, to involve the community and provide services without incarceration. Many of these are job skills training and substance abuse programs.

Table 4.1

Types of Felony Sentences Imposed by State Courts

Conviction Offense	Total	Percent of Felons Sentenced to: Incarceration Total	Prison	Jail	Probation
All offenses	100%	71%	46%	25%	29%
Violent offenses	100%	80%	59%	21%	20%
Murder	100	95	91	4	5
Rape	100	86	67	19	14
Robbery	100	90	73	17	10
Aggravated assault	100	72	45	27	28
Other violent	100	67	42	25	33
Property offenses	100%	66%	44%	22%	34%
Burglary	100	75	54	21	25
Larceny	100	65	40	25	35
Fraud	100	53	33	20	47
Drug offenses	100%	72%	43%	29%	28%
Possession	100	64	35	29	36
Trafficking	100	77	49	28	23

Source: Adapted from Bureau of Justice Statistics, National Judicial Reporting Program, 1990 *(Washington, D.C.: Government Printing Office, December, 1993), p. 6.*

Table 4.2

Average Length by Type of Felony Sentence Imposed by State Courts

Conviction Offense	Incarceration: Total	Prison	Jail	Probation
	Average Length of Sentence			
All offenses	52 mo	75 mo	8 mo	42 mo
Violent offenses	91 mo	119 mo	10 mo	46 mo
Murder	233	243	37	67
Rape	128	160	11	61
Robbery	97	115	12	50
Aggravated assault	52	78	9	43
Other violent	57	85	7	45
Property offenses	47 mo	65 mo	8 mo	44 mo
Burglary	61	80	9	48
Larceny	33	49	7	41
Fraud	40	58	6	43
Drug offenses	44 mo	66 mo	9 mo	42 mo
Possession	30	49	6	39
Trafficking	52	74	10	44

Source: Adapted from Bureau of Justice Statistics, Census of State and Federal Facilities, 1990 *(Washington, D.C.: Government Printing Office, December, 1993), p. 7.*

Incarceration Sentences

Sentences imposing a period of institutional commitment display marked variation in who controls the sentence selection and the length of incarceration. This results in diverse sentencing structures and a lack of standard definitions and terms. The following are types of **incarceration sentences:**

- Split sentences and **shock incarceration:** a penalty that explicitly requires the convicted person to serve a period of confinement in a local, state, or federal facility (the shock), followed by a period of probation. This penalty attempts to combine the use of community supervision with a short incarceration experience.
- **Incarceration:** the confinement of a convicted criminal in a federal or state prison or a local jail to serve a court-imposed sentence. Custody is usually within a jail, administered locally, or in a prison operated by the state or the federal government. In many states offenders sentenced to less than one year are held in a jail; those sentenced to longer terms are committed to the state prison. Increasing numbers are being held on death row.

Sentences imposing incarceration may be broadly classified as (1) nondiscretionary, (2) limited discretionary, and (3) discretionary. A nondiscretionary sentence makes a prison term a **mandatory sentence;** the incarceration is specified by statutory law and must be given upon conviction. Judges have no discretionary power to grant probation or suspend sentences, and in some jurisdictions, parole is not possible. This manner of sentencing is found in nearly all states for certain offenses.

Limited discretionary sentences usually allow for a suspended sentence, probation, or if a judge chooses, incarceration, a fixed amount of time that usually may be reduced by parole and/or good behavior time. This classification includes **determinate sentences.** A **presumptive sentence** also permits limited discretion. A judge is required to impose a sentence whose length is set by law for each offense or category of offense. When there are mitigating or aggravating circumstances, a judge may shorten or lengthen a period of incarceration within a specified minimum and maximum amount of time.

Discretionary sentences provide the most latitude. Judges may suspend sentence, grant probation, or if incarceration is chosen, fix the period of time to be served within a range from a low minimum to a high maximum. Commonly called **indeterminate sentences,** these sentences often give parole authorities substantial control over the length of time served.

Disparity in Sentencing

When offenders guilty of the same crime and with similar criminal histories are given different sentences that cannot be justified based on special circumstances, **sentencing disparity** exists. The existence of unjustified disparity in sentencing is generally recognized and has been well documented.[1]

Sentencing decisions that change the lives of offenders are handed down in a formal setting of courtrooms, such as this, throughout the nation.

Sentencing disparity exists for two main reasons. The first arises from the substantial amount of discretion permitted to officials of the criminal justice system. The second stems from a large array of personnel and institutions that have an impact on what an eventual sentence might be.

Legislatures affect sentencing by specifying penalties—usually fines and length of incarceration—for violating various laws they enact. Sometimes a comparison of penalties has revealed significant disparities. Although corrective measures are ongoing, penalties have often been prescribed without considering the relative seriousness of the offense and without considering penalties enacted for other offenses. As a result, legislative enactment has actually created disparity in some instances. An historical example of such a disparity was provided by the following account:

> . . . a recent legislatively sponsored inquiry revealed the following rather shocking provisions: One convicted of first degree murder must serve 10 years before he first becomes eligible for parole; one convicted of a lesser degree of murder may be forced to serve 15 years or more. Destruction of a house with fire is punishable by a maximum of 20 years; destruction of the same house with explosives carries a 10-year maximum (In Iowa burning of an empty isolated dwelling may lead to a 20-year sentence while burning of a church or school carries only a 10-year maximum). The Model Penal Code inquiry into burglary statutes revealed that in California a boy who broke into a passenger car to steal the contents of the glove compartment subjected himself to a maximum of 15 years; if he stole the entire car, he could only be sentenced to 10.[2]

In most jurisdictions, the judge is a powerful individual in terms of sentencing decisions. Often this power is virtually unreviewed, although appeals and parole decisions may provide a check in some cases. Although bound in a few instances to pronounce a mandatory sentence, the law usually provides considerable sentencing discretion but with little criteria or guidance. Thus, inconsistencies are likely to arise at this level. A central fact of importance is who pronounces the sentence. It has been stated of judges that:

> They are allowed to impose their beliefs, or biases, or prejudices, or reactions to particular defendants within the very wide margins afforded by our present statutory range of sentences. A natural result of this "nonsystem" is that similar offenders guilty of similar crimes commonly receive grossly disparate sentences.[3]

During one research effort a number of federal judges were given identical presentence reports for five defendants and asked to indicate what sentence they would impose if they were the sentencing judge. The results showed great disparity and lack of agreement as to what constituted an appropriate sentence. There were major differences regarding amounts of fines, whether or not to imprison the defendant, and lengths of probation. For example, in a case of bank robbery, one judge issued a five-year prison sentence with no fine while another imposed an eighteen-year prison term and a $5,000 fine. In another case involving theft and possession of stolen goods, one judge imposed four years of probation, and another judge issued a sentence of seven and one-half years of imprisonment.[4]

Correctional officials and parole boards affect sentencing, mainly the length of incarceration, by deciding on release dates. Although parole boards have no control over the decision on whether to incarcerate a defendant, they have great control regarding the length of confinement. Many sentences are of an indeterminate nature and parole boards have the authority to decide when to release individuals. Again, much discretion is allowed and these decisions involving whether and when to release offenders are generally not reviewed. Because guidelines and criteria by which parole decisions are made have been lacking, disparity has also arisen at this stage in the criminal justice process.

Changes in Sentencing

Given this history of disparity in sentencing, changes have been made by some jurisdictions in an effort to reduce this problem (changes in some other jurisdictions have not reduced problems of disparity). In a few instances, parole has been either abolished or changed to a supervisory function only, with no discretionary power regarding the length of imprisonment. Although approaches have varied, emphasis has been placed on trying to reduce the degree and range of discretion by moving away from indeterminate sentencing toward some form of sentencing structure with accompanying guidelines and criteria. No better example of this type of approach exists than the **sentencing guidelines** developed by the United States Sentencing Commission and authorized for use by the United States federal courts.

Federal Sentencing Guidelines

On November 1, 1987, the **federal sentencing guidelines** put into place a determinate sentencing system. Paroling authority was eliminated. An essential premise is that under previous law, which granted judges broad discretion and provided for release by parole, fragmented sentencing resulted. Sentencing was characterized by unwarranted disparity, a lack of fairness, and uncertainty. Judges previously imposed a wide variety of sentences on similar offenders convicted of similar crimes committed under similar circumstances. Moreover, convicted defendants did not know how long a prison term would actually be until the day of release. The guideline system seeks to cure these ills by generally requiring judges to sentence according to the guidelines and by eliminating parole so that offenders will now serve ''real time'' sentences.[5]

Stated policy objectives are to seek: (1) honesty in sentencing; (2) uniformity in sentencing; (3) proportionality in sentencing. The matter of honesty is approached by the abolition of parole and by making the court-imposed sentence the actual sentence that the offender will serve, thus avoiding confusion and implicit deception. A goal of uniformity is sought by narrowing the wide disparity in sentences imposed by different federal courts for similar criminal conduct by similar offenders. Proportionality is sought by imposing appropriately different sentences for criminal conduct of different severity.[6]

Although somewhat complex and technical, offenders are sentenced by taking into account two essential factors: (1) the **offense level** (how serious the offense is), and (2) the **criminal history category.** These two factors are utilized to develop a sentencing table or grid showing the months of imprisonment to be applied as provided in table 4.3.

Offenses are ranked by a point system with an increasing number of points as the seriousness increases. An example of a few selected crimes and the corresponding offense level follow.

Offense	*Offense Level*
First-degree murder	43
Assault	20
Sexual abuse	27
Air piracy	38
Drug offenses	6 to 43 (depending on specific offense)
Theft	4 to 17 (depending on amount)
Burglary	17 to 23 (depending on amount)
Robbery	18 to 24 (depending on amount)
Arson	6 to 24 (depending upon risk of death or injury to others)

Table 4.3

Federal Sentencing Table

Offense Level	Criminal History Category I	II	III	IV	V	VI
1	0–1	0–2	0–3	0–4	0–5	0–6
2	0–2	0–3	0–4	0–5	0–6	1–7
3	0–3	0–4	0–5	0–6	2–8	3–9
4	0–4	0–5	0–6	2–8	4–10	6–12
5	0–5	0–6	1–7	4–10	6–12	9–15
6	0–6	1–7	2–8	6–12	9–15	12–18
7	1–7	2–8	4–10	8–14	12–18	15–21
8	2–8	4–10	6–12	10–16	15–21	18–24
9	4–10	6–12	8–14	12–18	18–24	21–27
10	6–12	8–14	10–16	15–21	21–27	24–30
11	8–14	10–16	12–18	18–24	24–30	27–33
12	10–16	12–18	15–21	21–27	27–33	30–37
13	12–18	15–21	18–24	24–30	30–37	33–41
14	15–21	18–24	21–27	27–33	33–41	37–46
15	18–24	21–27	24–30	30–37	37–46	41–51
16	21–27	24–30	27–33	33–41	41–51	46–57
17	24–30	27–33	30–37	37–46	46–57	51–63
18	27–33	30–37	33–41	41–51	51–63	57–71
19	30–37	33–41	37–46	46–57	57–71	63–78
20	33–41	37–46	41–51	51–63	63–78	70–87
21	37–46	41–51	46–57	57–71	70–87	77–96
22	41–51	46–57	51–63	63–78	77–96	84–105
23	46–57	51–63	57–71	70–87	84–105	92–115
24	51–63	57–71	63–78	77–96	92–115	100–125
25	57–71	63–78	70–87	84–105	100–125	110–137
26	63–78	70–87	78–97	92–115	110–137	120–150
27	70–87	78–97	87–108	100–125	120–150	130–162
28	78–97	87–108	97–121	110–137	130–162	140–175
29	87–108	97–121	108–135	121–151	140–175	151–188
30	97–121	108–135	121–151	135–168	151–188	168–210
31	108–135	121–151	135–168	151–188	168–210	188–235
32	121–151	135–168	151–188	168–210	188–235	210–262
33	135–168	151–188	168–210	188–235	210–262	235–293
34	151–188	168–210	188–235	210–262	235–293	262–327
35	168–210	188–235	210–262	235–293	262–327	292–365
36	188–235	210–262	235–293	262–327	292–365	324–405
37	210–262	235–293	262–327	292–365	324–405	360–life
38	235–293	262–327	292–365	324–405	360–life	360–life
39	262–327	292–365	324–405	360–life	360–life	360–life
40	292–365	324–405	360–life	360–life	360–life	360–life
41	324–405	360–life	360–life	360–life	360–life	360–life
42	360–life	360–life	360–life	360–life	360–life	360–life
43	life	life	life	life	life	life

Source: United States Sentencing Commission, Sentencing Guidelines *(Washington, D.C.: October, 1987), p. 5.2.*

In a similar fashion, an offender is placed in a criminal history category which reflects past criminal behavior. The total points from A through E determine a criminal history category in the sentencing table.

A. Add three points for each prior sentence of imprisonment exceeding one year and one month.
B. Add two points for each prior sentence of imprisonment of at least sixty days not counted in A.
C. Add one point for each prior sentence not included in A or B, up to a total of four points for this item.
D. Add two points if the defendant committed the instant offense while under any criminal justice sentence, including probation, parole, supervised release, imprisonment, work release, or escape status.
E. Add two points if the defendant committed the instant offense less than two years after release from imprisonment on a sentence counted under A or B. If two points are added for item D, add only one point for this item.

The offense level (1–43) forms the vertical axis of the sentencing table. The criminal history category (I–VI) forms the horizontal axis of the table. The intersection of the offense level and criminal history category displays the guideline range in months of imprisonment. "Life" means life imprisonment. For example, the guideline range applicable to a defendant with an offense level of 15 and a criminal history category of III is twenty-four to thirty months of imprisonment.

Judges are to sentence within these guidelines, which significantly limits the amount of discretion they had prior to the guidelines' adoption. A judge must provide written justification for any sentence that is different from the guideline range. A sentence less severe can be appealed by the prosecution and a more severe sentence can be appealed by the defense.

The guidelines have been challenged on the basis of limiting the discretion of sentencing judges and by not permitting the individualizing of sentences. However, the constitutionality of these guidelines has been upheld. In 1989, *Mistretta v. United States,* the United States Supreme Court has ruled that these guidelines did not violate the division of power between the legislature and the judiciary and that they did not compromise the independence of the judiciary.[7] However, implementing the guidelines will require continued effort, and future cases will serve to clarify their operations.[8] Some judges and critics have also felt that this approach is too rigid and sometimes results in sentences that are far too severe.

State Sentencing Laws[9]

Sentencing laws enacted at the state level represent one of the most diverse aspects of criminal justice systems across the United States. Every state has a unique set of sentencing laws that undergo periodic and often substantial change. This results in a diversity that complicates classification of sentencing systems since approaches often blend into a hybrid system that overlaps boundaries between categories.

All states provide the option of nonincarceration sentences except in certain instances when a prison term is required (mandatory sentencing). With regard to incarceration, many states have adopted some form of determinate sentence, but most states still use indeterminate sentencing. Some states, following 1976 legislation in Maine have adopted entirely or predominantly determinate sentencing (California, Connecticut, Florida, Illinois, Indiana, Maine, Minnesota, New Mexico, North Carolina, Washington).

Mandatory sentencing laws for certain crimes are found in nearly all states. In twenty-five states incarceration is required for certain repeat offenders. In thirty states imprisonment is mandatory if a firearm was involved in committing a crime. In forty-five states prison is mandatory for specific offenses of categories of offenses, such as violent crimes and drug trafficking. In some instances habitual offenders must serve a mandatory minimum prison term.

Since beginning in the 1970s, about one-quarter of the states use sentencing guidelines. Typically, a range of sentences from probation through amount of time in prison is prescribed by employing a grid that considers offense severity and criminal history. For example, a prescribed sentence for a certain offense might be probation if the offender has no criminal history, a shorter prison term if an offender had one prior conviction, and an increasingly longer prison term as criminal history becomes more extensive. Sentencing guidelines at the state level are written into statutes in Florida, Louisiana, Maryland, Minnesota, New Jersey, Ohio, Pennsylvania, and Tennessee; are used statewide but not mandated by law in Utah; and are applied selectively in Massachusetts, Michigan, Rhode Island, and Wisconsin.

Changes in sentencing laws alter the manner in which state correctional systems operate. Determinate and mandatory sentences modify (or eliminate in some instances) parole decision-making. Some states develop parole guidelines, thereby reducing the use of ''good time'' and the use of discretionary decisions in determining release dates. For example, in one ten-year period, changes in sentencing laws and reduced release authorization lowered the percent of persons released from state prisons by a parole board decision from 72 percent in 1977 to 41 percent in 1987.[10] The number and category of offenders under correctional supervision are obviously influenced by sentencing laws. Although disparity may have been reduced, changes in sentencing have lead to an unprecedented expansion of prison populations and overcrowding.

Changes in sentencing laws may also bring about new administrative requirements. Activities such as collection of victim restitution funds, operation of community service programs, levying fees for probation supervision, and other services change traditional correctional practices.

Summary

Sentencing is a major process by which society deals with those found guilty of a crime. Corrections is that component of the criminal justice system responsible for implementing sentencing decisions. Differing philosophies and discretionary justice account for a wide range of sentencing practices and periodic changes.

Sentences can be generally categorized into those that require incarceration and those that do not. Sentences imposing incarceration are often classified as mandatory, determinate, and indeterminate.

The existence of well-documented disparity in sentencing has brought change, some of which is in the direction of limiting discretion by developing sentencing guidelines to be followed. A good example is the set of guidelines being used by United States federal courts.

Changes in sentencing and the development of guidelines bring continual controversy and often result in calls for and actual implementation of additional change. Long-term outcome is unclear, especially when considering the impact of crime rates. Recent consequences for corrections have been more clear: an increased number of persons requiring supervision; unprecedented growth in prison populations; and overcrowding of prisons.

Discussion Topics

1. Has anything beneficial been accomplished by recent changes in sentencing laws?
2. What is society trying to express through sentencing?
3. Is disparity in sentencing something our nation should be concerned about?
4. Do federal sentencing guidelines place too much restriction on judges?
5. Should attempts be made to reduce the diversity of sentencing laws between the states?
6. Divide into small groups, read an identical PSI and determine a sentence. Was there agreement or disagreement on the sentences?

Additional Readings

Doble, John et. al. *Punishing Criminals.* New York, New York: The Edna McConnell Clark Foundation, October, 1991.

Frankel, Marvin E. *Criminal Sentences.* New York: Hill and Wang, 1973.

Shane-DuBow, S., A. P. Brown, and E. Olsen. *Sentencing Reform in the United States.* Washington, D.C.: United States Department of Justice, 1985.

Notes

1. See Marvin E. Frankel, *Criminal Sanctions* (New York, N.Y.: Hill and Wang, 1973) and Willard Gaylin, *Partial Justice* (New York, NY: Knopf, 1974).
2. American Bar Association Project on Standards for Criminal Justice, *Standards Relating to Sentencing Alternatives and Procedures* (New York: Office of the Criminal Justice Project, 1968), 49.
3. Pierce O'Donnell, Michael J. Churgin, and Dennis E. Curtis, *Toward a Just and Effective Sentencing System* (New York: Praeger Publishers, 1977), 3.

4. Anthony Partidge and William B. Eldridge, *Second Circuit Sentencing Study: A Report to the Judges of the Second Circuit 9* (1974).
5. U.S. Department of Justice, *Prosecutors Handbook on Sentencing Guidelines* (Washington, D.C.: 1 November 1987), i.
6. United States Sentencing Commission, *Sentencing Guidelines* (Washington, D.C.: October, 1987), 1.2.
7. *Mistretta v. United States,* 109 U.S. 647 (1989).
8. Stephen G. Breyer and Kenneth R. Fainberg, ''The Federal Sentencing Guidelines: A Dialogue,'' *Criminal Law Bulletin* (January/February 1990): 5–37.
9. Adapted from United States Department of Justice, *Report to the Nation on Crime and Justice,* 2d ed. (Washington, D.C.: Bureau of Justice Statistics, March, 1988), 90–91.
10. Ibid.

Unit 2

Incarceration

The most serious and dramatic form of correctional supervision is incarceration. Numerous subjects and management problems associated with shorter-term confinement in local jails to longer-term sentences in state and federal prisons are presented in this group of chapters.

Chapter 5 Jails
Chapter 6 Prison Facilities and Populations
Chapter 7 Inmate Life and Prisoner Rights
Chapter 8 Prison Programs and Management
Chapter 9 Women in Prison

Chapter 5

Jails

Learning Objectives

After studying this chapter students should be able to:

1. Provide a definition for jails.
2. Present a brief history of jails.
3. Describe various characteristics of jail populations.
4. Discuss the volume of activity associated with the nation's jails.
5. Identify needs of jail residents.
6. Recall basic operational procedures found in jails.
7. Identify and discuss a variety of management concerns jail administrators need to be aware of.
8. Identify some conditions that have resulted in some jails being placed under court order to improve.
9. Explain the variety of alternative pretrial release options available.
10. Contrast new generation jails which incorporate direct supervision with older jails.

Key Terms

Alternative release options
Bail
Bondsperson
Chemically dependent
Chronic offenders
Classification
Contraband
Convicted status
Court appearances
Detoxification
Direct supervision
Essential services
Federal Bail Reform Act
Fee system
Financial bond
Incarceration
Inmate rights
Intake
Jail
Jail census
Jail operational procedures
Jail standards
Local control
Lockups
Orientation
Pretrial release program
Programs
Recognizance (ROR)
Regional jail
Release
Socioeconomic characteristics
Special needs
Supervised release
Three generations of jails
Turnkey fee
Unconvicted status
Weekend jail

Overview

Incarceration is to confine someone; to imprison them; to lock them up. The most widely experienced form of incarceration is a stay in **jail.** Whether as a pretrial detainee (an unconvicted suspect) or as a sentenced offender, a total of several million persons are confined each year in the jails of the nation. Incarceration in the local jail represents the first contact, and for some the only contact, with what is commonly considered a part of corrections. Therefore, it is appropriate to begin our examination of incarceration by examining the jail.

Historical Development of Jails

The word jail is Latin in origin and an alternate form of the word, gaol (pronounced the same way), which has a Norman/French origin. Whatever the spelling or linguistic origin, it is clear that places of confinement have existed for centuries. Ancient Egypt, Greece, and Rome—all had designated confinement facilities. It appears that as societies evolve from an extended family or tribal system into a more complex social and political system, there arises a need for local detention. Early confinement areas were characterized by ''unscalable pits, dungeons, suspended cages, and sturdy trees to which prisoners were chained pending trial.''[1]

Jails in England

Rudiments of a criminal justice system emerged under the Anglo-Saxon feudal system in England. The countryside was divided into governmental units known as shires. In each jurisdiction there was one official charged with enforcing the law. This official (shire-reeve or sheriff) was appointed by the crown. The sheriff, acting as the king's or queen's representative, had a variety of law enforcement duties and was also charged with responsibility for maintaining a jail. In 1166, King Henry II required each sheriff to establish a place to secure offenders until the next appearance of the king's court in that area. It should be noted that detention facilities were also operated by municipal and city governments as well as various religious orders.

Two important precedents were established in these early years; jails were the responsibility of local units of government, and were administered by law enforcement personnel. These two features of early English jails remain common elements of present-day United States facilities.

Jails of the sixteenth through the eighteenth centuries were often places of filth and disease. A form of typhus known as jail fever was common in England. Causes of jail fever were linked to the unsanitary conditions found in local jails.

Jails were poorly supported by local governments. Jailers were often appointed for life and earned their living from a **fee system.** This system provided for payment of a fee to the jailer by jail residents or their families to maintain the facility. The fee system often brought abusive practices such as some residents being charged more than others and some being required to pay an additional fee just prior to being released. Jailers sold items such as food, clothing, and toiletries to residents. It was common practice for jailers to hire out jail residents for manual work in the community.

The deplorable conditions in these early jails, coupled with the large number of citizens who were incarcerated, led to cries for change. Three important legislative acts of the seventeenth century established some basic rights for the incarcerated:

1. 1628 Petition of Right assured the right to freedom before trial;
2. 1679 Habeas Corpus Act provided a remedy for improper incarceration;
3. 1689 the English Bill of Rights outlawed the imposition of excessive bail.[2]

These same rights later reappeared in the Constitution and Bill of Rights of the United States.

Jails for a time were the only detention facilities in many communities and thus were utilized to house all categories of individuals. As governments recognized a need for a comprehensive system of incarceration, three basic types of institutions were developed: (1) the common jail, (2) houses of corrections, and (3) workhouses.

Jails became facilities primarily used for pretrial detention, although many individuals who were unable to pay their fines were also incarcerated there. Houses of corrections, or bridewells as they were known, started in 1553 and served as training schools for delinquent youth, provided housing and support for older and poorer persons, and detained vagrants. Workhouses were for sentenced criminals and provided a center for inmate labor.

All three of these facilities were the responsibility of local government. These facilities shared responsibility for detaining individuals passing through the criminal justice system. Although modified forms of each of these facilities were to be developed in colonial America, the jail in most cases became solely responsible for detaining criminals and those accused of crime.[3]

John Howard (1726–1790), a local sheriff turned jail reformer, became a major leader in the movement to improve conditions in English jails. Howard's concern for jail change grew out of his experience as a sheriff and his personal experience as a prisoner of war. His untiring efforts helped to bring an end to the frequently abused fee system when in 1774 Parliament provided for salaries for jailers.

His efforts have remained a model for generations to follow. His name has in fact been adopted as the symbol for a modern-day prison improvement group. The John Howard Association, a Chicago-based organization, was founded in 1901 and has functioned throughout the century as an advocate of improved prison and jail conditions.

American Colonial Jails

The system of jails that developed in colonial America were basically a modification of the English system. The jail remained a responsibility of local government with the sheriff commonly serving as administrator. The fee system was retained, and many deplorable conditions found in English jails were recreated in the colonies. Although information on early jails is limited, it seems apparent that a wide variety of facilities and conditions existed. The "public gaol" of Williamsburg, Virginia, described in figure 5.1, was a model for its time and provides an excellent example of early American jails. The restored facility, which can be seen at colonial Williamsburg, included separate areas for criminals and debtors. The practice of providing housing for jailers within the facility is also reflected in figure 5.1. Most early facilities were operated by a husband and wife who lived at the jail.

Early Changes

Early jails served as all-purpose detention facilities. Many factors came together in the eighteenth century to bring about a major change in the use of incarceration in response to criminal behavior. A general humanitarian mood was reflected in efforts by John Howard in England and William Penn in America. These efforts began to focus attention on alternatives to then existing practices of incarceration. Groups such as the Philadelphia Society for Alleviation of the Miseries of Prisons called for jail change. Various groups were often directed by local civic and religious leaders and had a great deal of influence. There was also a reaction in the newly independent United States to the harsh English legal system and its extensive use of capital punishment.

These factors coalesced in 1790 and resulted in the conversion of Philadelphia's Walnut Street Jail into a state penitentiary. The penitentiary movement, which developed and spread throughout the United States, left the jails behind. Throughout the nineteenth century, jails were generally ignored by reformers who had focused their attention on the prisons. Jails were viewed as temporary facilities for pretrial detention and incarceration of some misdemeanants. Jails remained the responsibility of local officials, while prisons were placed under the direction of state government. This decentralized control coupled with limited public attention left jails to develop in a hodgepodge fashion in a rapidly expanding nation.

The old English fee system was modified and became the **turnkey fee.** This new system was a per-resident fee paid by the local or state government to cover the total cost of jail operation. In the South the prison movement was not as active, and jails retained a much wider range of residents. By the late nineteenth century, chain gangs and other public and private work details were common practice in jails in the South and other regions of the country. Work

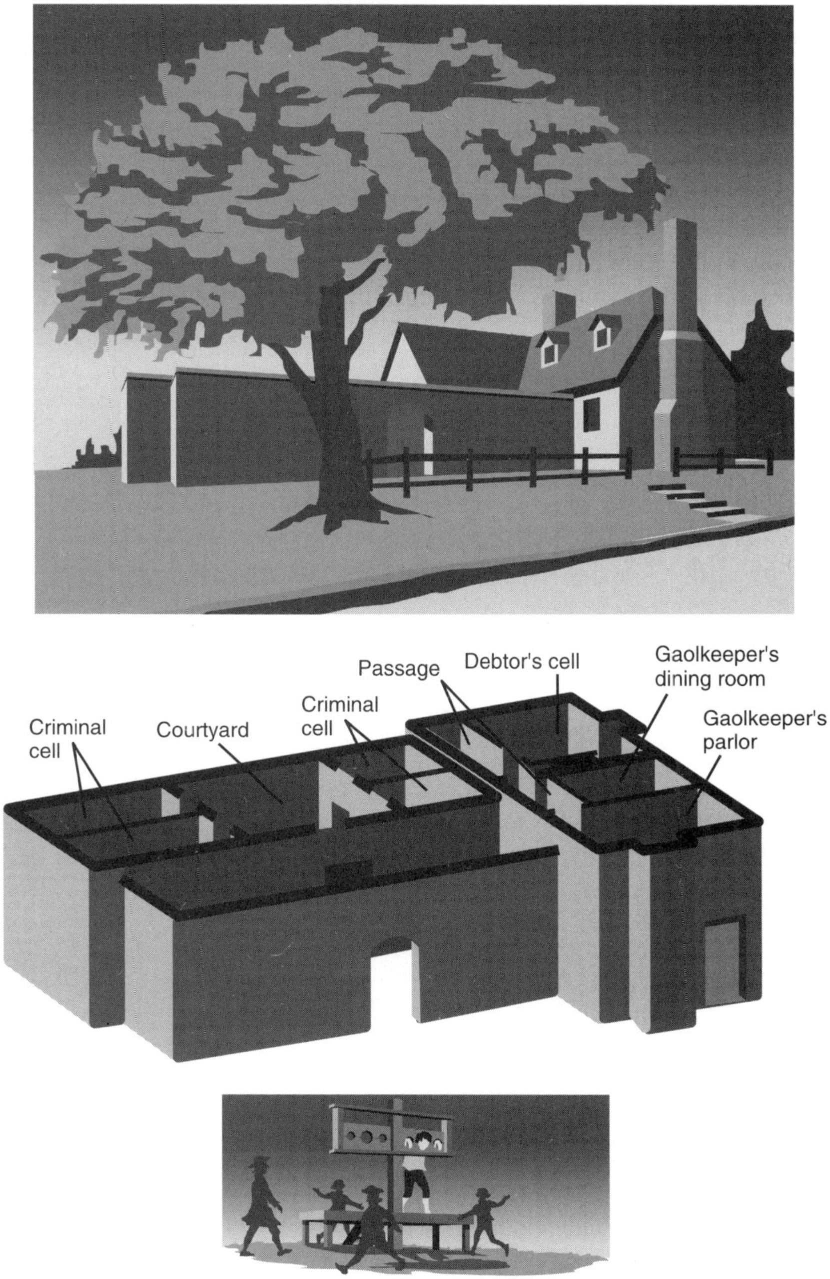

Figure 5.1 The public gaol

crews helped defray costs of jail operations, relieved the enforced idleness which plagued most jails, and also supported a prevailing notion of the deterrent value of hard labor.

Changes that did take place during the nineteenth century resulted in special populations being removed from jails, but seldom did change result in an improvement of the often deplorable conditions that existed in local facilities. By the mid–1800s the practice of placing debtors in jail had significantly decreased. Special institutions were being developed for long-term incarceration and care of juveniles. In 1825, New York City House of Refuge became the first facility designed solely to house juveniles.

Upon entry into the twentieth century, jails were facing many of the same problems that plagued them throughout the 1800s. During the twentieth century, there has been an increased interest in jail change. The first national survey of jails was conducted in 1931. The study by Hasting H. Hart[4] identified 11,000 jails and lockups in the United States. The report formed the basis of jail research and knowledge for the next forty years.

Jails in the United States

Definition

Part of the difficulty in studying jails is understanding terminology associated with the field. Jails are usually locally controlled, and since there are many different facilities, a variety of terms and labels have developed. For example, jails in Chicago and Detroit are called houses of corrections, and in New York City and Miami, they are referred to as houses of detention.

A jail can be defined as a confinement facility authorized by local or state law to hold individuals for periods in excess of forty-eight hours.[5] This is not to say that jails cannot hold individuals for less than forty-eight hours, but that they have the authority for longer periods of confinements. This distinction is important when comparing jails to **lockups** (temporary holding facilities) which by law or practice can incarcerate an individual for less than forty-eight hours. Typically, lockups are found in police stations, city halls, courthouses, and other noncorrectional facilities. It is common practice for individuals held in a local lockup to be transferred to a jail following a preliminary screening of an individual case.

It is important to differentiate between the jail and the other most common incarceration facility, the prison. The most distinguishing factor is that jails hold pretrial detainees and prisons do not. Prisons are administered at the state or federal level and not locally controlled. Jails typically incarcerate individuals with sentences of one year or less, while prisons have been designed primarily for criminals to be confined for more than one year. These characteristics most clearly differentiate jails from prisons.

Although championed by some, critics point out that **local control** of jail facilities is the root problem of many inadequate jails. Local political leaders are often very conservative, unwilling to implement programs and to provide sufficient funding for an adequate jail. As a result some jails become so deficient that they have to be closed. Others are unsafe and unsanitary places to reside and work. In addition, public safety is jeopardized since some buildings cannot be adequately secured and escapes occur.

Facilities

The approximately 3,300 jails presently in operation represent the most widely used type of confinement in the United States. Despite the proliferation of jails, very little was known about their actual operation, organization, and characteristics until the 1970s. Complete enumerations of the nation's jails are now conducted every five years. Information is collected from a sample of jails in each of the four years between complete censuses. These surveys of the nation's jails provide definitive looks at local incarceration. Some information reported in this chapter has been generated from these **jail census** reports. An increasing amount of research, writing, and interest have been focused on conditions in our nation's jails during recent years.

Many of the nation's jails are very aged facilities, constructed long before any consideration was given to correctional standards by either the courts or the profession. Jails are located and used throughout the country and vary greatly in size and structure. Many are overcrowded for various reasons. Thousands sentenced to state prisons are held in jails because there is no room at the prison. Increasing numbers of substance abusers and mentally ill persons find themselves in jail. Stricter laws and a lack of release options also contribute to facility overcrowding.

Jail Populations

Jail populations are measured by several approaches. We must remember that jail populations change very quickly, constantly increasing or decreasing during every day as people are admitted and released. Such a situation presents some methodology challenges, but nevertheless, reasonable estimated measures are provided in data obtained by the Bureau of Justice Statistics jail surveys. Figure 5.2 shows recent massive increases that have occurred in both jail populations and the rate of jail incarcerations per 100,000 United States residents. Local jail populations increased from 209,582 in 1982 to 444,584 in 1992, an increase of 112 percent. Similarly, the rate of jail incarcerations per 100,000 United States residents increased from 90 in 1982 to 174 in 1992, an increase of 93 percent. From 1984 to 1992 the rate of increase of white inmates was from 68 to 109, an increase of 60 percent, while the rate of increase of black inmates was from 339 to 619, an increase of 83 percent.

Table 5.1 provides information on the number and proportions of persons in jail on a daily basis.

Figure 5.2 Number and rate of jail incarceration, 1982–1992

Source: Adapted from Bureau of Justice Statistics, Jail Inmates 1992 *(Washington, D.C.: Government Printing Office, August, 1993), cover page.*

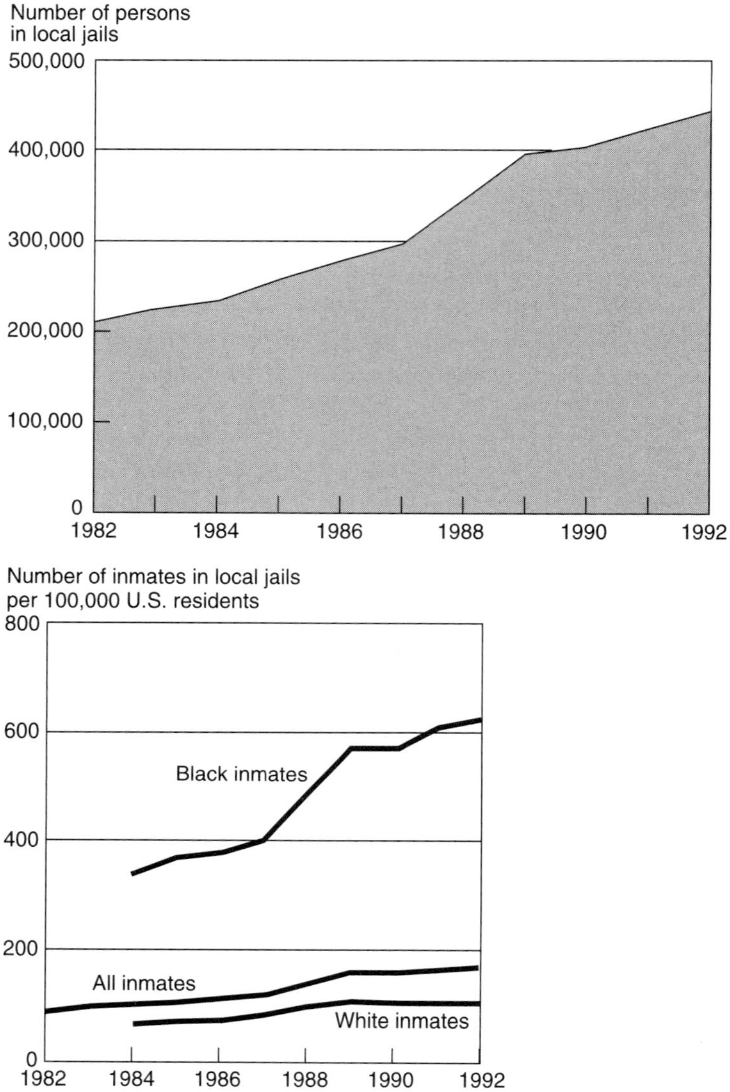

Legal Status

It is important to realize that some jail residents have been convicted while others have not. Those in a **convicted status** are those awaiting sentence, those serving a sentence, and those returned because of a probation or parole violation. Those in an **unconvicted status** are awaiting arraignment, awaiting trial, or are on trial. The proportion of jail residents in a convicted and unconvicted legal status by sex is shown in figure 5.3. Slightly over one-half of jail residents are unconvicted. Some will subsequently be convicted, but others will not. This is but one of many measures that shows the diverse nature of jail residents and

Table 5.1

Jail Population: One-Day Count

	Number	Percent
One-day count		
All inmates	444,584	100
Adults	441,781	99
Male	401,106	90
Female	40,674	9
Juveniles	2,804	1

Source: Adapted from Bureau of Justice Statistics, Jail Inmates 1992 *(Washington, D.C.: Government Printing Office, August, 1993), p. 2.*

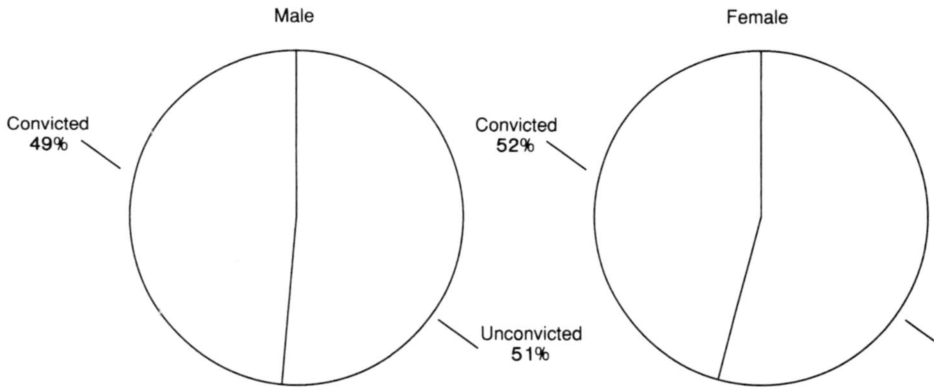

Figure 5.3 Legal status of jail residents by sex

Source: Adapted from Bureau of Justice Statistics, Jail Inmates, 1992 *(Washington, D.C.: Government Printing Office, August, 1993), p. 2.*

highlights the important role that the management of inmates and effective classification should play.

Annual Activity

The total volume of activity at the nation's jails is staggering. As shown in table 5.2, the combined admissions and releases is moving toward exceeding 20 million. A significant proportion of this total is accounted for by **chronic offenders** who have multiple admissions and releases, which are counted each time they enter and leave.

Socioeconomic Characteristics of Inmates

All segments of the United States population are represented in the makeup of our jails; however, the **socioeconomic characteristics** of the typical inmate does not resemble that of the average citizen as indicated by tables 5.3 and 5.4.

Table 5.2

Annual Jail Admissions and Releases, by Legal Status and Sex	
Total admissions	10,064,927
Adults	10,005,138
Male	8,894,706
Female	1,110,432
Juveniles	59,789
Male	51,226
Female	8,563
Total releases	9,870,546
Adults	9,811,198
Male	8,723,872
Female	1,087,326
Juveniles	59,348
Male	50,913
Female	8,435

Source: Adapted from Bureau of Justice Statistics, Correctional Populations in the United States, 1990 *(Washington, D.C.: Government Printing Office, July, 1992), p. 9.*

A typical jail resident is poor, undereducated, and unemployed. Approximately 45 percent of jail residents were unemployed at the time of arrest. Residents have an earned median income that would place them at the poverty level. Many have not completed high school. Over half the males and nearly three-fourths of females have dependent children.

Research suggests that this population contains fewer dangerous criminals than commonly thought and that a different purpose, which is to manage society's rabble, may be served. John Irwin studied this population in depth and concludes that:

> . . . beyond poverty and its correlates—under-education, unemployment, and minority status—jail prisoners share two essential characteristics: detachment and disrepute. They are detached because they are not well integrated into conventional society, they are not members of conventional social networks, and they are carriers of unconventional values and beliefs. They are disreputable because they are perceived as irksome, offensive, threatening, capable of arousal, even prorevolutionary. . . . I shall refer to them as the rabble, meaning the "disorganized" and "disorderly," the "lowest class of people."
>
> I found that it is these two features—detachment and disrepute—that lead the police to watch and arrest the rabble so frequently, regardless of whether or not they are engaged in crime, or at least in serious crime. (Most of the rabble commit petty crimes, such as drinking on the street, and are usually vulnerable to arrest.)
>
> These findings suggest that the basic purpose of the jail differs radically from the purpose ascribed to it by government officials and academicians. It is this: the jail was invented, and continues to be operated, in order to manage society's rabble. Society's impulse to manage the rabble has many sources, but the subjectively perceived "offensiveness" of the rabble is at least as important as any real threat it poses to society.[6]

Table 5.3

Demographic Characteristics of Jail Inmates	
Characteristic	***Percent of Jail Inmates***
Total	100%
Sex	
Male	91%
Female	9
Race	
White	51%
Male	46
Female	5
Black	47%
Male	43
Female	4
Other	2%
Male	1
Female	—
Hispanic origin	
Hispanic	14%
Male	13
Female	1
Non-Hispanic	86%
Male	78
Female	8

Source: Adapted from Bureau of Justice Statistics, Correctional Populations in the United States, 1990 *(Washington, D.C.: Government Printing Office, July, 1992), p. 9.*

Table 5.4

Selected Social and Economic Characteristics of Jail Inmates	
Education	—41% did not complete high school
Unemployment	—45%
Income	—Median income at poverty level
Married	—21%
Have dependent children—males	—54%
Have dependent children—females	—71%
Ever used any drug	—75%

Source: Adapted from Bureau of Justice Statistics, Report to the Nation on Crime and Justice, *2d ed. (Washington, D.C.: Government Printing Office, March, 1988), pp. 48–50.*

Needs of Residents

Social Service Needs

The social service needs of inmates can be divided into two categories: short-term and long-term. Short-term social service needs generally relate to problems associated with confinement and to legal issues. Inmates who find themselves incarcerated may need to make arrangements for child care or other temporary assistance.

Long-term social service needs address issues that may have led to incarceration. Any maladaptive behaviors demonstrated by jail inmates commonly require more extensive services than are available in most jails. Often long-term services focus on identification of problem areas and making appropriate referrals upon an individual's release.

Special Needs

It should not be surprising that individuals with a variety of **special needs** enter our jails. Law enforcement officers frequently are able to screen out special needs populations and divert them to alternate facilities, such as mental health centers, hospitals, and drug and alcohol centers. On many occasions, however, law enforcement personnel may be unaware of an individual's special need, there may be no alternate facility available, or the nature of the person's criminal activity may make an alternate placement unadvisable.

Whatever the reason, it is apparent that many individuals with special needs are incarcerated in the nation's jails. It is the responsibility of the jail staff to properly screen and identify any special needs of inmates. The following categories represent some of the most common special needs of jail residents.

Medical Needs

When citizens are incarcerated, they become wards of the state, and responsibility for medical care rests with the governmental unit that has jurisdiction over the jail or prison. In numerous cases, such as *Estelle v. Gamble*[7] and *Mills v. Oliver*,[8] federal courts have held that the local or state government must assume responsibility for reasonable medical care of the incarcerated.

Common medical problems include injuries related to the crime that resulted in arrest and a variety of chronic problems such as diabetes, epilepsy, and tuberculosis. Because most jails are small and are unable to afford a full-time medical staff, the medical needs of residents pose particular difficulties.

Mental Health Needs

Increasingly, many jail residents are in need of the services available from mental health workers. Often these needs are short-term in nature and related to their incarceration, but many inmates have experienced chronic problems relating to their mental health. Limitations of most jails prohibits employment of a full-time mental health staff. Two common approaches to providing mental health services to jail residents have been to establish referral agreements with community mental health agencies and to provide special training in crisis intervention for the local jail staff. All too often, however, these services are lacking, and increasing numbers of mentally ill have become ''street people'' in urban areas.

The need for mental health services for jail residents is illustrated by the incidence of jail suicide. Jail residents have one of the highest rates of suicide in the United States. Over 400 inmates commit suicide each year in the nation's jails.[9]

The jail suicide victim would most likely be a young, single male arrested for a crime relating to alcohol and/or be under the influence of alcohol. The typical victim takes his life by hanging within three hours of incarceration.[10] Prevention of the jail suicide calls for short-term intervention into a crisis situation. The jail staff needs to be aware of the potential for suicide and be prepared to intervene.

Jails must recognize the need for and provide mental health services for inmates. The high cost of all health services places a strain on the budget of many jails. Cooperative arrangements with community agencies seem the most efficient and cost-effective alternative for most jails.

Chemical Dependence

A **chemically dependent** individual is one who is physiologically and psychologically dependent on alcohol, opium derivatives, and synthetic drugs with morphinelike properties, stimulants, and depressants.[11] A significant portion of the jail population has a history of substance abuse and chemical dependence.

The chemically dependent inmate and problems associated with this condition are major concerns for all jails. No other single problem is as pervasive among jail populations. Two primary issues arise for the jail in dealing with chemically dependent inmates: detoxification and treatment.

Detoxification is the process by which an individual who is physiologically and/or psychologically dependent upon a drug is brought to a drug-free state.[12] Ideally, the detoxification process is accomplished gradually through the use of substitute medication, although the process can occur ''cold turkey'' with no medically approved withdrawal assistance. Unassisted withdrawal from chemical dependence can be hazardous. Withdrawal from alcohol commonly results in delirium tremors ''DTs.'' DTs can result in death. Withdrawal from some drugs may increase the risk of suicide.[13]

To assist adequately in the detoxification process, the jail staff must be trained to screen individuals in need of special care. Provisions have been made in many jails to provide special detoxification units for the chemically dependent; however, placement in a ''drunk tank'' remains the most common response in many jails.

Many jurisdictions have sought to remove simple chemical dependence from the realm of criminal offenses. Other efforts have focused on diversion of the chemically dependent from the criminal justice system. When community detoxification facilities are available, law enforcement and jail personnel often divert the chemically dependent away from jail and into these alternate facilities. Unfortunately, many communities do not have detoxification centers and the chemically dependent remain in jail. Providing adequate detoxification and treatment programs for the chemically dependent can be viewed as a major problem in United States jails.

Female Jail Residents

As indicated in table 5.3, the female population of the nation's jails represent approximately 9 percent of the total. Three-fourths of these women have dependent children, which presents special problems seldom addressed. Although some large urban areas have a sufficient female arrest rate to justify a separate facility, most jails are faced with the problems of housing both males and females within the same building. A presence of female inmates in a predominately male facility requires special arrangements. Female correctional officers are generally utilized for body searches and other direct inmate contact activities where questions of propriety might be raised. Female inmates are generally kept in areas of the jail away from the vision of male residents. Because of the widespread ban against sexual relations between inmates, consideration must be given to the supervision of any group activities.

Because of their limited numbers, special services and programs for female inmates have been limited or nonexistent in most jails. The female jail resident has been largely ignored in the United States.

Jail Operations

Because jails vary so greatly in size and organization, it is difficult to make any definitive statements about how every jail operates. However, there are some basic **jail operational procedures** that must be addressed by all facilities regardless of size or location.

Basic operational procedures of jails may be divided into five categories: intake, classification, orientation, court appearance, and release. Each of these procedures is discussed in the following sections.

Intake

Intake or receiving is the initial point of entry into the jail and the official entry into the criminal justice system. The intake process consists mainly of transferring responsibility for the arrestee from the custody of law enforcement personnel to the jail. The official booking process is conducted, and the case is logged into jail records. Arrestees are taken through a standard procedure of verification of the arrest warrant, assignment of a booking number, property check, fingerprinting, being photographed, and other preliminary identification processes. This is also the time when an individual is given an opportunity to place the traditional "one phone call."

The initial phase may vary depending on the condition of the arrestee at the time of arrival. The intake process is considered a very high-stress period in jail operations. The arrestee is frequently agitated and uncooperative and under the influence of alcohol or other substances. The police want to process the individual and move on, while the jail staff may be flooded with a number of new arrivals at the same time. A well-planned and smoothly operated intake process can help to assure a professional jail operation.

Classification

Classification is an ongoing formal process concerned with identification, categorization, and assignment of inmates to various levels of security, programs, and work. The initial classification is concerned primarily with housing and any special needs which an arrestee may have. Common factors considered in classification decisions are sex, age, category of offense charge, prior criminal history, special medical needs, and available space.

Nine different types of housing units have been identified as common in jails: temporary holding area, single cell, multiple cell, dormitory, sobering unit, observation cell, segregation cell, hospital ward or room, and individual room.[14] Again it must be pointed out that not all jails have this full range of cells, but the variety of housing units illustrates the importance of the classification decision.

It is important to note that classification decisions often must be made on very limited information. The jail staff must rely upon their own records and the data provided by law enforcement personnel and the arrestee. The ideal system would base classification decisions upon both the instant offense and also the individual's history of criminal activity. Because background information is generally not available as rapidly as needed, it is common practice to treat all new arrivals as maximum security residents until additional information becomes available. The initial classification decision can have a grave impact on the safety and security of jail residents. Of particular concern are individuals with medical or mental health needs. It has been suggested that the United States jail suicide rate could be significantly reduced with improved classification procedures.[15]

Orientation

Although many arrestees have been in jail before, it is important to provide an **orientation** for all new arrivals. Confinement in jail places individuals in a dependent status. The jail is obligated to inform new residents of jail rules and procedures. An effective orientation process can result in reduced conflict and, in the long run, can decrease the need for repeated questions. Many jails have standard informational handouts for new arrivals. Whatever the technique used, an orientation process is useful in all facilities.

Court Appearances

Jail residents make repeated and frequent **court appearances.** The first such appearance takes place within twenty-four hours or at the next court session. Because of the close working relationship with the court, most jails are located in the general proximity of the courthouse. Safe and timely delivery of prisoners to the court often requires detailed planning and a significant number of staff to supervise the process.

Release

The process of the **release** of residents from jail, although not as hectic as admission, is also an important process. Major considerations in the release process are whether the type of release is on bail, completion of sentence, dismissal of charges, transfer to another institution, or some other form of release. It is also crucial to make positive identification of the releasee and to return any personal property.

Management of Jails

Essential Services

Because jail inmates are wards of the state and therefore dependent upon their keeper for basic needs, there are many **essential services** that must be provided. The jail is responsible for providing food, clothing, facility maintenance, and visiting and interviewing procedures and facilities. Other services include mail, sanitation, and the storage of personal property.

Many of these services are constitutionally guaranteed and have been the subject of court rulings. These services can act as forms of passive control over inmate behavior. Adequate food, decent housing facilities, and reasonable visiting procedures can work to reduce the resistance of jail residents to custodial detention. It is important to remember that inmates retain the right to have basic services provided, and jail personnel are not permitted to deny individuals their basic needs. The courts have allowed certain modification of services, such as restrictions on visiting or mail, based on institutional security. For instance, it is acceptable to limit visiting hours within a jail, but it has not been acceptable to prohibit visiting entirely as a form of punishment. Pretrial residents are more likely restricted to closed visits, while those serving short-term sentences may be allowed open visits. A great deal of staff time and energy is involved in maintaining and supervising the delivery of essential services.

Security

A primary responsibility of jails is the secure detention of the jail population. The role of security (or custody) is to assure that inmates are detained in a fashion that assures the safety of both the prisoners and the public.

Because jail populations change so rapidly, it is necessary that security personnel be particularly vigilant. Jail security staff members often do not have the luxury of time to get to know inmates. They have the responsibility to protect jail residents from assaults, and the potential for violence and sexual attacks can be decreased by direct staff intervention along with proper planning and facility design. Correctional officers in a jail must work with many unknowns and therefore face a very difficult task.

Controlling Contraband and Fires

Two areas of special concern in jail security are **contraband** and fires. Control of contraband in jails is made difficult by a high level of traffic within the jail and varying categories of inmates confined within a single facility. Inmates regularly receive visits from lawyers, family members, and associates. In addition, inmates routinely leave the jail to attend court or for other activities. Many jails have community work programs that also remove residents from the facility. All of these factors make the control of contraband a major concern for jails.

Most fires in jails are intentionally set by inmates. Fires commonly involve ignition of clothing, mattresses, and padded wall coverings. Because of the risk of loss of life to both inmates and staff in the restricted jail setting, constant attention must be given to fire safety.

Some modern jails have been equipped with elaborate electronic control systems to monitor inmate activity and to assist in the security process, but the importance of well-trained and efficient custody personnel remain the key to jail security.

Programs

Programs in jails are ongoing activities directed at meeting social and rehabilitative needs of the residents. Such programs are only available in some jails. Typically, programs might include education and work release, counseling programs, substance abuse treatment, and social services.

There appear to be two crucial factors in determining whether a particular jail provides programs for residents. The two factors are the size of the facility and the average length of stay of residents. The larger the facility and the longer residents are there, the more likely one is to find jail programs available to inmates. It would appear that both factors relate to economies of scale, however.

Many smaller facilities often do not provide programs because of the high cost per inmate of such activities.

The problem of idleness in jails remains. In 1913 Wisconsin enacted the landmark Huber Law which allowed for work release. Provisions of this law allowed jail residents to pursue employment in the community while still incarcerated. This Wisconsin law formed a model for similar legislation and programs in most other states. Although work release is typically only available to a limited number of short-term honor status inmates, it does allow for a reduction in enforced idleness in jails.

Another attempt to use incarceration more constructively has been the **weekend jail** program available in many jurisdictions. Weekend jail sentences are a compromise between incarceration and probation. Convicted persons are allowed to live and work in the community through the week but must report to the jail on weekends to serve their sentences. These programs do present some problems for jail administrators who are concerned with the regular pattern of weekend population increases, which can be compounded by such special sentences. Maintaining jail security and limiting the flow of contraband are also relevant issues of concern in jails utilized for weekend sentencing. This plan provides for incarceration without requiring inmates to sever family ties or to terminate employment.[16] Weekend sentences can prove to be a means of reducing both sentence length and the expense of incarceration.

Staffing and Funding

Jail employees are confronted daily with a wide range of individuals. The jailer must communicate with and supervise individuals who are often uncooperative, afraid, angry, intoxicated, and at times, dangerous. The attitude and manner jail employees assume toward inmates have been found to be a key factor in directing and determining prisoners' behavior.

Jails and other correctional institutions are frequently criticized for functioning on the basis of custodial convenience. Custodial convenience describes a method of operation in which the overriding principle or goal is directed at what is most acceptable to the demands of security and the correctional staff. It has been pointed out that many procedures in jails are designed to make the job easier for employees, while only secondary consideration is given to overall benefits for inmates or the institution.

Although styles of operating vary greatly among jail facilities, problems of funding are shared by nearly all jails. In most jurisdictions, the funding of jails is a responsibility of local government. Funds needed to operate jails compete with requests from schools, highway departments, and other community services. Because jails lack a voting constituency with significant clout in most communities, funding for their operations is generally not given a high priority. Many jail administrators are faced with increasing demands for services but have only limited or even reduced funds to support their programs.

Table 5.5

Subject of Court Orders among Jails with Larger Populations		
	Number	**Percent**
Total	503	100
Subject of court order:		
Limit population	131	26
Crowded living units	118	23
Recreation facilities	62	12
Medical facilities or services	57	11
Staffing patterns	53	10
Library services	49	9
Totality of conditions	41	8
Inmate classification	40	8
Grievance procedures or policies	38	7
Visitation practices or policies	37	7
Disciplinary procedures or policies	37	7
Food services	29	6
Education or training programs	25	5
Fire hazards	22	4
Administrative segregation procedures and policies	21	4
Counseling programs	18	3
Other	14	2

Source: Adapted from Bureau of Justice Statistics, Jail Inmates 1992 *(Washington, D.C.: Government Printing Office, August, 1993), p. 5.*

Inmate Rights

Jail residents retain many basic constitutional rights. Those imprisoned awaiting trial retain all the rights of free citizens, except those necessarily suspended by the fact of their confinement. Inmates who have been convicted of criminal activity, although not entitled to the same freedoms as those held pending trial, nonetheless, are provided some **inmate rights,** covered more extensively in chapter 7.

Inmates have two basic courses of action if they feel that any violations of constitutionally guaranteed rights have occurred. They can bring a court action under the Civil Rights Act of 1871. This action would bring the inmates' complaints before the federal courts, rather than state or local courts, and could result in free legal services and possible assessment of civil damages. Inmates could also initiate a separate civil court action to recoup any losses they might have suffered as a result of the violation of rights.

Although federal court intervention into jails has not been as common an occurrence as reviews of prison conditions, the judiciary has had an impact on the operation of jails. As shown by table 5.5, a survey of 503 larger jails indicates that in one year 131 jails, or 26 percent, were under court order to reduce their inmate population. Forty-one jails, or 8 percent, were cited for a totality of conditions; that is, several conditions taken together that resulted in a cumulative negative effect. This table also shows an extensive list of conditions that were the subject of court orders. The most frequently cited conditions were:

- 23 percent crowded living conditions;
- 12 percent inadequate recreational facilities;
- 11 percent medical conditions;
- 10 percent inadequate staffing.

The Bail System

The operation of a system to administer **bail** within a particular jurisdiction has a major influence on local jails. Bail is a general term that identifies any one of several means of pretrial release from detention of accused individuals. Bail was designed to assure a future appearance in court. Most defendants are eligible for release pending trial. It has been reported that nearly nine of every ten unconvicted jail inmates have had bail set.[17] Whether by **financial bond** or some alternative release option, most defendants are not detained for long prior to trial. However, some are unable or are not permitted to post a financial bond or to participate in an alternative release option and, therefore must stay in jail.

The present United States bail system can be traced to England where in 1275 the Statute of Westminster I first established rules governing bail. This statute defined bailable offenses and assigned responsibility for bail decisions to local judges.[18]

The bail system in the United States is governed by state and federal law. Protection against excessive bail is provided in the Eighth Amendment to the Constitution. In many jurisdictions, bail is available for most individuals accused of noncapital offenses. The traditional means of obtaining release from jail was to place a security bond in the form of cash or property with the court to assure future appearance. Upon appearance before the court at a designated time, the bond was returned. Failure to appear before the court resulted in the forfeiture of the bond.

Many individuals accused of crime lacked the resources to make their bond and were forced to remain in jail or seek assistance from others. A professional **bondsperson** began to provide guaranteed bail bond for a fee. These individuals who were generally licensed by the state would guarantee the amount of bond established by the court for a fee of 10 or 15 percent. If an accused person had a bail bond set at $1,000 and sought the services of a bondsperson with a fee of 15 percent, the cost to the incarcerated individual would be $150. For this fee the bondsperson would guarantee to pay the full bond should the accused not appear in court on the assigned date. The fee paid to the bondsperson was nonrecoverable regardless of whether or not the accused appeared before the court. Various means of posting financial bond are shown in table 5.6.

Table 5.6

Financial Bonds and Alternative Release Options	
Financial Bond	***Alternative Release Options***
Fully secured bail—The defendant posts the full amount of bail with the court. *Privately secured bail*—A bondsman signs a promissory note to the court for the bail amount and charges the defendant a fee for the service (usually 10% of the bail amount). If the defendant fails to appear, the bondsman must pay the court the full amount. Frequently, the bondsman requires the defendant to post collateral in addition to the fee. *Deposit bail*—The courts allow the defendant to deposit a percentage (usually 10%) of the full bail with the court. The full amount of the bail is required if the defendant fails to appear. The percentage bail is returned after disposition of the case, but the court often retains 1% for administrative costs. *Unsecured bail*—The defendant pays no money to the court but is liable for the full amount of bail should he or she fail to appear.	*Release on recognizance (ROR)*—The court releases the defendant on the promise that he or she will appear in court as required. *Conditional release*—The court releases the defendant subject to his or her following specific conditions set by the court, such as attendance at drug treatment therapy or staying away from the complaining witness. *Third-party custody*—The defendant is released into the custody of an individual or agency that promises to assure his or her appearance in court. No monetary transactions are involved in this type of release. *Citation release*—Arrestees are released pending their first court appearance on a written order issued by law enforcement personnel.

Source: Adapted from Bureau of Justice Statistics, Report to the Nation on Crime and Justice, *2d ed. (Washington, D.C.: Government Printing Office, March, 1988), p. 76.*

Problems and Change

There has been a great deal of criticism of this private bail-bonding system. Abuses of this system have been noted by many. The accused persons could be forced to pay what might be a high fee to obtain their release. Many poorer jail residents might be unable to raise even the 10 percent required for the bonding fee and thus be required to remain incarcerated. Other individuals who seemed to be likely to appear before the court were denied bonding services. There was the continuing dilemma of individuals being incarcerated prior to formal adjudication and sentencing. Such pretrial detainees would be unable to work or participate in community programs and would be separated from their families prior to conviction. Finally, because of the large sums of money generated from bail bonding, there have been many allegations of bribes and payoffs to local officials to assure preferential treatment for a particular bondsperson.

Because of these and other problems, movement to change the bail system developed. Organizations such as the New York City-based Vera Institute of Justice's Manhattan Bail Project began in 1961 advocating changes in the bail system to release people on their own **recognizance (ROR).** It was emphasized that a major purpose of pretrial detention and bail was to assure appearance of the accused before the court. Reformers maintained that a presumption of innocence until proven guilty, which is a cornerstone of our legal system, dictated that only high-risk individuals remain incarcerated pending trial.

Efforts of this bail reform movement resulted in many changes in the nation's pretrial release procedures, including the passage of the 1966 **Federal Bail Reform Act.** This federal legislation became an early model for bail alternatives throughout the country by instructing judges to use the likelihood of court appearance as a basis for release decisions and to use the least restrictive alternative. The Bail Reform Act of 1984 substantially changed the 1966 Act by instructing judges to include public safety as a factor to be considered in release and detention decisions.[19] Factors that might indicate danger to the community include the use of a firearm and injury to the victim or arresting officer.

The Act includes provisions for pretrial detention of suspects charged with violent crimes, crimes with a possible life or death sentence, major drug offenses, and felonies when a suspect has a history of serious crime. Since this Act has been in effect, both the percent of defendants held before trial and the average time held have increased.[20]

Alternatives to Bail: Pretrial Release Options

Hundreds of pretrial **alternative release options** operate throughout the United States.[21] Table 5.6 illustrates basic pretrial release options used today.

Release on personal recognizance, or ROR, is the most widely used form of pretrial release. Recognizance means a personal pledge or promise, in this case a promise to appear in court. In most jurisdictions that use ROR, there is a screening process that reviews a number of salient factors to determine eligibility for ROR.

Supervised release is granted when a third person agrees to assume responsibility for an accused. Frequently a third person is a representative of a service agency that provides counseling and other programs for the accused. For instance, commonly, supervised release is often granted to an individual who agreed to participate in a drug or alcohol treatment program pending trial.

As an example, Kentucky has established a **pretrial release program.** The local pretrial service officer, working in conjunction with the court, screens jail residents to determine their eligibility for release and then makes recommendations to the court. This program has proven very effective as nearly three-fourths of the custodially arrested individuals have been found eligible for ROR. Of those released through the pretrial service program, only about 4 percent deliberately fail to appear before the court at a designated time.

Pretrial service officers in Kentucky use a point system to evaluate eligibility for release, as shown on the interview form in figure 5.4. The system results in a recommendation of release on recognizance for any eligible individual who has a positive score of eight or more.

Figure 5.4 Interview form

Source: Courtesy State of Kentucky, Administrative Office of the Courts.

AOC-PT-21
Rev. 2-92 AKA: ____________ Social Security Number: ____________

NAME: ____________ DOB: __/__/__ Age: ____ Sex: ____ Race: ____
Last First Middle/Maiden

DATE OF ARREST: ____________ CHARGE: ____________
COURT: ____________
VERIFIED COURT DATE: ____________
Yes No

PRESENT ADDRESS: ____________
Street/Apt. No. City State Zip Code

LENGTH OF RESIDENCE: Present: Yrs.____ Mos.____ Area: Yrs.____ Mos.____ Phone: ( ) ____________
☐ Own ☐ Rent ☐ Other ____________

ALTERNATE/PRIOR RESIDENCE: ____________
Street/Apt. No. City State Zip Code
With Whom: ____________ Phone: ( ) ____________ Lgth. of Res.: Yrs. ____ Mos. ____

LIVES WITH: ☐ Alone ☐ Spouse ☐ Parents ☐ Grandparents ☐ Children ☐ Brother/Sister ☐ Other Relatives ☐ Other ____________
MARITAL STATUS: ☐ Single ☐ Married ☐ Divorced ☐ Widowed ☐ Common Law ☐ Separated No. of Children: ____
SPOUSE'S NAME: ____________ SPOUSE'S SOURCE OF INCOME: ____________
FAMILY IN AREA: Name: ____________ Relationship: ____________
Address: ____________ Phone: ( ) ____________

☐ EMPLOYED ☐ UNEMPLOYED How Long Yrs. ____ Mos. ____
☐ Full-time ☐ Part-time ☐ Seasonal ☐ Welfare ☐ Unemployment ☐ Disability ☐ Retirement ☐ Other ____________
EMPLOYER: ____________ Job Position: ____________
EMPLOYER'S ADDRESS: ____________
Street City State Zip Code
Phone: ( ) ____________ Can Contact: ☐ Yes ☐ No Who: ____________
PRIOR/OTHER SOURCE OF INCOME: Source: ____________ Job Position: ____________
Address: ____________
Phone: ( ) ____________ Length of Employment: ____________
ATTENDS SCHOOL: ☐ Yes ☐ No School: ____________
Address ____________
GED or High School Degree: ☐ Yes ☐ No Can Contact: ☐ Yes ☐ No Phone: ( ) ____________

PRIOR ARREST: ☐ Yes ☐ None If yes, Where: ____________
PENDING CHARGES: ☐ Yes ☐ No How Released: ____________ Next Court Date: ____________
Where: ____________ Charges: ____________
ON PROBATION/PAROLE: ☐ Yes ☐ No Probation/Parole Officer's Name: ____________
Address: ____________
Phone: ( ) ____________

DRIVER'S LICENSE NUMBER: ____________ ☐ Kentucky ☐ None ☐ Other: ____________

IN COURT:	Name	Address	Relationship	Phone
1.				
2.				
3.				
4.				

COMMENTS:

Court Dates: 1. ____________ 2. ____________ 3. ____________ 4. ____________

ELIGIBLE INELIGIBLE ____________ Points	COURTS INITIAL DECISION/ JUDGE Date/Time	24 Hour Review: Yes No Courts Decision:	HOW RELEASED:

Figure 5.4 *(continued)*

WARNING

This interview form will be used by the Judge or Trial Commissioner to set bail. It will also be used for personal identification, future bond reviews, service of warrants, and sentencing if found guilty. The Judge may allow your attorney or probation/parole officer to review the information. Except for these situations, any information you provide will be confidential and not released without your written consent or court order. You do not have to say anything, and can stop answering questions at any time. Signing this form means you want to be interviewed.

DECLINED INTERVIEW OR REFUSED TO SIGN AFTER BEING WARNED: ______________________
S/Defendant

Witnessed By: ______________________
Date and Time: ______________________

Witnessed By: ______________________
Interviewer: ______________________
Date and Time: ______________________

Circle only one number for each category of criteria except Miscellaneous and Previous Criminal Record

RESIDENCE

- +5 Has been a resident of the commonwealth for more than one year.
- +3 Has been a resident of the commonwealth for less than one year but more than three months.

PERSONAL TIES

- +4 Lives with spouse, grandparents, children, parents, and/or guardian.
- +3 Lives with other relatives.
- +2 Lives with non-related roommates.

ECONOMIC TIES (Double length of employment if part-time)

- +5 Has held present job for more than one year OR is a full-time student.
- +4 Has held present job for less than one year but more than three months.
- +3 Is dependent on spouse, parents, other relatives, or legal guardian, unemployment, disability, retirement, or welfare compensation.
- +2 Has held present job for less than three months.

MISCELLANEOUS

- +3 Owns property in the Commonwealth
- +1 Has a telephone.
- +1 Expects someone at arraignment.

PREVIOUS CRIMINAL RECORD (+)

- +3 No convictions on record (excluding traffic violations) in last two years.

(A) ____________ **TOTAL POSITIVE POINTS**

PREVIOUS CRIMINAL RECORD (-) (FTA must be verified by court records or personal knowledge)

- -3 AWOL on record (current military personnel only).
- -5 Probated or paroled after felony conviction in last two years.
- -5 FTA on traffic citation in last two years.
- -10 FTA on misdemeanor charge in last five years.
- -15 FTA on felony charge at any time.
- -15 Violation Conditional Release while case is pending and active.

(B) ____________ **TOTAL NEGATIVE POINTS**

____________ **TOTAL PRETRIAL RELEASE POINTS**
("A" minus "B")

DATE	OFFENSE	COURT	DISPOSITION

Alternative release options have helped to reduce problems of overcrowding in jails. A major impact has been felt in many large urban areas from the rapid release of the pretrial accused. It should be emphasized that these reforms have not been without their critics. Two major questions have arisen: how many fail to appear in court? How many of those released are rearrested before trial? There is a need to protect the public from criminals and, at the same time, protect the rights of all citizens by not incarcerating individuals who are presumed innocent until proven guilty unless absolutely necessary.

Available information suggests that upwards of nine in ten appear at all court sessions with those charged with more serious offenses more likely to appear. Failure to appear or return by force does not exceed 4 percent of cases. Rearrest rates are between 10 and 20 percent.[22]

Concern for public safety has resulted in the incorporating of conditional provisions into pretrial release in many states. One or more of the following may apply:

- Exclusion of certain crimes from automatic bail eligibility
- Exclusion of release conditions
- Limitations for those previously convicted
- Revocation of pretrial release if a new crime is committed
- Provisions for pretrial detention to ensure safety[23]

New Facilities

Over half of the nation's jails have been built since 1950. Although there has been a great deal of jail construction in recent years, many local facilities, particularly in rural areas, are aged and in need of renovation or replacement. Often, conservative local politicians have not wanted to spend money for adequate jails, with the result that decisions to build new facilities have often been forced by problems of overcrowding, inadequate security, and court intervention. At a minimum, it often requires two to three years to plan and construct a new facility.

Better jails can be built. **Three generations of jails** have evolved with different architectural designs and philosophies of supervision. Some newer model jails have incorporated advanced architectural design with **direct supervision** of residents.[24] Direct supervision is best understood by contrasting it with earlier styles of supervision and designs, as shown by figure 5.5. The design and management of most jails have provided intermittent or remote surveillance. With intermittent surveillance, cells are lined up in rows with staff unable to observe all living areas from one location. Officers and staff are separated by bars. With remote surveillance, observation is improved by inmates, and officers are isolated from each other with little or no opportunity for interaction. With direct supervision, there is no enclosed officer booth, and officers spend time in the living area of inmates actively supervising and interacting more naturally with residents. Reported experiences with direct supervision

Some antiquated jails have been replaced by second- and third-generation structures.

indicate that cost savings are possible, maintenance costs are reduced, working conditions are improved, and violence is reduced.[25]

Future of Jails

Regional Jails

There has been some attempt to move toward **regional jail** facilities. Regionalized jails have been proposed as an alternative to the numerous small facilities located in local communities. A regional jail concept includes cost sharing among jurisdictions to construct new facilities with a large inmate capacity, thus making feasible the expansion of jail services. A regionalized approach has proven successful in some areas, but the adoption of such a plan generally encounters much resistance.

Regionalized jails have been opposed by some because they are seen as eroding local control. Other drawbacks include the problems of transportation for law enforcement, legal counsel, and families. Local people often resist the removal of a valuable source of revenue from their community. Some advocates of community reintegration view regionalization of jails as a barrier to the reentry of inmates into society.

Despite its many advantages, the regional jail concept has not gained widespread support. A primary reason for the limited use of regionalized jails seems clearly to be the reluctance of individual jurisdictions to cooperate on such ventures. There are also legislated and constitutional prohibitions against such consolidations in some states.

Figure 5.5 Three generations of jails

Source: W. Raymond Nelson, Cost Savings in New Generation Jails: The Direct Supervision Approach *(Washington, D.C.: National Institute of Justice, July, 1988), pp. 2–3.*

A. Intermittent surveillance

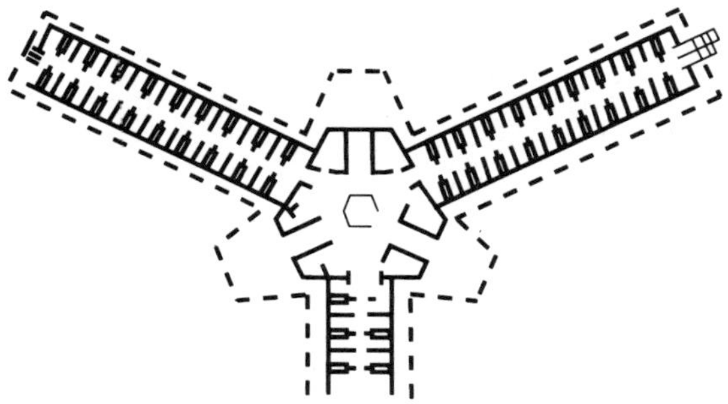

First generation

B. Remote surveillance

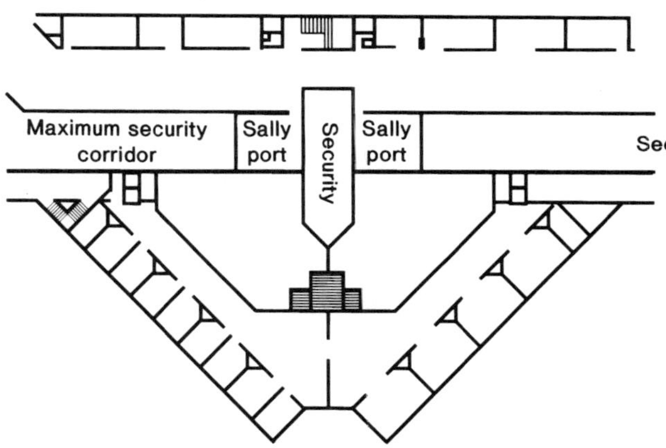

Second generation

C. Direct supervision

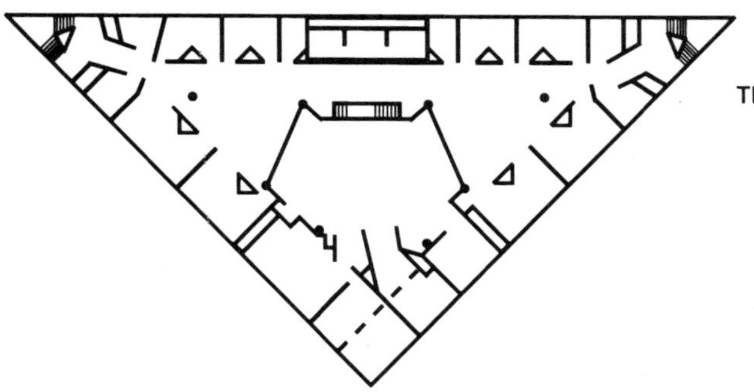

Third generation

Jail Standards and Court Intervention

Although correctional standards have existed for many years, it was not until the 1970s that such guidelines for jails came into prominence. There is presently in existence an array of standards that apply to jails, including ones sponsored by the American Correctional Association, American Bar Association, National Sheriff's Association, National Fire Prevention Association, American Public Health Association, American Psychological Association, and United Nations. Most states have specific standards that apply to jails.

Compliance with and enforcement of these standards vary greatly depending upon the set of standards as well as upon local and state laws. In many jurisdictions, the courts have become the chief enforcers of **jail standards.** In an increasing number of cases, the courts have turned to various standards as the yardstick by which to measure the adequacy of jail programs and facilities. Areas of particular interest to the courts have been medical services, recreational opportunities, cases relating to overcrowding, and issues revolving around a jail's failure to adequately protect inmates from danger.

Continuing Role of Jails

Jails remain chiefly concerned with the secure incarceration of pretrial and posttrial inmates. The overriding importance of this secure detention role will remain, but techniques and approaches taken may be altered. With the introduction of sophisticated electronic equipment, better design, and increased emphasis upon training of jail staffs, the security operations within the nation's jails should continue to be strengthened.

What is yet to be seen is what role jails will play in the diagnosis of and provision of services for inmates. Jails may remain basically temporary local detention units that have only minimal concern for the identification and remediation of inmate problems, or they may expand their role to become more comprehensive community resource centers. Future directions of jails depend upon community interest and support, professional standards, and mandates provided by the courts.

Summary

More citizens experience incarceration in jail than any other part of corrections. Some of these persons are unconvicted suspects, while others are serving short-term sentences as a result of conviction. Millions are confined each year for some period of time in the nation's jails.

Local places of confinement have existed for centuries. Such practices existed in ancient civilizations and became institutionalized in England under the Anglo-Saxon feudal system. Early American jails mirrored many English practices.

Jails are authorized to hold people in excess of forty-eight hours which distinguishes them from lockups. Although jails incarcerate people as do prisons, major distinctions are that jails are usually locally controlled, jail residents have a shorter length of stay, and the jail population includes many who are in an unconvicted legal status.

Jail populations change every day as people are admitted and released. About one-half of jail residents are in an unconvicted legal status. Many inmates have long-term, deeply rooted problems that go beyond the immediate reason for their incarceration. Many are undereducated, unemployed, poor, with alcohol or other drug usage. Some have mental health problems. The high suicide rate among jail residents is but one illustration of the need for special training of management and staff in the mental health area.

After spending a short time in jail, most unconvicted residents are released on some form of bail or alternative release option. In many jurisdictions full financial bail has been replaced by release on personal recognizance or supervised release. Outcomes from alternative forms of release have generally been favorable. Overcrowding has been reduced, most defendants appear in court, and rearrest rates while on release have been low.

Many jails are antiquated but efforts to improve jails continue. Model jails have been developed; increasing inspection and adherence to standards are evident in some jurisdictions; and some localities have constructed so-called new generation, state-of-the-art facilities.

Discussion Topics

1. It would be better if jails operated with less local-level and more state-level control. Do you agree or disagree? Why?
2. Do you think the number of jails should be reduced by establishing more regional jails? Why or why not?
3. Visit a local jail. How well does management classify and separate convicted from unconvicted residents?
4. To what extent do you agree with John Irwin's contention that jails exist to manage society's rabbles?

Additional Readings

American Journal of Criminal Justice. Fall, 1989. Entire issue devoted to topic of jails.

Irwin, John. *The Jail*. Berkeley: University of California Press, 1985.

Zupan, Linda L. *Jails: Reform and the New Generation Philosophy*. Cincinnati, OH: Anderson Publishing Company, 1991.

Notes

1. Hans W. Mattick, "The Contemporary Jails of the United States: An Unknown and Neglected Area of Justice," *Handbook of Criminology*, ed. Daniel Glaser (Chicago: Rand McNally, 1974), 782.
2. Ronald Goldfarb, *Jails* (Garden City, NJ: Anchor Books, 1976), 36–37.
3. J. M. Moynahan and Earle K. Stewart, "The Origin of the American Jail," *Federal Probation* (December 1978): 41–50.
4. Hasting H. Hart, "Police Jails and Village Lockups," Report on Penal Institutions, Probation and Parole, no. 9 (Washington, D.C.: National Commission of Law Observance and Enforcement, 1931), 330–345.
5. Moynahan and Stewart, "The Origin of the American Jail," 41.

6. John Irwin, *The Jail* (Berkeley, CA: University of California Press, 1985), 2.
7. *Estelle v. Gamble,* U.S. 97 S. CT. 285, (1976).
8. *Mills v. Oliver,* 366 F. Supp. 77, E.D. VA., (1973).
9. Lindsay M. Hayes and Joseph R. Rowan, *National Study of Jail Suicides: Seven Years Later* (Alexandria, VA: National Center on Institutions and Alternatives, 1988), 11.
10. Lindsay M. Hayes, *And Darkness Closes In . . . National Study of Jail Suicide* (Washington, D.C.: The National Center on Institutions and Alternatives, 1982), 2.
11. American Medical Association, *Guide for the Care and Treatment of Chemically Dependent Inmates* (Chicago: AMA, 1980), 1.
12. Ibid.
13. Dorynne Czechowicz, *Detoxification Treatment Manual* (Washington, D.C.: United States Government Printing Office, 1978), 29–32.
14. Frances O. Jansen and Ruth Johns, *Management and Supervision of Small Jails,* (Springfield, IL: Charles C. Thomas Publisher, 1978), 61.
15. Hayes, *And Darkness Closes In . . .* 18.
16. Edgar May, ''Weekend Jail: Doing Time on the Installment Plan,'' *Corrections Magazine,* (March 1978): 28–38.
17. Bureau of Justice Statistics, *Report to the Nation on Crime and Justice* 2d ed. (Washington, D.C.: Government Printing Office, March, 1988), 76.
18. Goldfarb, *Jails,* 36.
19. Bureau of Justice Statistics, *Pretrial Release and Detention: The Bail Reform Act of 1984* (Washington, D.C.: Government Printing Office, February, 1988).
20. Ibid, 5.
21. National Association of Pretrial Services, ''Pretrial Program Directory'' (Washington, D.C.: July 1986).
22. Bureau of Justice Statistics, *Report to the Nation,* 77.
23. Ibid.
24. W. Raymond Nelson, *Cost Savings in New Generation Jails: The Direct Supervision Approach* (Washington, D.C.: National Institute of Justice, July, 1988).
25. Ibid.

Chapter 6

Prison Facilities and Populations

Learning Objectives

After studying this chapter students should be able to:

1. Briefly describe the development of prisons and reformatories in the United States.
2. Explain the differences between the congregate system and the separate system.
3. Describe differences in custody levels.
4. Present some specific cost estimates to build and operate a prison.
5. Understand trends and projections in the number and rate of incarceration in the nation.
6. Recognize problems presented by prison inmates with AIDS and describe some approaches to managing this disease in a prison setting.
7. Reflect an increased awareness of elderly inmates and provide some recommendations for handling.
8. Understand the presence and problems presented by gangs and persistent felons in many prisons.

Key Terms

AIDS
Classification
Congregate system
Construction costs
Elmira Reformatory
HIV
Incarceration rate
Institutional classification
Maximum (or close) custody prisons
Medium custody prisons
Minimum custody prisons
New Gate Prison
Operating costs
Penitentiary
Persistent felons
Prison design
Prison population
Reclassification
Reformatory
Separate system
Systemwide classification

Introduction

A sentence of incarceration in a state or federal prison for convicted offenders is a most dramatic sanction. This is probably the sanction uppermost in people's minds throughout society when they think of corrections. We very easily visualize a sentence being pronounced by a judge in the courtroom and the offender being taken to prison. In reality, this has become an ever increasing event in American society. As we move into the twenty-first century, more people are in state and federal prisons than ever before; many new prisons have recently been built or are under construction. Overcrowding is found in many facilities. Expenditures for prisons are at an all time high. Through various media channels, the visibility of prisons has greatly increased and debates over incarceration have intensified. This chapter reviews the goals of corrections and then examines prison facilities and the numbers who go to prison.

From the discussion in chapter 3, you will recall that we discussed several aims or goals of corrections by utilizing four basic models: custodial, justice, least restrictive alternatives, and rehabilitation. We can ask, what goals does imprisonment serve? Obviously, punishment is being administered. One must leave home, leave the community and friends, give up freedoms, and live in a very structured environment. A custodial function is also served. An offender is incapable of further illegal acts against society while in prison. Little or no emphasis is placed on changing the offender's behavior.

Goals of the justice model may also be served. A debt is being paid to society for wrongful behavior. Offenders deserve just punishment in an amount relative to the seriousness of their crime. This goal is less concerned with changing behavior than protecting society through just means.

Rehabilitation is geared toward one learning or relearning appropriate behavior. Correctional efforts assume a strong change agent and teaching role, but this is very difficult or even impossible to achieve in prison settings.

Reintegration is directed toward the community helping offenders to be productive citizens and, therefore, has been accorded little direct relevance to imprisonment. Incarceration may have a short-term role in some instances so long as a least restrictive alternative rationale applies. Given the nature of prisons, it is reasonably clear that most prisons perform primarily a custody function. Some elements from other models may be found occasionally in individual institutions, and even then only secondarily to goals of security and custody.

Scholars should also be ever mindful that research and information about prisons has concerned mainly male prisons and inmates, neglecting female prisons and inmates. While women prisoners have faced some of the same experiences as males, many other experiences have been significantly different and often discriminatory. Some materials in chapters 6 through 8 apply to both male and female prisoners and facilities. However, because of fundamental differences the topic of women prisoners and facilities is also examined separately and in more detail in chapter 9.

The emergence of a prison system on a nationwide and massive scale was developed in the United States. This development is examined in the next section.

Development of Prisons in the United States

The concept of long-term incarceration of offenders in prisons and other types of correctional facilities is often associated with the United States. Many factors and influences came together to encourage the development of prisons in this country. As was previously discussed, the British had developed a system of institutions to handle many of those individuals who did not readily fit into society. The English gaol, workhouse, and poorhouse were all forerunners of jails and prisons in the United States.

Colonial America was inhabited by many individuals seeking a reformation of traditional social practices. There was great resentment of the harsh European criminal codes, which frequently prescribed death, disfigurement, or banishment as the punishment for a wide variety of crimes. As was noted, many early colonists came to America not as adventurous explorers, but as criminals either sentenced by the courts or fleeing prosecution. Because many citizens had experiences with harsh punishments and wanted to move away from traditional approaches, there was strong support for changing the system used to punish criminals.

Early American immigrants also brought with them very strong religious beliefs. Belief in the Judeo-Christian ethic calling for obedience to God's law and hard work was strong in colonial America. There was also a belief in the reformation of man and the ability of individuals to change and to seek forgiveness for their past sins.

Economic considerations also influenced the development of correctional institutions. America was a young nation in need of many improvements. Cheap and available labor to build roads and help support the fledgling nation's industries was always in short supply. This new nation also lacked economic strength to support large numbers of idle individuals who might find their way into criminal activity.

Economic and fiscal considerations relating to the development of penal institutions have been linked in an evolutionary sense to slavery.[1] The relationship between punishment and slavery can be traced to ancient civilizations, such as the Greeks. Enslavement of defeated soldiers and captured populations was an accepted practice in many nations, including the United States. Demands of the labor market for northern industrial workers and southern agricultural laborers can be viewed as influential factors in the establishment of a penal system in the United States.

These factors combined in the United States and provided the impetus to develop a system of correctional institutions. The prisons helped end widespread use of capital punishment, provided a lasting symbol of deterrence, and established a valuable community resource (inmate labor). Early attempts were scattered and varied, but the growth and development of prisons has been a nearly continuous process since the American colonial era.

In 1632, the Massachusetts Bay Colony established the nation's first prison. The colony's leadership erected a small wooden building in Boston. This facility served as the only place of incarceration in the colony for the next eighteen years.[2] Similar local and colonial facilities were constructed throughout America. The first state prison in the United States was established in Connecticut. The **New Gate Prison,** which was named after the infamous London institution, was first opened in 1773.* The prison consisted of a series of buildings constructed atop an abandoned vertical shaft to a copper mine (figure 6.1). Early residents of the prison were housed in the various chambers of the mine. Until its closing in 1825, the mine and prison provided both industry and housing for the inmates. Despite the wretched conditions that were known to exist in the prison, it was touted as an improvement over earlier practices of corporal punishment and execution.

Early Prison Changes

Very early in the nation's history, a movement developed to change the system of incarceration that was developing. It was recognized by many leaders of the time that two evils plagued prisons: congregate confinement and idleness. Writing of his visits to America, Alexis de Toqueville described what he called "popular truth" in the United States: "Whoever has studied the interior of prisons and the moral state of their inmates, has become convinced that communication between these persons renders their moral reformation impossible. . . ."[3] He also recognized the pains of idleness and commented that labor was the comfort for solitude.[4]

Although there was general agreement on what the problems were, there was not the same harmony when it came to addressing these issues. The debate over how best to operate a prison was framed into two schools of thought, known as the congregate and separate systems.

Congregate System

The **congregate system** began in New York in 1819 with the opening of Auburn prison. This system provided for confinement in individual cells at night but brought the inmates together into congregate workshops. Inmates were absolutely prohibited from talking with one another. Individuals were to labor in

*The New Gate Museum in East Granby, Connecticut, is now operated by the Connecticut Historical Commission as an exhibit and tour facility. For more information, contact the Commission at its headquarters in Hartford, Connecticut, 06106.

Figure 6.1 New Gate at East Granby, Connecticut, an abandoned copper mine, used as a state prison beginning in 1773
Courtesy of the New Gate Museum

silence for their transgressions. "A punitive philosophy predominated in the prison and found expression in mass treatment, rigid repression, regimentation, silence rules, unsanitary, poorly lighted cells, and a lack of anything but the most rudimentary efforts at rehabilitation."[5]

Auburn Prison was under the leadership of Elam Lynds beginning in 1830. In Auburn, "He put the prisoners to work in small strictly supervised units in workshop and out-of-doors during the daytime—to be locked up in individual cells at night. Complete silence was to be observed. A breach of this rule was punished by flogging."[6] Although verbal silence was enforced, it is well-known that inmates developed an effective nonverbal system of communication. Prisoners were held in small individual cells, often stacked several high, as depicted in figure 6.2.

Figure 6.2 The Auburn plan

Courtesy of Federal Bureau of Prisons

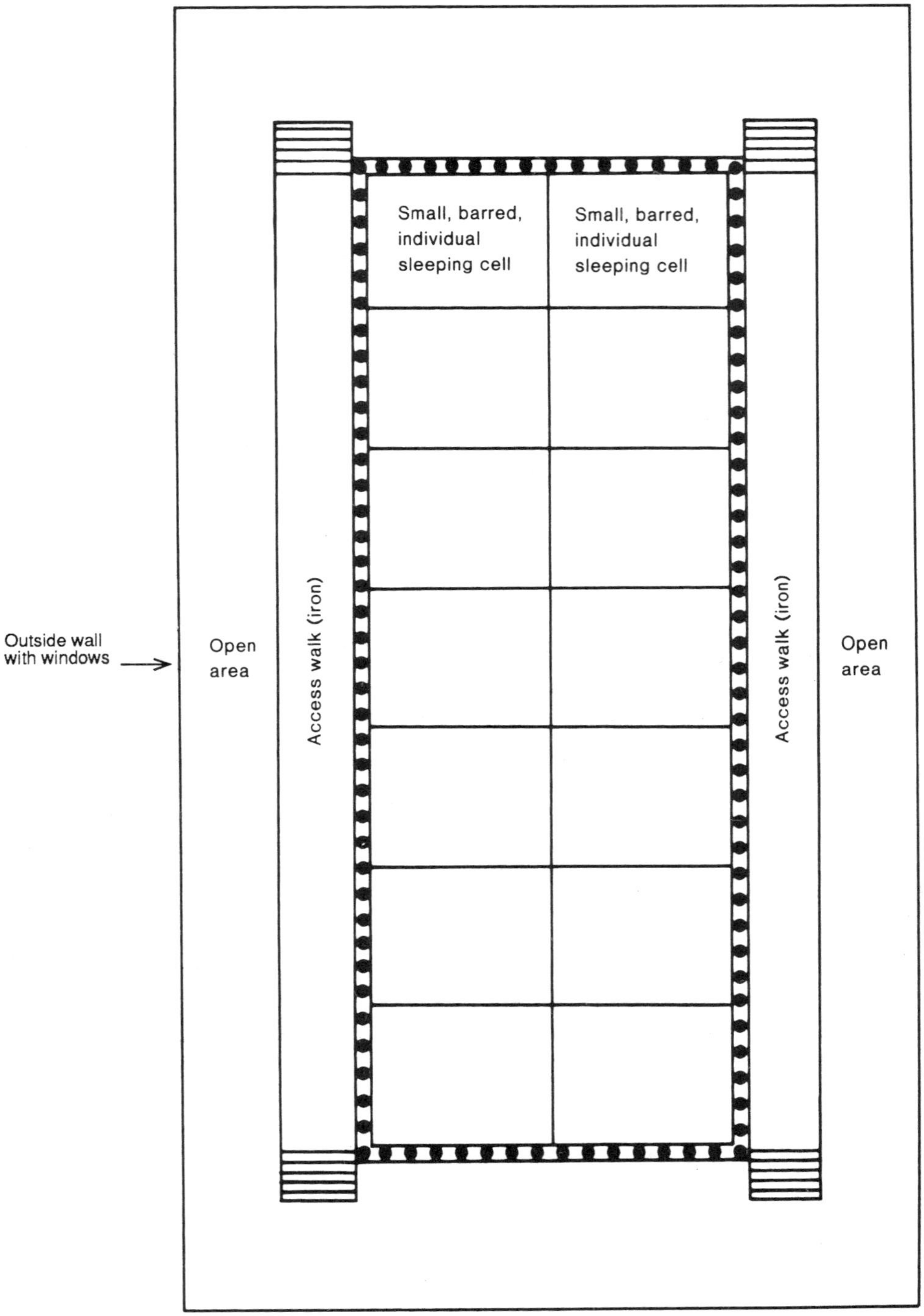

Separate System

The **separate system** was best exemplified by the Eastern Penitentiary near Philadelphia. Residents in this system lived, ate, and worked in separate cells, isolated and out of contact with other inmates, having only necessary contact with prison officials. Thus, the **penitentiary** was designed for prisoners to do penitence for their wrongdoing. After some period of time they would come to see the error of their ways, repent for their crime, and lead law-abiding lives upon release.

Cell blocks in this system were arranged in a radial fashion like spokes of a wheel from a central hub. Individual cells were larger than at Auburn, plus each cell had an attached, private outside exercise area as shown in figure 6.3.

Resolving a Debate

During the middle and latter 1800s, the nation, as well as visitors from European countries, became swept up in a great debate about whether to build prisons according to a congregate or separate system. Proponents on both sides expounded at length about the relative merits of each. In the end, economics would be the deciding factor. As prison development accelerated following the Civil War, the congregate system grew in favor primarily because of the economic advantages that could be realized from congregate work and cheaper prison construction as compared to the separate system.

Building Reformatories

The penitentiary system was well established by the time of the United States Civil War. The aftereffects of this war, as has been the case with many subsequent wars, were to be seen in the nation's prisons. The social and economic upheaval of those years of internal conflict left the nation and many of its citizens scarred. Crimes became a major concern following the war. There was again a need for cheap labor to rebuild the nation's industries and transportation systems that had been destroyed or allowed to deteriorate during the war.

In the South, the prison system took on many of the characteristics of the recently prohibited practice of slavery. Many blacks and poor white prisoners were leased out by the states to private businesses and individuals to serve their protracted sentences in a form of penal slavery.[7]

In the North and Midwest, the post-Civil war period became an era of correctional reform and major innovation in the history of prisons. The ideas of correctional reformers, such as Sir Walter Croften and Alexander Maconochie, sparked changes in the United States. Alexander Maconochie, whose efforts at penal reform on the small Norfolk Island located 930 miles northeast of Sydney, Australia, did not go unnoticed. Maconochie established two axioms for the operation of prisons:

"Brutality and cruelty not only humiliate those who suffer them, but also a society that deliberately practices or tolerates them." "The convict shall be punished for the past and trained for the future."[8]

Figure 6.3 The Pennsylvania plan

Courtesy of Federal Bureau of Prisons

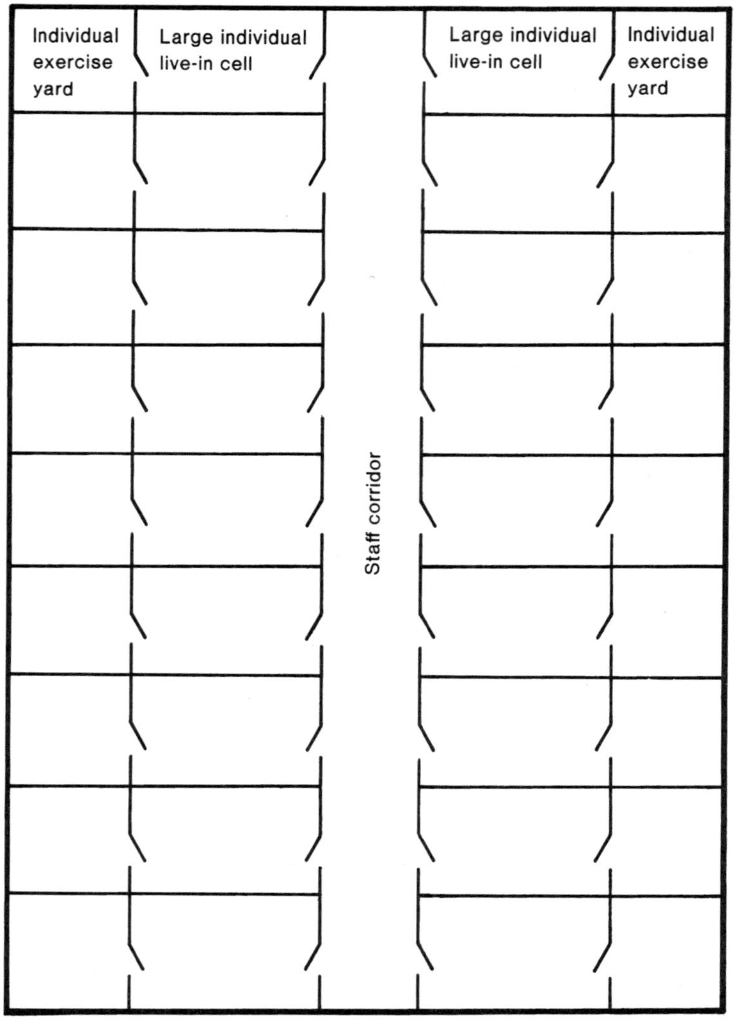

This spirit of reform led to the establishment of **Elmira Reformatory.** Elmira, under the leadership of Zebulon R. Brockway, became the new model for the nation. Prisons throughout the nation were designed on the Elmira model. These facilities were for younger, first-time offenders who had generally committed less serious crimes. Education was stressed, and prison officials decided when to release an individual based on their behavior and performance. Upon release, volunteers known as guardians provided supervision in the community, forerunners of parole and indeterminate sentences. The real promise of the **reformatory** was never completely realized as practices at the institutions reverted to previous harshness and punitive environments. Many of these facilities simply became designated as medium security prisons.

The Elmira Reformatory, known for the innovative approaches of Zebulon Brockway, superintendent during the 1800s.

Prison Facilities

Prisons generally operate to hold persons sentenced to incarceration for more than one year. The number of state and federal prisons is approaching 1100. These facilities display variations in major characteristics as shown in table 6.1. Most house only males; a small percentage house females or are coed. Some are near or beyond 100 years old. A substantial percent are relatively new, indicating recent increases in new construction. About 39 percent of facilities are less than twenty years old. Prisons also vary greatly in size, holding from several hundred to several thousand inmates.

Custody Levels

Prison facilities are designed to provide different levels of security, usually designated at the State level as maximum, medium, or minimum custody level.

- **Maximum (or close) custody prisons** are typically surrounded by a double fence or wall (usually eighteen to twenty-five feet high) with armed guards in observation towers. Fewer facilities have razor wire and electronic sensing devices. Such facilities usually have large interior cell blocks for inmate housing areas. About one in four State prisons are classified as maximum security, and about 44 percent of the nation's inmates are held in these facilities.

Table 6.1

Characteristics of State Prisons		
Characteristics	*Percent of Prisons*	*Percent of Inmates*
Total	100%	100%
Region		
Northeast	15	17
Midwest	20	20
South	48	44
West	17	19
Size		
Less than 500 inmates	65	22
500–1,000	20	27
More than 1,000	15	51
Custody level		
Maximum security	25	44
Medium security	39	44
Minimum security	35	12
Sex of inmates housed		
All male	88	91
All female	7	3
Coed	5	5
Age of facility		
Over 100 years	5	12
50–99 years	16	23
25–49 years	22	18
15–24 years	14	13
5–14 years	23	20
5 years or less	20	15
Not known	—	—

Source: Adapted from Bureau of Justice Statistics, Report to the Nation on Crime and Justice, *2d ed. (Washington, D.C.: Government Printing Office, March, 1988), p. 107.*

- **Medium custody prisons** are typically enclosed by double fences topped with barbed wire. Housing architecture is varied, consisting of outside cell blocks in units of 150 cells or less, dormitories, and cubicles. About 39 percent of all prisons are medium security and 44 percent of the nation's inmates are held in such facilities.
- **Minimum custody prisons** typically do not have armed posts but may use fences or electronic surveillance devices to secure the perimeter of the facility. More than a third of the nation's prisons are minimum security facilities, but they house only about one of eight inmates. This is indicative of their generally smaller size.[9]

Figure 6.4 Prison designs

Reprinted courtesy The Edna McConnell Clark Foundation, New York, NY

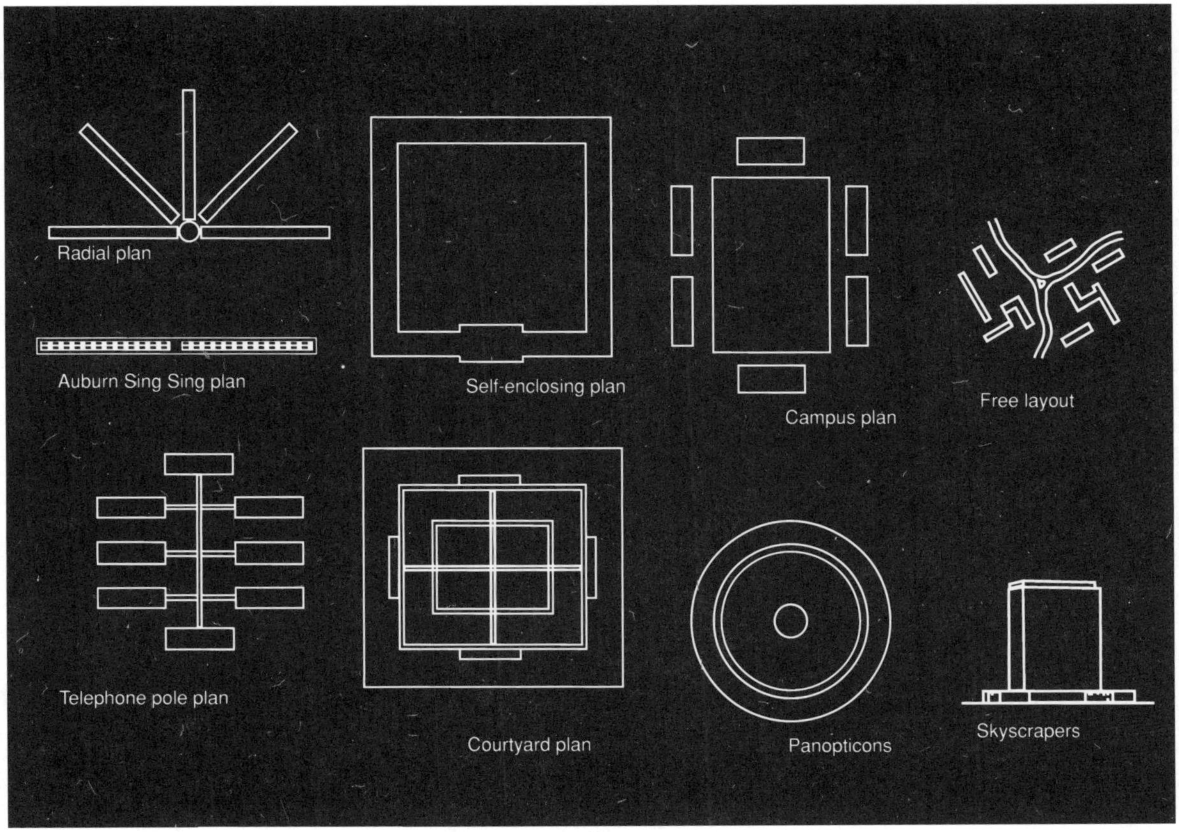

Design

Prison design and architecture has been very imaginative and varied. This is illustrated in the various diagrams in figure 6.4. Not surprisingly, prison design is often related to providing a planned level of security and custody. The Auburn and telephone pole designs have often been planned to provide maximum security. On the other hand, minimum security facilities often utilize a campus-type or free layout plan.

Prison Costs

As we learn in this section, the costs of building a prison are very high; the costs of operating a prison are also very expensive.

Construction costs are usually expressed as a cost per prison bed, arrived at by dividing total costs by the number of available beds. Application of this calculation yields the following construction cost estimates per prison bed:

Security Level	***Range***	***Average***
Maximum	$21,525–$155,300	$70,768
Medium	$16,000–$125,000	$53,360
Minimum	$7,000–$112,842	$29,599[10]

As we move into the twenty-first century these average costs could advance to $116,000 (maximum), $88,000 (medium), and $49,000 (minimum).

Across the United States a wide range in construction costs is evident. Reasons for such variation include:

- Construction costs vary by region
- Land costs vary
- Costs may be offset by using prisoner labor
- Some accounting methods may include architect fees, insurance, financing, and additional costs while others do not
- Variation in amount of space per prisoner[11]

Construction cost estimates usually do not include finance charges. This is applicable for some prisons because they are built with borrowed money. In some instances interest paid on the borrowed money could triple the original construction costs. For example, over a twenty- to twenty-five-year period, an initial construction cost of $70,000 per bed could reach in excess of $200,000 per bed when interest payments are included.

These construction costs are only the beginning. Long-term **operating costs** are staggering. Depending on such items as salary variations, security level, utility costs, and use of prison labor, yearly operating costs per prisoner have been estimated to range from about $8,600 up to $30,000.[12] Over a thirty-year period, for example, these costs far exceed construction costs. If we assume a midrange cost of $15,000 per year in a 500-bed prison, the total estimated one-year operating costs are $7.5 million ($15,000 × 500). Without even factoring in for inflation this becomes an estimated total of $225 million for thirty years of operating costs (7.5 × 30). Therefore, estimated total costs for thirty years would be:

Construction Costs	$ 35,000,000	13%
Operating Costs	$225,000,000	87%
	$260,000,000	100%

Obviously, prison bed space is extremely expensive to taxpayers. Such sobering costs should raise questions about whether or not society is benefiting

Figure 6.5 United States prison population

Source: Adapted from Diana N. Travisono, ed., Juvenile and Adult Correctional Departments, Institutions, Agencies and Paroling Authorities Directory *(American Correctional Association, Laurel, MD, 1991), p. XL, and Bureau of Justice Statistics,* Prisoners in 1993 *(Washington, D.C.: Government Printing Office, June, 1994), p. 2.*

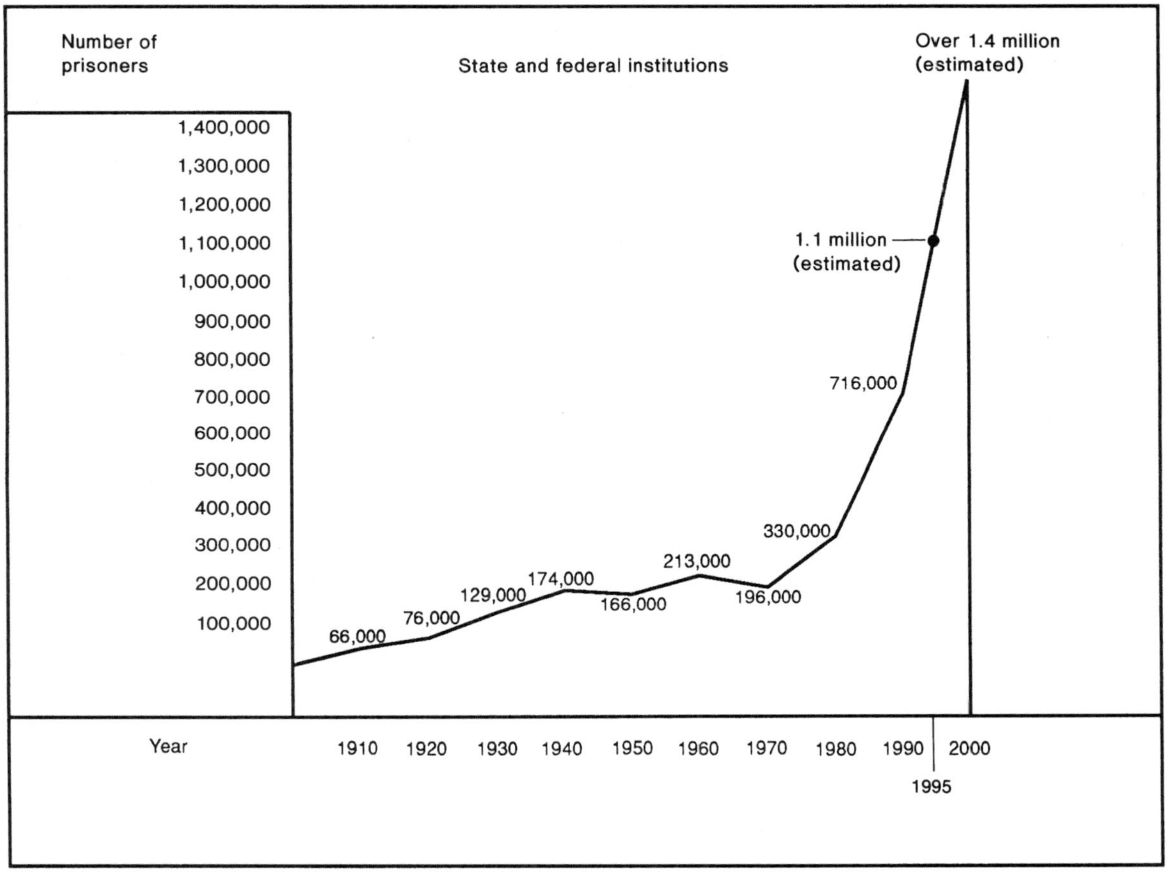

from such expenditures. Are there alternatives? Does everyone sentenced to prison need to be there? Does prison improve people? Could some prisoners serve shorter sentences? Is crime being reduced?

Prison Population

Characteristics of the **prison population** that strike us immediately are increasing absolute numbers and increases per 100,000 in the rate of incarceration. These characteristics are presented in figures 6.5 and 6.6. There have been dramatic increases in the number of individuals incarcerated in the United States. For three decades (fifties, sixties, and seventies) following World War II, the state prison population averaged under 200,000. By 1980, this figure climbed to over 300,000, approaching 800,000 by 1990, an increase of over

Figure 6.6 United States incarceration rates

Source: Adapted from Diana N. Travisono, ed., Juvenile and Adult Correctional Departments, Institutions, Agencies and Paroling Authorities Directory *(American Correctional Association, Laurel, MD, 1991), p. XL, and Bureau of Justice Statistics,* Prisoners in 1993 *(Washington, D.C.: Government Printing Office, June, 1994), p. 2.*

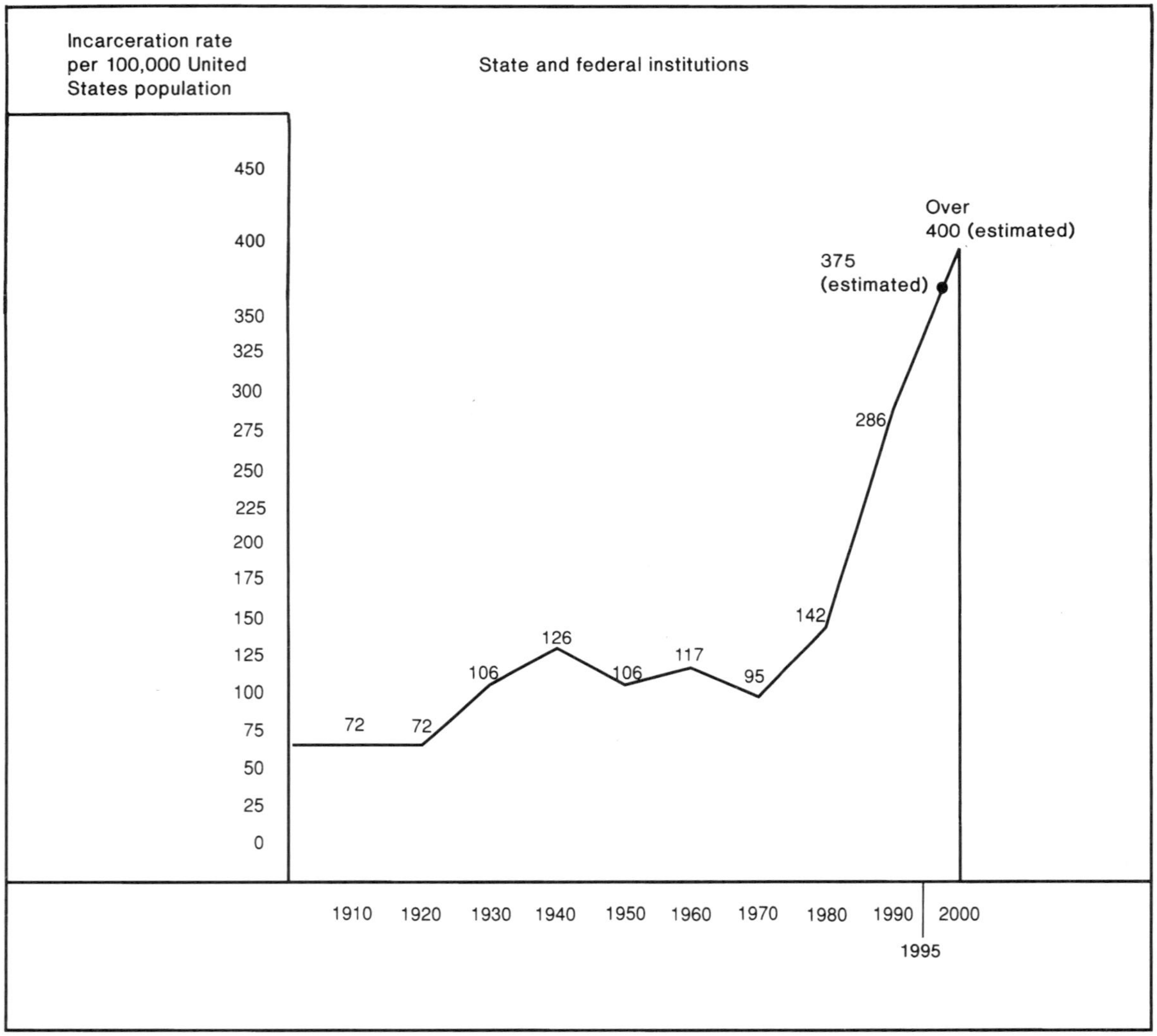

Figure 6.7 Rate (per 100,000 civilian population) of sentenced prisoners incarcerated in state institutions, by jurisdiction

Source: Adapted from Bureau of Justice Statistics, Prisoners in 1993 *(Washington, D.C.: Government Printing Office, June, 1994), p. 2.*

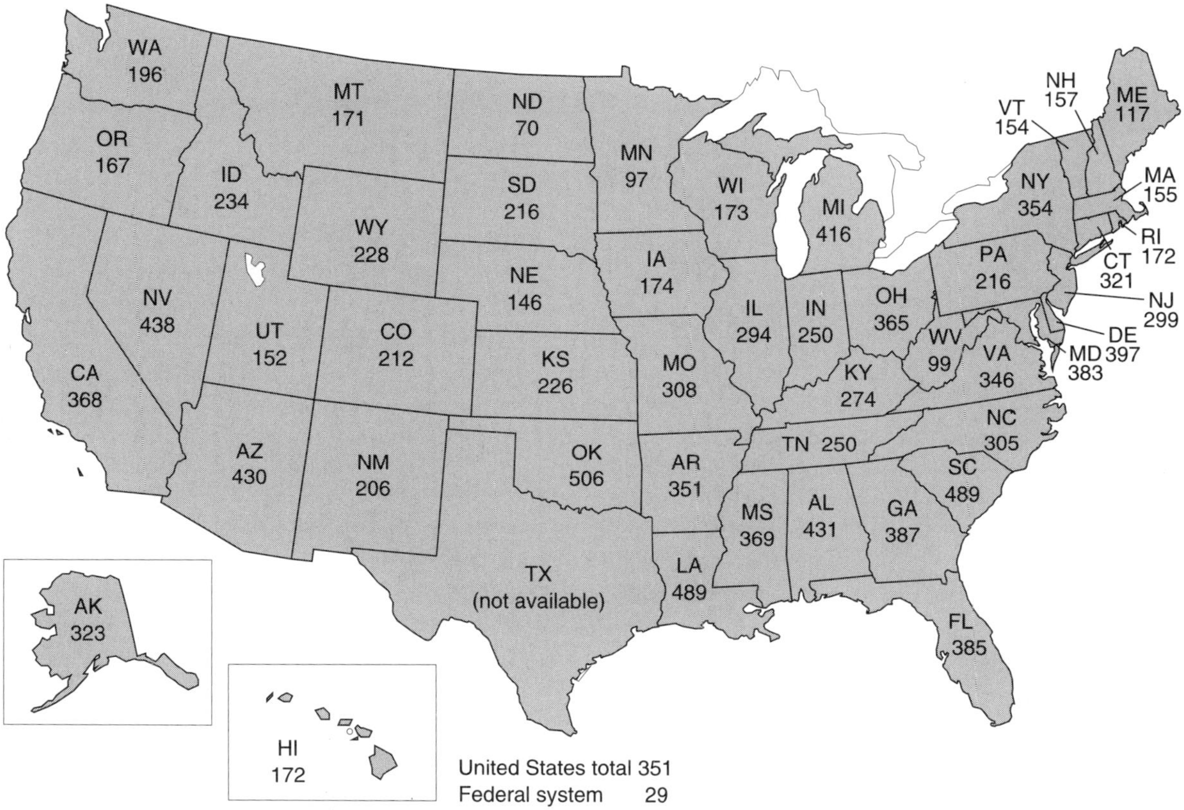

166 percent in just ten years. The number of state and federal prisoners currently exceeds 1 million. In a free society, such numbers are unprecedented. In recent years the size of the prison population has been growing by over 200 people *every day*. It is, therefore, not surprising that most prisons are overcrowded and operating beyond capacity.

During the same time period, dramatic increases were also occurring in the **incarceration rate,** which averaged around 100 during the three decades following World War II. By 1990, this rate increased 2.5 times to nearly 300, and could extend toward 400 as we enter the twenty-first century.

There are major differences in incarceration rates among the states as shown in figure 6.7. Each state makes its own policy decisions about whom to send to prison and for how long. As a result, dramatically different incarceration rates occur which do not necessarily correlate with a state's violent crime rate or with per-capita costs of corrections. For example, at the time of a recent study, Florida ranked number one in the violent crime rate and twelfth for the incarceration rate; New York ranked second in the violent crime rate but twentieth for the incarceration rate. Delaware was ranked fourth for the incarceration rate and twenty-second in the violent crime rate; South Carolina ranked second for incarceration rate and tenth for violent crime rate.[13]

Classification of the Prison Population

Classification may be thought of as a process whereby a prison population is divided into subgroups on the basis of security and program needs. This provides a mechanism for correctional officials to review and identify their population, to group them into categories, and to work with them in an efficient and effective manner toward the goal of operating a safe and orderly prison.

The term classification describes a number of distinct yet interrelated processes that operate within a correctional institution and throughout a correctional system. The generic use of the term applies to all formal decision-making processes regarding an inmate. The classification scheme of a particular correctional department encompasses three subfunctions of classification.

Systemwide classification refers to the initial decisions that are made by a corrections department regarding the degree of custody, level of supervision, and institutional assignment of a newly sentenced offender. This initial process generally takes place at a central reception-orientation-classification center.

Institutional classification is the first formal process the inmate goes through upon arrival at the assigned correctional institution. The institutional classification committee assigns specifics tasks, housing, and activities for the inmate within the parameters established earlier by the systemwide classification process. The major components of the inmate's official life in prison are established and subsequently monitored through institutional classification.

Reclassification is a subunit of the institutional classification process. Reclassification is the monitoring and adjustment phase of institutional classification; it reviews and formalizes changes in the inmate's official prison status. Thus the term classification describes a variety of activities that significantly influence and direct the incarcerated individual's life in prison.

Ideally, classification seeks to match inmate needs with correctional resources. However, in reality, treatment needs are usually outweighed by determining an appropriate custody and security level for each prisoner. After review and assessment, state inmates are typically assigned to a maximum, medium, or minimum security level prison. In some instances prisoners with special needs such as AIDS and other medical problems, mental health problems, or educational needs are classified into subgroups for appropriate handling.

Security Level Designations

Prisons operate mainly according to a custodial model. Decisions in prison are based largely on public safety and custody or security needs, not on the needs of inmates. The level of custody granted a prisoner is considered to be crucial by most inmates. The application of security designations is illustrated by the following example from Oregon:

> Maximum security is reserved for active and extreme escape risks; individuals who are a continuing source of agitation; and inmates who pose a threat of actual or potential physical violence toward others. Maximum security is only assigned after a special administrative hearing which considers such factors as disciplinary reports, the length and number of stays in disciplinary isolation, prior history of rules violation and also individual inmates' requests for maximum security confinement. Individuals under maximum security are provided with special housing and are only permitted out of their cell/room in the custody of a staff member.
>
> Close security is reserved for inmates whose trustworthiness has not been demonstrated. The designation serves as an alert to staff of the potential security risk of the so designated inmates. Close security inmates are permitted free movement within designated secure areas.
>
> Medium security designations are given to inmates who have demonstrated some degree of trustworthiness. The inmate's previous observed behavior and/or verified history of conduct are used to make such a determination. These inmates are allowed out of the perimeter security of the institute without constant supervision, but under the direction of a staff member.
>
> Minimum security is assigned to inmates who appear from all indications to be trustworthy. Inmates sentenced for the crime of aggravated murder are not eligible for minimum security, until they have served their minimum sentence. These inmates are allowed to participate in work and programs beyond the perimeter security area without supervision.[14]

Special Populations

There are some special groups of individuals who are incarcerated. These special populations differ significantly from typical inmates. Three examples of special groups selected for additional discussions are inmates with AIDS (Acquired Immune Deficiency Syndrome), prison groups, and elderly inmates. As shown by table 6.2, as we move further into the 1990s, a total of 11,565 confirmed AIDS cases have been reported among inmates across the nation in federal, state, and larger city, county correctional systems. This is a cumulative total since the epidemic began. In addition, a cumulative total of 3,474 inmates have died of AIDS while in custody.[15]

AIDS and Prison Inmates

AIDS disease was first recognized in 1981 and is caused by a virus called **HIV** (Human Immunodeficiency Virus). This virus damages the body's immune system, causing the body to lose its ability to fight off germs; these germs then go on to cause certain diseases that do not occur in people with healthy immune systems, such as PCP pneumonia and certain cancers. Such diseases are usually fatal.

Table 6.2

Cumulative Total AIDS Cases among Correctional Inmates and the Population at Large, U.S., 1985–1992/1993

	Correctional Cases[a]		*Cases in Total U.S. Population*[b]	
	Number of Cases	*% Increase from Preceding Survey*[c]	*Number of Cases*	*% Increase from Preceding Survey*[d]
November 1985	766	N/A	14,519	N/A
October 1986	1,232	61%	26,002	79%
October 1987	1,964	59	41,770	61
October 1988[e]	3,136	60	73,621	76
October 1989[f]	5,411	72	110,333	50
October 1990[g]	6,985	29	152,231	38
November 1992–March 1993[h]	11,565	66	249,199	64

[a]The figures in this and other tables represent inmate AIDS cases in the federal prison system, all 50 state prison systems, and a *sample* of 28–37 city and county jail systems (depending on the year of the NIJ Survey).

[b]Adult/adolescent cases only. Pediatric cases excluded.

[c]In most cases, the column presents the difference between the number of cases in the given year and the number in the prior year. The exception is November 1992–March 1993.

[d]As with the correctional cases, the percent increase reflects the change from the preceding survey. In most cases, this is a one-year interval. However, the percent increase for November 1992 through March 1993 reflects a two-year interval.

[e]Figures for 1988 include 28 city/county jail systems.

[f]Figures for 1989 include 32 city/county jail systems.

[g]Figures for 1990 include 27 city/county jail systems.

[h]Figures for 1992/1993 include 31 city/county jail systems.

Sources: CDC, AIDS Weekly Surveillance Reports—U.S., *November 4, 1985, October 5, 1986, October 5, 1987, October 3, 1988; CDC,* HIV/AIDS Surveillance Report, *November 1990, February 1993 (cases reported through 1992); NIJ/CDC Questionnaire Responses.*

All people who test positive for HIV do not have full-blown AIDS. However, infected people without symptoms can spread the infection even if they never develop full-blown AIDS. Doctors must do a series of blood tests to determine if a person has HIV/AIDS infection.

For HIV to spread, infected blood must enter the bloodstream of another person. This may occur through heterosexual or homosexual contact and blood-to-blood contact, such as sharing drug needles and passage from an infected mother to her child during pregnancy. HIV is not transmitted through casual contact.

Recent estimates concerning HIV and AIDS among the nation's prison and jail inmates include the following:

- over 17,000 (2.2%) inmates were infected with HIV
- during a one-year period, 528 inmates (28%) in state prisons were attributable to AIDS

- during a one-year period 24% of all deaths in jails were AIDS related
- during a one-year period 51% of state prison inmates indicated being tested for HIV and knowing the results[16]

Such a disease carries several implications for managing inmates with AIDS, including: (1) preventing the spread of AIDS to other prisoners; (2) legal rights of AIDS infected prisoners; and (3) employee concerns.[17]

Until a vaccine or cure is found, preventive education is the most effective means to help prevent the spread of AIDS to others in the prison population. Offenders and staff alike generally view educational efforts positively. Failure of a prison to provide such education might be harmful if an inmate in custody filed such a suit.

A more controversial approach is whether all inmates should submit to mandatory testing for HIV. All states, the federal Bureau of Prisons and the District of Columbia test inmates on some basis for HIV. As shown by table 6.3, a variety of policies have evolved. Seventeen jurisdictions test all prisoners, either at admission, release, or during custody. The other thirty-five jurisdictions test only some prisoners.

Inmates diagnosed with AIDS present special treatment challenges. Appropriate treatment should be available and be dictated by the medical needs of the patient. In this regard, staff education is essential. Thought needs to be given to adequately preparing for the intense medical needs of AIDS-infected inmates. Failure to do so could pose great legal risks.

Some inmates may test positive for HIV but present no symptoms. The question arises as to who may be notified without violating the offender's right to privacy. In fact, some states prohibit the disclosure of test results without consent of the person. Such individuals pose little significant risk to employees, but medical staff should be informed. Written policy for notifying medical personnel should be developed and conform to applicable laws of confidentiality and disclosure.

Another issue is whether those with HIV/AIDS should be segregated from the rest of the prison population. In some facilities a decision has been made to segregate, while in others segregation has been avoided. In general, the courts have upheld both approaches as long as the discretionary decision of correctional officials was based on legitimate health, safety, and institutional security considerations. For example, segregation policies were upheld in *Cordero v. Coughlin,*[18] and nonsegregation policies were upheld in *LaRocca v. Dalshein.*[19]

A central issue is consideration of similar circumstances. It is not permissible to arbitrarily refuse one person's privileges that are granted to others in similar circumstances. However, if an inmate's situation differs from those granted a privilege, there is no basis for legal complaint. A trend away from blanket segregation seems to be continuing.[20] Isolating every inmate appears to be counterproductive and overrides classification schemes. To date there has been no documented case of job-related HIV infection or AIDS among criminal justice (including corrections) personnel. However, employees need to be

Table 6.3

Testing Policies for the Antibody to the Human Immunodeficiency Virus That Causes AIDS

All Incoming Inmates	*All Inmates Currently in Custody*	*All Inmates at Time of Release*	*High-Risk Groups*	*Upon Inmate Request*	*Upon Clinical Indication of Need*	*Upon Involvement in Incident*	*Random Sample*	*Other*
Alabama	Rhode Island	Alabama	Arkansas	Alaska	Alaska	California	Arkansas	Hawaii
Colorado	Utah	Federal	Connecticut	Arizona	Arizona	Florida	Dist. of Col.	Illinois
Georgia	Wyoming	Missouri	Dist. of Col.	Arkansas	California	Hawaii	Federal	Mississippi
Idaho		Nevada	Illinois	California	Colorado	Kentucky	Maryland	New Jersey
Iowa		Wyoming	Indiana	Colorado	Connecticut	Louisiana	Massachusetts	New Mexico
Michigan			Kentucky	Connecticut	Delaware	Maryland	New York	North Carolina
Mississippi			Minnesota	Delaware	Dist. of Col.	Massachusetts	Wyoming	Oregon
Missouri			New York	Dist. of Col.	Federal	Michigan		South Carolina
Nebraska			North Carolina	Federal	Florida	Minnesota		Tennessee
Nevada			Ohio	Florida	Georgia	Missouri		Washington
New Hampshire			South Carolina	Georgia	Hawaii	New Hampshire		Wisconsin
North Dakota			South Dakota	Hawaii	Illinois	New Jersey		
Oklahoma			Tennessee	Indiana	Indiana	New Mexico		
Rhode Island			Texas	Kansas	Kansas	New York		
Utah			West Virginia	Kentucky	Kentucky	Ohio		
Wyoming				Louisiana	Louisiana	South Carolina		
				Maine	Maryland	Tennessee		
				Maryland	Massachusetts	Texas		
				Massachusetts	Michigan	Virginia		
				Michigan	Minnesota	Wyoming		
				Minnesota	Mississippi			
				Missouri	Missouri			
				New Hampshire	Montana			
				New York	New Hampshire			
				North Carolina	New Jersey			
				Ohio	New Mexico			
				Oregon	New York			
				Pennsylvania	North Carolina			
				Rhode Island	Ohio			
				South Carolina	Oregon			
				South Dakota	Pennsylvania			
				Tennessee	Rhode Island			
				Texas	South Carolina			
				Vermont	Tennessee			
				Virginia	Texas			
				Washington	Virginia			
				West Virginia	Washington			
				Wisconsin	West Virginia			
				Wyoming	Wisconsin			
					Wyoming			

Source: Adapted from Bureau of Justice Statistics, HIV in U.S. Prisons and Jails *(Washington, D.C.: Government Printing Office, September, 1993), p. 3.*

educated and trained to understand the disease in order to maintain safe and healthy working conditions and to be in a position of strength should lawsuits arise. Documentation of training and precautionary procedures is important.

Strong educational programs aimed at corrections workers can help counter unfounded rumors, misinformation and work disruptions. Table 6.4 presents some frequent concerns accompanied by an appropriate action message.

Table 6.4

Educational and Action Messages for AIDS-Related Training of Law Enforcement and Criminal Justice Personnel

Issue/Concern	Educational and Action Messages
Human bites	• The person who bites is typically exposed to the victim's blood, rather than the reverse; therefore, the victim is at extremely low risk for HIV infection. • HIV transmission through saliva is highly unlikely because the virus has only been isolated in extremely low concentrations in saliva. • If bitten by an individual who has tested seropositive, one should allow the wound to bleed, wash the area thoroughly, and seek medical attention.
Spitting	• Viral transmission through saliva is highly unlikely. • CDC no longer recommends "universal precautions" for saliva.
Urine/feces	• HIV has been isolated in only very low concentrations in urine and not at all in feces. • There have been no cases of AIDS or HIV infection associated with either urine or feces. • CDC no longer recommends "universal precautions" for urine or feces.
Cuts/puncture wounds	• Use caution in handling sharp objects and searching areas hidden from view. • Needle-stick studies show risk of infection is very low.
CPR/first aid	• Use masks/airways to eliminate the minimal risk of HIV transmission associated with CPR. • Avoid blood-to-blood contact by keeping wounds covered and wearing gloves when in contact with bleeding wounds.
Body removal	• Observe the crime scene rule: do not touch anything. • Those who must come into contact with blood or other body fluids contaminated with visible blood should wear gloves in accordance with official policy and CDC guidelines.
Casual contact	• No cases of AIDS or HIV infection are attributed to casual contact.
Contact with blood or body fluids	• Wear gloves if contact with blood, semen, or body fluids containing visible blood is likely. • If contact occurs, wash thoroughly with soap and water; clean spills with 1:10 solution of household bleach and water.
Contact with dried blood	• The drying process inactivates the virus. Laboratory studies showing persistence of AIDS virus for 3 days in dried samples used viral preparation 100,000 times more concentrated than that found in normal blood samples.

Source: Theodore M. Hammett, AIDS and HIV: Training and Education in Criminal Justice Agencies *(Washington, D.C.: National Institute of Justice, August, 1989), p. 3.*

Elderly Inmates

It has been suggested that elderly inmates are "forgotten people."[21] Research data about this group are limited and only recently has much attention been given to studying this group. The literature reflects some differences in the lower age of elderly or geriatric inmates. Age fifty is sometimes considered as the lower cutting point, while others have considered fifty-five or higher as necessary for inclusion into their group. In many systems, the term geriatric is applied to inmates age fifty-nine or older.[22]

Systematic and nationwide information regarding the numbers trends and rates of incarcerated elderly inmates is not readily available. Nevertheless, there is every indication that the numbers are increasing and will continue to do so. For example, it was reported that during one year in Florida, inmates age fifty-one and older increased 51 percent, while the total prison population grew by 25 percent.[23]

Whatever the actual numbers may be, there are significant numbers of elderly inmates incarcerated in the nation's prisons who need a variety of special care considerations.

Chief among these is medical attention. It has been reported that "nearly every geriatric inmate has some long-term debilitation that requires frequent medical attention."[24] Many inmates suffer from such illnesses as diabetes, pulmonary diseases, circulatory problems, arthritis, and even Alzheimer's disease. Some are chronically ill and die in prison.

It must also be considered that as the body ages nutritional requirements change. Type and composition of food required will differ from those of the young. Various digestive problems may be more pronounced.

Housing of elderly inmates also requires special consideration. Some may require housing segregated from the general population in order to reduce the risk of victimization, to gain access to buildings, and to deal with adjustment problems related to living in prison.

The list of special needs for this population is long, complex, many times expensive, and are just beginning to be addressed. One group of recommendations includes the following:

1. Expanding and making accessible educational, rehabilitation, and recreational programs
2. Structuring programs that do not compete with younger inmates
3. Incorporating arts and crafts geared for the elderly
4. Providing appropriate levels of recreational activities
5. Greater consideration given to housing elderly inmates in separate facilities[25]

Prison Gangs

It has been estimated that about 3 percent of the prisoner population are gang members, and most prison systems have reported the existence of prison gangs.[26] Their presence has become significant in prisons in many states and especially so in Illinois, California, and Pennsylvania.

Gang members usually reflect gangs found outside prisons across the nation. Such groups are often organized along racial, ethnic, and geographical location. The names of several prominent inmate groups reflect these deep racial divisions, names such as La Nuestra Familia (Latin), Black Guerrilla Family, Mexican Mafia, and Aryan Brotherhood. Obviously, such divisions carry a great potential for inmate violence, which sometimes occurs in gang-related assaults and killings. Gangs often continue to network with their counterparts on the street and in other prisons.

It has been suggested that:

> Inmates join gangs for many reasons, often to pursue illegal activities, but some reasons are legitimate: companionship, a sense of belonging, protection, and to provide direction for their lives. Leaders and hard-core members form the basis for operations by writing bylaws, designing organizational structure, maintaining correspondence, and developing a means of gaining strength by enticing or forcing others to join. Many in the criminal justice system believe the group is more dangerous than the individual.[27]

Gangs seek to control the illegal operations of the prison. This would include drug trafficking, extortion, gambling activities, and prostitution.

Although small by a percentage basis, the presence of gangs can be a major disruptive force in a prison population. Corrections has been slow in developing an awareness of this problem, training staff to respond, and developing an overall strategy. It has been noted that "corrections officials can make confronting this issue easier, more consistent, and less slipshod by dealing with it more professionally and emphasizing prevention rather than reacting to events."[28] Toward that end, recommendations have included the following:

1. Developing policy and control strategies
2. Early detection and tracking
3. Adopting successful control models
4. Developing a national clearinghouse of information
5. Developing expert task forces
6. Maintaining communication with gang members[29]

Persistent Felons

Persistent felons comprise an identifiable special population within prison. One study estimated that in some states approximately 45 percent of inmates could be classified as persistent felons.[30] As more states enacted legislation providing for long-term incarceration, the number of such inmates increased.

Persistent felon inmates have been convicted of repeated felony violations. In many states, an individual who has been convicted of two or three felonies may be considered a persistent felon. Generally, the criminal statutes provide for a sentence enhancement or a separate (additional) sentence for such offenders. The lengths of these additional mandatory sentences vary, but most range from three to ten years in length (many of which do not allow for any form of early release).

Some factors associated with persistent criminals include lower socioeconomic status, limited employment and education, and a history of substance abuse. Race is also a factor: "Blacks account for slightly less than 12 percent of the American population . . . but they account for 52 percent of all arrests for violent crime."[31] The profile of a persistent felon greatly resembles the "state raised youth" described by John Irwin.[32] A young offender with a history of criminal behavior becomes a persistent felon who is starting to dominate the nation's prisons.

Persistent criminals, because of longer sentences that prohibit early release, are a larger proportion of United States prison populations. These individuals present special considerations for both custody and program staffs. For example, they may require a higher level of supervision, thus creating a greater demand for maximum security bed space. However, these do not appear to become involved in a significantly greater amount of institutional violence than other inmates,[33] and they seem to participate at about the same level as other inmates in work and treatment programs.[34] However, longer sentences make it more difficult to plan programs because they may not seek out change-oriented activities until well into their terms. For example, it makes little sense to train an inmate in a vocational skill that he cannot practice on the streets until some distant time. Developing innovative programs that address the needs of persistent felons presents special challenges.

Summary

A prison sentence is dramatic and easily visualized in people's thoughts. More people than ever before are being incarcerated in the nation's prisons, which are serving primarily a custodial function.

Americans rejected harsh corporal and capital punishment methods, establishing incarceration on a large-scale basis, which at the time was considered less harsh and economically beneficial by providing free labor. Major debates ensued about whether to develop prisons according to a congregate or separate system. Eventually, mainly for economic reasons, most were built according to a congregate system.

State prisons generally hold inmates whose sentence of incarceration exceeds one year. There are approximately 800 such facilities, some very old and others very new. Different prisons are designed to hold persons with maximum, medium, or minimum levels of custody.

Prisons are very expensive for taxpayers to construct and operate. The per-bed construction costs reach into the tens of thousands of dollars and longer term operating costs per facility go into the hundred of millions of dollars.

Both the number of people in prison and the incarceration rate per 100,000 civilians are at all time highs. In recent years the prison population has been growing by over 200 every day. Most prisons are overcrowded.

Prisoners are placed in various categories by a classification process. The major element in this process is to determine in which custody level—maximum, medium, or minimum—to house the inmate.

Numerous special groups of people are incarcerated. Three examples selected for discussion are inmates with AIDS (Acquired Immune Deficiency Syndrome), elderly inmates, and prison gangs. Each of these groups present different and difficult challenges, which have prompted various recommendations for prison management.

Discussion Topics

1. What goals should prisons serve? Why?
2. How might prison costs be reduced?
3. What could be done to reduce prison overcrowding?
4. Why do state incarceration rates and crime rates show differences?
5. Are there ways that the classification process could be improved?
6. Should elderly inmates be housed in separate units? Why or why not?

Additional Readings

Colvin, Mark. *The Penitentiary in Crisis.* Albany, NY: State University of New York Press, 1992.

McKelvey, Blake. *American Prisons: A History of Good Intentions.* Montclair, NJ: Patterson-Smith, 1977.

Petersilia, Joan and Paul Honig. *The Prison Experience of Career Criminals.* Washington, D.C.: U.S. Department of Justice, 1976.

Smolowe, Jill. ''. . . And Throw Away the Key.'' New York, NY: *Time,* February 7, 1994, 54–59.

Notes

1. Thorsten J. Sellin, *Slavery and the Penal System* (New York: Elsevier, 1976).
2. Robert M. Carter et al, *Correctional Institutions* (Philadelphia: J.B. Lippincott, 1972), 19.
3. Gustave de Beaumont and Alexis de Toqueville, *On the Penitentiary System in the United States and Its Application in France* (New York: Augustus M. Kelly Publishers, 1970), 21.
4. Ibid., 23.
5. Carter et al., *Correctional Institutions,* 26.
6. Torsten Iriksson, *The Reformers* (New York: Elsevier, 1976), 50.
7. Sellin, *Slavery and the Penal System,* 145.
8. Eriksson, *The Reformers,* 83.
9. Bureau of Justice Statistics, *Report to the Nation on Crime and Justice,* 2d ed. (Washington, D.C.: Government Printing Office, March, 1988), 107.
10. Ibid., 124.
11. Bruce Cary and Stephen Getlinger, *Time to Build? The Realities of Prison Construction* (Edna McConnell Clark Foundation, New York, NY, 1984), 15–19.
12. Bureau of Justice Statistics, *Census of State and Federal Correctional Facilities* (Washington, D.C.: Government Printing Office, May, 1992), 17.
13. *Corrections Compendium,* ''Ranking the States'' (Lincoln, NE, February, 1987), 9–10.
14. Thomas G. Toomles, ''A Matrix System for Custody Classification and Supervision,'' in *Classification* (College Park, MD: American Correctional Association, 1981), 14.
15. Theodore M. Hammett, et al., *1992 Update: HIV/AIDS in Correctional Facilities* (Washington, D.C.: National Institute of Justice, January, 1994), 14–15.
16. Bureau of Justice Statistics, *HIV in U.S. Prisons and Jails* (Washington, D.C.: Government Printing Office, September, 1993), 1–6.
17. Marianne Takos and Theodore M. Hammett, *Legal Issues Affecting Offenders and Staff* (Washington, D.C.: National Institute of Justice, May, 1989).
18. 607 F. Supp. 9 (S.D.N.Y., 1984).
19. 467 N.Y.S. 2d 302 (App. Div. 1983).
20. Moini and Hammett, *1989 Update,* 53.
21. Gennaro F. Vito and Deborah G. Wilson, ''Forgotten People: Elderly Inmates,'' *Federal Probation* (March, 1985): 18.
22. A. W. Kelsey, ''Elderly Inmates: Providing Safe and Humane Care,'' *Corrections Today* (May, 1986): 56.

23. Richard L. Dugger, ''The Graying of America's Prisons,'' *Corrections Today* (June, 1988): 28.
24. Kelsey, ''Elderly Inmates . . . ,'' 56.
25. Dugger, ''The Graying of America's Prisons,'' 31–34.
26. United States Department of Justice, *Prison Gangs: Their Extent, Nature and Impact on Prisons* (Washington, D.C.: Government Printing Office, 1985), vii.
27. Steve Daniels, ''Prisons Gangs: Confronting the Threat,'' *Corrections Today* (April, 1987): 66.
28. Ibid., 162.
29. United States Department of Justice, *Prison Gangs,* xix.
30. Joan Petersilia and Paul Honig, *The Prison Experience of Career Criminals* (Washington, D.C.: U.S. Department of Justice, 1980), xii.
31. John P. Conrad, ''It is Very Hard to Predict, Especially the Future,'' in *Confinement in Maximum Custody,* eds. David A. Ward and Kenneth F. Schoen (Lexington, MA: Lexington Books, 1981), 6.
32. John Irwin, *Prisons in Turmoil* (Boston: Little, Brown and Company, 1980), 51.
33. Petersilia and Honig, *Prison Experience of Career Criminals,* 81.
34. Ibid., xvi.

Chapter 7

Inmate Life and Prisoner Rights

Learning Objectives

After studying this chapter students should be able to:

1. Explain how the concept of total institutions is often useful in describing life in prison.
2. Describe traditional and more contemporary views regarding the nature of the inmate subculture.
3. Discuss some everyday problems faced by many in prison.
4. Understand that life in a prison in the United States is designed to be a life of sexual deprivation.
5. Describe the hands-off doctrine.
6. Identify when and how prisoners began to win some constitutional rights.
7. Identify some major sources of rights.
8. Recall four major categories of rights.
9. Identify several specific cases within each category and the decision rendered.
10. Evaluate various outcomes resulting from court intervention.

Key Terms

Access to courts
Administrative segregation
Civil rights
Conjugal visits
Contempt powers
Court intervention
Cruel and unusual punishment
Deprivation model
Disciplinary actions
Eighth Amendment Rights
Equal protection clause
First Amendment Rights
Formal reward systems
Fourteenth Amendment Rights
Furlough programs
Good time
Hands-off doctrine
Hustling
Importation model
Increased privileges
Inmate code
Inmate subculture
Lowered security levels
Mushfake
Niches
Ordered segmentation
Prisoners' rights
Prison homosexuality
Section 1983
Sexual deprivation
Significant others
Slave of the state
Solitary confinement
Total institutions
Writ of certiorari
Writ of habeas corpus

Introduction

A condition of incarceration affects inmates in different ways. Individuals who are imprisoned go on with their lives and adjust to the rigors of congregate living in confinement. There are, however, a number of factors that influence the lifestyle of inmates and their level of acceptance of life in prison. An individual's personal characteristics and ability to adjust are key determinates of the lifestyle adopted in prison. Another primary consideration in determining lifestyle in prison includes the security level of the institution. The variance in living conditions and activity options is tremendous among prisons. For example, an inmate housed in an open dormitory in a maximum security institution has many adjustments to make that are not required of the prisoner in a single room in a minimum security facility. The size and composition of the inmate population can greatly affect the stresses that are placed upon the incarcerated individual. In general, the larger the correctional facility, the greater the stress placed upon each inmate. Some prison populations are homogeneous in terms of criminal background, but most have a very heterogeneous population that can make for added stress and possible confrontation for the resident.

Total Institutions

Erving Goffman, in his classic *Asylums,* illustrated the dehumanizing effects of **total institutions.** Those who live and work in such environments are directly and sometimes dramatically affected by the experience.

> The central feature of total institutions can be described as a breakdown of barriers ordinarily separating . . . spheres of life. First, all aspects of life are conducted in the same place and under the same single authority. Second, each phase of the member's daily activity is carried on in the immediate company of a large batch of others, all of whom are treated alike and required to do the same thing together. Third, all phases of the day's activities are tightly scheduled, with one activity leading at a prearranged time into the next, the whole system of explicit formal rulings being imposed from above by a system and a body of officials. Finally, the various enforced activities are brought together into a single rational plan purportedly designed to fulfill the official aims of the institution.[1]

Total institutions generally have barriers to social integration with the outside world built into their structure. Sometimes these barriers are formal and take on strict prohibitions against outside contact. In other institutions, the barriers are informal and merely develop over time. The total institution typically provides physical restrictions, in that residents are maintained within a certain identified area. Generally, all aspects of life are conducted in this identified space and under a single authority or command.

In total institutions, most activities are conducted in groups, and the residents' time is scheduled, generally by some central authority or supervisor.

Institutions considered total in nature are generally under a bureaucratic managerial system with clear lines of separation between residents and staff.

Most prisons in the United States fit the description of total institutions, especially those designated as maximum and medium security levels, although some conform more closely than others. Prisons today are more open to public scrutiny, and inmates have gained some rights that have improved their life and conditions. Nevertheless, many prisons remain to a great degree as total institutions.

Throughout the following discussions of life in prison, the role and behaviors of prison residents are examined. It is important to keep in mind that the special environment of a total institution has profound effects on the everyday lives of these individuals. The artificial and unnatural conditions created in a correctional institution can shape and influence day-to-day and future behaviors. The confined quarters, an authoritarian system of control, and a forced confinement combine to create a unique experience known as prison life.

Although many prisons remain, to a great degree, total institutions, some efforts have been made to make the prison environment more normal. Operations designed to increase normalization have been reported by Miles D. Harer; examples include the use of a human relations approach to managing inmates, employing female correctional officers, developing inmate grievance procedures, expansion of furlough and visitation programs, and support for prison educational programs.[2]

Inmate Subculture: Traditional Views

Early studies of **inmate subculture** began with a description of prisons as self-contained and isolated institutions. Focus was placed on prisoners as being subject to numerous restrictions within a very coercive environment, cut off from the world outside. In an effort to manage their life in this generally deprived situation, an accommodating subculture was generated by the inmates inside the prison. Such explanations are commonly identified as a **deprivation model** because the origin of this subculture was considered to be from inside the prison in response to negative and forced living conditions found therein.[3]

Deprivation Model

This model assumes that the key elements in the development of prison social systems are the depersonalization and stigmatization associated with incarceration and the adjudication process of the criminal justice system. This model views the prison subculture as an unintended negative outcome of the organization and operation of prisons.

The inmate is removed from his or her home (normal environment) and is thrust into a highly stratified social order. The new inmates have hit ''rock bottom'' and now must fend for themselves in an alien environment. Placing many individuals with similar problems together in a confined and restrictive environment yields a unique social order. Thus goes the logic for a deprivation model that maintains that the subculture arises to provide self-esteem for the individual inmate through resistance to and manipulation of the prison's organizational structure. The cohesion and solidarity among inmates facing common problems helps reduce the pains of imprisonment.[4] Inmates and the prison officials are on opposite sides, and the inevitable conflict that occurs directly affects the development and maintenance of the inmate subculture.

Importation Model

Other scholars challenged a deprivation model proposing that inmate values were brought into the prison.[5] It has been pointed out that the deprivation model is far too restrictive in its view of the prison social organization and the various influences upon it. The model has also been criticized because it does not account for particular subcultural reactions to positive change-oriented reward systems that exist in many correctional institutions.[6]

The **importation model** focuses on preprison and extraprison influences on the development of the inmate social system. Without discounting the importance of issues identified by a deprivation model, the importation model considers the prior life experiences of the individual inmate. According to this view, prisoners bring with them a set of norms and group allegiances they held on the street, changing them to accommodate to life in prison. The importation model is also concerned with the feelings of alienation and powerlessness among inmates. This model views structurally generated powerlessness as a significant determinant of adaptation to prison life.

Inmate Subculture: Contemporary Views

More recent analyses have criticized both the deprivation and importation models as unrealistic and no longer applicable to today's prison conditions.[7] One such concern focuses on the racial composition of prison populations that contain a high proportion of blacks. If it is assumed that an inmate code does exist, it would appear unlikely that ". . . black and white prisoners were equally committed to each of its tenants or that members of each racial group applied the code uniformly to themselves and to members of the other group."[8] Prison populations contain a variety of racial and ethnic groups and are not as homogeneous as assumed by earlier models and, for this reason, are likely to contain multiple inmate subcultural codes. A reality found among contemporary prison populations is the formation of small friendship groups or small cliques.[9] For example, prisoners who share similar interests or backgrounds, have mutual interests, or have been friends on the outside are likely to form such groups. These groups display a range of solidarity and function. As Jacobs has stated: "The cliques range from loosely committed groups of friends who share leisure hours together to tightly knit organizations in which the members cooperate in rackets, thefts, and violence, share each other's materials, protect each other from attacks, and avenge each other when attacked. In between these extremes there are a wide range of small to large and more or less cohesive groups."[10]

These types of relationships have been identified as **ordered segmentation.**[11] The norms of this inmate subculture are often based on a low level of trust and an intense loyalty to one's own group. Few prisoners maintain friendships among different groups. There is generally a hostile relationship with members of different groups, and in some instances, "gang" violence erupts between members of different groups.

Whatever the origins of an inmate subculture, relationships in prison are governed by inmate-generated norms and values known as the **inmate code.** This code provides rules of conduct by which inmates expect other inmates to live. One paramount example is not to inform on another inmate. Failure to conform to the inmate code brings sanctions and alienation from other prisoners.

Daily Life and Problems in Prison

As a means of control, inmates are put through a very routinized process of life in prison. Their day is structured for them with a schedule of events and activities dictated by the rules and procedures of the institution. The following is a typical daily schedule for the inmates of a maximum security facility.

General Population (Week Days)	6:30 A.M.—wake up
	7:00—breakfast and count
	7:45—to work and program
	12:00–12:45—lunch and count
	4:30—end workday and count
	5:00—dinner
	6:00—evening activities
	9:00—in room and count
	Midnight—count
Work/Program	Education at institution level—7:45 A.M. to 4:30 P.M.
	Work assignment—7:45 A.M. to 4:30 P.M.
	Visiting—6:00 to 9:00 P.M.
	Recreation in cluster
	Recreation in complex yard—six days per week
	Recreation in main yard—once per week
	Education in complex—6:00–9:00 (ongoing when in room)
	Education 6:00 to 9:00—institutional level

Making It in Prison

Though some inmates experience tangible stress, which is destructive and disabling for them, most inmates somehow adapt to prison and remain comparatively healthy and sane behind walls.[12] When nonincarcerated persons contemplate serving time in prison, they may well wonder if they could make it through the experience. Yet, most individuals are able to cope with the experience. Making it in prison requires more than just effective emotional coping mechanisms. Most inmates become involved at some level with the sub-rosa economy of the prison. **Hustling,** or illegitimate economic activities,[13] range from the exchange of cigarettes for extra food or preferential treatment at the laundry to the purchase of drugs and weapons. The majority of inmates only participate in the hustling system to obtain goods or services that will make their stay in prison less burdensome. Inmates often ''mushfake'' their way through prison. A **mushfake** is a prison-made copy of something that was available on the streets.[14] Residents of correctional institutions often attempt to recreate as many of the comforts of home as they can within the prison.

There are inmates in any correctional institution who are hustling for the high profits and exploitation of others. There are numerous cases of drug dealing, loan sharking, selling protection, and other illicit practices to be found in most correctional facilities. Although these activities occur, the typical inmate is only interested in getting by as comfortably as possible and getting out of prison as soon as possible. Many inmates want the good things hustling can provide but avoid full participation in the sub-rosa system because of accompanying dangers.

Prison, because it is a total environment, does not readily allow for common relaxation of tensions, or escapes from everyday routine. There are no vacations from prison, no corner bars, and only limited access to one's family. Inmates must seek other ways or places to feel safe or relaxed. In examining inmate behavior, Hans Toch has determined that many inmates find **niches** within the institution that afford them a needed escape from stress. ''A niche is a functional subsetting containing objects, space, resources, people and relationships between people.''[15] The niche can be a room, a regular visit, a job setting, or a formalized program offered by the correctional system. It matters less what the niche is so long as it exists. Often seemingly small, insignificant actions or places can be crucial to the emotional survival of an inmate.

For the most part, inmates learn to cope with the stress of incarceration and are able to maintain their physical and mental health. There are, however, threats to individual safety that are beyond the control of the prisoner.

Violence and Victimization in Prison

The nature and extent of violence in prison is well documented.[16] Most of the violence in prison is inmate versus inmate and does not involve correctional personnel. Although staff is seldom involved in the violence, it has been shown that increased levels of supervision can lessen the levels of victimization found in a prison.[17]

The type and composition of the institution has a great deal to do with levels of violence. In a maximum security institution, imprisoned street gang members may continue their battles over turf and status, and levels of violence may be high. The type of institution can intensify the occasions for violence, but in most correctional facilities, there are common situations that lead to confrontation. Some of the common occasions for violence include disputes over homosexual relationships, sexual assaults, and the general sense of terror

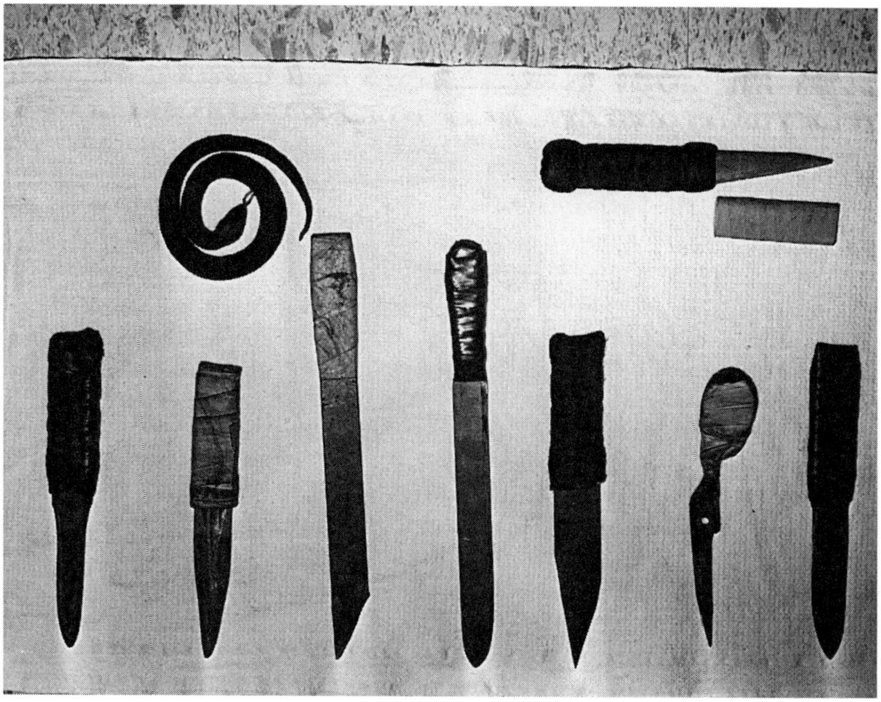

Contraband weapons seized in a maximum security institution shakedown.

that may result in defensive actions. Other situations include exploitation and gambling, retaliations against informers, and the general ''macho'' character of the prison subculture.[18]

Violence and victimization, in some prisons more than others, remains a problem. In large measure, the responsibility for reducing prison violence rests with correctional administrators and staffs. Inmates alone cannot reduce the level of violence in prisons because of the predatory nature of many of the residents. Institutional personnel, through leadership, classification, and education techniques must create an environment to decrease the levels of violence and victimization.

Sexual Deprivation

Life in prison in the United States is designed to be a life of **sexual deprivation.** Sexual relations are forbidden and situations which carry a potential for sexual contact are closely monitored. Prisons provide a one-sex environment but are populated to a great extent by younger persons with an active sexual drive. Not surprisingly, therefore, the subject of homosexuality is often linked to prison life, a subject that tends to be fraught with various myths, stereotypes, and emotionalism.

It is difficult to accurately measure the levels of homosexual activities in prison just as it is difficult to determine the same behavior in the general population. Changing societal attitudes toward homosexual behavior may have altered some traditional views, but the practice in prison continues to be unacceptable. The reasons for homosexual behavior in prison are diverse, but include the fact that some inmates' preprison preference was homosexual. There are also economic factors, cases of coercion and assault, and a situational factor wherein heterosexual relations are not possible.

In her study of **prison homosexuality,** Alice Popper notes that there is really very limited knowledge of the prevalence of homosexual behavior (both assaultive and consensual) in prison. Institutional rules discouraging physical contact and reliance on self-reports of questionable accuracy make any accurate determination of the level of homosexuality in prison nearly impossible.

According to Popper, homosexuality in prison may be a learned response to cope with situational stresses and needs.[19] The prison homosexual situation has numerous facets ranging from the atrocities of homosexual rape to consensual sexual acts between inmates. Because of the lack of accurate information on the subject and current societal attitudes toward the practice, many myths and misperceptions have developed regarding prison homosexuality. Two of the most common myths regarding prison homosexuality are that (1) most inmates have a homosexual experience in prison, and (2) inmates approve of prison homosexuality.

Most studies have revealed that less than 50 percent of inmates surveyed reported a prison homosexual experience. These studies also report that inmates express opposition to homosexuality the same as do young people outside correctional institutions.[20]

It has been suggested that the most effective and plausible means of reducing prison homosexuality would be to:

- decrease the length of sentences;
- establish more furlough and conjugal visiting programs;
- provide sexual education programs for inmates; and
- make available crisis counseling for victims of sexual assault.[21]

In the United States only very limited attempts have been made to deal with sexual deprivation utilizing conjugal visitation. **Conjugal visits** provide for unsupervised and private visits between inmates and spouses at or near the prison. Only a handful of jurisdictions have authorized such visits. Such practice violates the sexual mores of many in the general public and is limited to married inmates, which often comprise a smaller proportion of prison populations.

In most jurisdictions, **furlough programs** that allow for inmate visits in the community are available. Furloughs are generally granted only to specially identified minimum security inmates. A typical furlough allows inmates to visit in their home community for several days without supervision. Although severely limited, furlough programs provide an additional avenue for incarcerated individuals to maintain relationships to lessen some of the controversial issues that surround the use of conjugal visits.

Significant Others

Obviously normal family relations cannot be continued while a family member or close friend is in prison. Efforts to assist inmates in maintaining ties with **significant others** in the community have been instituted in many correctional facilities. Although many marriages and family relationships may be permanently severed during incarceration, a significant number of families regularly visit or maintain contact by letters and phone.

For years, visiting in correctional institutions was considered a privilege and was severely limited. A number of court challenges in the early 1970s altered the visiting regulations in most correctional institutions. Correctional administrators can limit the number and length of visits, but a justifiable rationale must be provided. In most correctional facilities, visits are given some priority, and even inmates housed in isolation are granted visits by family and friends.

Visitors face many problems, including the expense of travel. Many correctional institutions are located in rural areas miles away from urban centers where the families of inmates live. Federal prisons may well be located in another state or even a different part of the country. Many inmates have children who, along with the natural fear of prison, pose an additional stress for a couple who must conduct their family affairs around a visiting table in a room crowded with many other people, or separated by a glass partition.

It is important for inmates to try to maintain their ties with significant others. Inmates with a family are often less of a behavior problem within the institution and more motivated to change behavior patterns and actively participate in prison programs. Having community ties keeps the inmate involved in the world outside prison, and it is hoped that this helps lessen the negative effects of institutionalization. Inmates who have a positive view of their life chances upon release appear to be less affected by the process of prisonization, thus enhancing their opportunity to successfully return to free society.[22] Helping inmates maintain community and family ties may be one of the most important tasks facing our correctional institutions.

Inmates' families may also suffer the pains of imprisonment. Families may experience emotional frustrations, economic hardship, and considerable difficulty adjusting to the incarceration of a family member. The emotional reaction to the incarceration of a spouse has been compared to the grieving process associated with the death of a husband or wife.[23]

Prison Discipline

A basic concern in every correctional facility is maintaining effective control over inmates. An institution with a well-established and effective system of control and discipline is safer, and it provides a more tolerable living and working environment. Correctional institutions and their employees utilize a great deal of discretion in dealing with inmates. The effective correctional employee has the knowledge and ability to use the disciplinary process only when necessary and not to rely upon punitive measures to control inmate behavior.

It has long been recognized that there are patterns of behavior in individuals who are incarcerated. In a now classic study, Wheeler determined that short-term offenders tended to exhibit a U-shaped pattern of violations or anti-institutional behavior. These inmates tended to commit the greatest number of violations during the middle portion of their incarceration. Wheeler's supposition was that this was the time when they were farthest from the free world and its influences.[24] It appears that long-term inmates do not have a discernible pattern of misconduct and, in fact, are far less likely to have significant violations of institutional rules.

There are numerous forms of discipline used in prison. **Disciplinary actions** include verbal reprimands, loss of privileges, solitary confinement, changes in security status, loss of accumulated good time, and even imposition of new criminal charges. Prior to court intervention, the disciplinary actions taken by prison officials were sometimes harsh, often arbitrary, and inhumane. With the advent of court intervention, such as *Wolf v. McDonnell*,[25] came the application of some limited due process protection for inmates, discussed later in this chapter.

The most common punishments for individual violations of prison rules are the loss of privileges and placement in **solitary confinement.** Countless movies and books have depicted the horrors of being placed in the ''hole'' for extended periods of time and the ill effects that result from such treatment. Current usages of solitary confinement, although far from a pleasant experience, are seldom akin to the theatrical image that has been created. The use of solitary confinement is not illegal per se, but the conditions of confinement are not to be shocking or debasing as in the absence of food or water. Inmates held in solitary confinement are regularly attended to, and placement there is frequently reviewed. A typical stay would be five to ten days depending upon the violations and the particular correctional system involved. Other troublemakers may be placed in **administrative segregation,** a separate section of the prison which is very controlled and separate from the general population.

Other punitive actions include the raising of an inmate's security status which could result in a transfer to a different institution. In the case of criminal offenses, the correctional officials generally turn the case over to local law enforcement authorities who can then proceed with criminal proceedings against the inmate involved.

Disciplinary Processes

Generally, disciplinary processes begin by a correctional officer or other employee who completes a violation report (often called a ticket). The inmate at this time receives a verbal notification of the violation, and the written form is submitted for disciplinary action. In many institutions, a hearing officer reviews the report, investigates the case, and calls in the accused inmate for an interview. In the case of a minor violation, the inmate may be assigned an additional work detail or other lesser punishment by the hearing officer. In the case of more severe violations, a formal disciplinary hearing is scheduled. Table 7.1 provides an example of minimum and maximum penalties possible for various offenses.

Following the filing of a disciplinary report, inmate notification of the violation, and at times a screening of the case, the disciplinary action is ready to enter the formal hearing stage. One important decision that must be made prior to the hearing is whether or not the inmate will be detained in administrative isolation pending a formal determination of guilt or innocence. Although there is a great deal of discretion available to correctional officials in determining who will be held in prehearing detention, as a general rule only the more serious violations of institutional regulations (such as assault or fighting) result in administrative isolation. It must be noted that there have been numerous cases of abuse relating to prehearing detention.

The hearing is generally held within one week of the violation but may take longer if there is a need for extensive investigation, or if there is a request by the inmate for additional time to prepare his defense. Typically a board of staff members conducts the hearing, determines the validity of the report, and assigns the punishment. The board is typically made up of a representative of the custody staff, the program staff, and the prison administration (a very similar composition to that found on the classification committee). Generally the disciplinary hearing is tape-recorded or transcribed to provide a record for future review.

The formality of the hearing varies among jurisdictions, but in general, the hearing tends to be a rather short and informal process. Most jurisdictions do not permit the inmate to be represented by legal counsel at the disciplinary hearing, but a number of states do allow the accused inmate to be accompanied by a fellow resident. The inmate is given an opportunity to confront the officer who filed the report, and, in some cases, to question other individuals who have provided evidence against him or her. This opportunity to confront one's accusers is a privilege, not a right, and in disciplinary hearings is subject to the approval of the hearing committee. An effort is made by prison officials to protect inmate informers who might provide information in such cases. Revealing the identity of informers in a prison would certainly lead to serious reprisals by other inmates and most probably the serious injury or death of the informers.

Once the two sides in the case have been presented, the hearing committee makes a determination of the legitimacy of the report. If the report is determined to be invalid, or is at least unsubstantiated through the hearing, then the inmate is cleared of any wrongdoing. If, however, the report is determined to be valid, then a final disposition is ordered. The disposition can range from a simple warning in less serious first violations to loss of good time or a recommendation of transfer to another level of security. One of the most common dispositions is the assignment of the inmate to some period in solitary confinement. In many

Table 7.1

Inmate Offense Categories and Penalties		
Offense	***Minimum Penalty***	***Maximum Penalty***
Category I (Minor Violations)		
1. Feigning illness	1	4
2. Improper or unauthorized use of or possession of state equipment or materials	1	4
3. Illegal possession of canteen tickets, or possession of money more than $20 in excess of amount authorized	1	4
4. Littering	1	4
5. Improper or unauthorized use of telephone system	1	4
6. Improper use of a pass	1	4
7. Illegal possession of any item or quantities not on an authorized property list	1	4
8. Failure to have and display I.D. card as required by institutional policy	1	4
9. Failure to abide by any published institutional schedule or documented rule	1	4
10. Violation of institutional dress code	1	4
11. Carrying food from the dining room and kitchen	1	4
12. Abusive or vulgar language	1	4
Category II (Minor Violations)		
1. Possession of contraband (For definition of contraband and types of items included therein see CPP 9.6)	2	5
2. Disruptive behavior	2	5
Category III (Major Violations)		
1. Interfering with an employee in the performance of his duty	2	7
2. Refusing to obey a direct order	2	7
3. Refusing to work	2	7
4. Violation of mail or visiting regulations	2	7
5. Breaking and entering into another inmate's locker, room, cell or living unit	2	7
6. Unexcused absence from assignment	2	7
7. Failure to carry out work assignment	2	7
8. Bucking an inmate line, wherever formed	2	7
9. Involvement in the writing, circulating or signing of petitions which could lead to disruption of institutional operations	2	7
10. Failure to clean bed area	2	7
11. Unauthorized changing of bed	2	7
12. Fighting	2	7
13. Inflicting injury to self	2	7
14. Charging another inmate for any services	2	7
15. Violation of the Furlough Code of Conduct	2	7
16. Being in a restricted or unauthorized area	2	7
17. Unauthorized communication with or from inmates	2	7
18. Forgery of any type	2	7
19. Violating a condition of an outside work detail	2	7
20. Failure to abide by penalties imposed by Adjustment Committee or Adjustment Officer or Unit Hearing Officer	2	7
21. Abusive or vulgar language directed toward an employee, visitor, or non-inmate	2	7
22. Lying to an employee	2	7
23. Unauthorized communication with any member of the public or staff	2	7
24. Conviction for any Category I or II Offense twice within 90 days	2	7

Table 7.1 (continued)

Offense	Minimum Penalty	Maximum Penalty
Category IV (Major Violations)		
1. Assaulting another inmate but with no injury occurring	2	8
2. Unauthorized use of drugs or intoxicants	2	8
3. Refusing to submit to a drug urinalysis test, breathalyzer, or search	2	8
4. Interfering with the taking of a drug urinalysis test, breathalyzer, or search	2	8
5. Smuggling of contraband items into, out of, or within the institution	2	8
6. Engaging in extortion or blackmail or making threatening statements	2	8
7. Confusing or missing a count	2	8
8. Nonviolent demonstration or inciting a nonviolent demonstration that could lead to a disruption of institutional operations	2	8
9. Unauthorized absence from the institution	2	8
10. Destruction, alteration, or defacing of state, personal, or community property of less than $100 in value	2	8
11. Obtaining money, goods, privileges, or services under false pretenses	2	8
12. Inappropriate sexual behavior	2	8
13. Gambling or possession of gambling paraphernalia	2	8
14. Stealing or possession of stolen personal, state, or community property under $100	2	8
15. Unauthorized buying, selling, giving, trading, loaning, or borrowing	2	8
16. Possession of tattoo paraphernalia	2	8
17. Tattooing self or others	2	8
18. Unauthorized use of medication	2	8
Category V (Major Violations)		
1. Destroying, altering, or defacing of state, personal, or community property valued at $100 or more	4	9
2. Destroying or tampering with life safety equipment	4	9
3. Destroying or tampering with locking or security devices	4	9
4. Running from or resisting apprehension by an official	4	9
5. Loan sharking, collecting, or incurring debts	4	9
6. Stealing or possession of stolen personal, state, or community property over $100	4	9
7. Bribing a correctional employee	4	9
Category VI (Major Violations)		
1. Inciting to riot or rioting	6	10
2. Escape	6	10
3. Deliberately setting a fire	6	10
4. Possession or promoting of dangerous contraband (as defined in CPP 9.6)	6	10
5. Possession of money $20 or more in excess of authorized amount where possession of money is authorized	6	10
6. Possession of tokens or money where not authorized	6	10
7. Possession of staff uniform clothing or uniform related items	6	10
8. Robbery	6	10
Category VII (Major Violations)		
1. Assaulting an employee or non-inmate	11	11
2. Assault resulting in the death or serious injury of another inmate	11	11
3. Sexual assault	11	11
4. Assault resulting in the death or injury of an employee or non-inmate	12	12

(continued on next page)

Table 7.1 (continued)

Inchoate Offenses

1. A person may be found guilty of an offense listed in this policy if he:
   a. attempts to commit the offense;
   b. solicits another or others to commit the offense;
   c. conspires with another or others to commit the offense; or
   d. facilitates the action of another or others in committing the offense.

Penalty Code—General Principles

1. Two penalties may be assessed for each offense so long as one penalty is penalty 1 through 5.
   a. Disciplinary segregation may be ordered to be served consecutively for each offense.
   b. Any time spent in detention shall be credited against any subsequent sentence imposed.
   c. If any two (2) minor violations are committed within ninety (90) days the penalty range for the second offense may be increased from 1–4 to 2–7.

   Conviction of any two minor violations may include the conviction of the same offense twice.

Penalties

1. Reprimand and warning.
2. Restriction of privileges not to exceed six (6) months.
3. Extra duty assignment for a specific period of time not to exceed forty (40) hours.
4. Restitution
   a. the Adjustment Committee, Adjustment Officer or Unit Hearing Officer may order restitution in cases of: destruction, injury, improper use, removal or theft of property of the state, employees, visitors or other inmates.
   b. self-infliction of injury or feigning illness.
   c. infliction of injury to others.
   d. obtaining money, goods, privileges or services under false pretenses.
5. Loss of privileged housing or meritorious living conditions.
6. Assignment to disciplinary segregation for a maximum of fifteen (15) days, each offense.
7. Loss of up to sixty (60) days good time, each offense.
8. Loss of up to sixty (60) days good time and assignment to disciplinary segregation for a maximum of forty-five (45) days, each offense.
9. Loss of up to ninety (90) days good time and assignment to disciplinary segregation for a maximum of sixty (60) days, each offense.
10. Loss of up to one hundred eighty (180) days good time and assignment to disciplinary segregation for a maximum of ninety (90) days, each offense.
11. Loss of up to two (2) years non-restorable good time and assignment to disciplinary segregation for a maximum of one hundred eighty (180) days, each offense.
12. Loss of up to four (4) years non-restorable good time and assignment to disciplinary segregation for a maximum of one (1) year, each offense.

Reduction in Sentence to Disciplinary Segregation

The Classification Committee may recommend a reduction of disciplinary segregation time to the Warden as provided in CPP 10.2.

1. Criteria for reduction shall be contained in CPP 10.2.
2. The warden, or institutional duty officer, may reduce disciplinary segregation time in emergency situations where cell space is needed.

Suspension of Sentence

A sentence or any part of a sentence may be suspended for a period of up to six (6) months as provided in CPP 15.6.

Notification to Inmates and Staff

All inmates and staff shall be notified of changes in this policy. All inmates shall be notified of the changes as part of the orientation process upon reception at all institutions. Copies of changes shall be posted in areas accessible to inmates and staff.

Source: Reprinted by permission of the Commonwealth of Kentucky Department of Corrections, Frankfort, KY.

jurisdictions, any disposition that involves a change in the inmate's institutional status may need approval by the institutional classification committee.

Appeal and Review

Although the recommendation of the disciplinary committee is generally followed by the classification committee, there is at least an opportunity for a review of the more serious cases at this level. Inmates who choose to appeal the decision of the disciplinary committee are generally informed of the formal process. The appeal is frequently administrative in nature, with the hierarchical process involving the institutional superintendent or designate and one or more central office personnel. Inmates who have exhausted the correctional system's appeal processes can seek redress through the courts.

Rewards versus Punishments

Despite the fact that institutional incarceration and the disciplinary processes that have been established within these facilities are punitive in nature, there are also **formal reward systems** that have been developed to encourage socially acceptable behavior. The three most common forms of reward systems operating within correctional facilities are **good time, increased privileges,** and **lowered security levels.**

In many correctional institutions privileges such as type of housing unit, access to entertainment, and extra commissary use are granted to inmates who adhere to the institutional rules. The granting of increased privileges or freedoms is a key program component in many juvenile facilities.

Good time is a system for shortening one's sentence through appropriate behavior. The mechanisms for granting good time differ among jurisdictions, but in most systems, inmates are credited periodically with a certain amount of time. There are established formulas for any reduction or restoration of an individual's sentence based on periods of time without any significant rules infractions.

Inmates can also become eligible for lowered security levels based upon good behavior. Many correctional systems consider the disciplinary record of an inmate when reviewing eligibility for a reduction in security level. Honor status or other designations are given to select groups of inmates who have exemplary institutional records. Honor status inmates frequently live and work outside the prison compound and are given the greatest freedom of any incarcerated individuals.

Both earning good time and a reduction of security levels to encourage desirable behavior effect the term of an inmate's prison stay. Good time directly reduces the time of an inmate's life in prison. The lowering of an individual's security status enhances chances for release consideration. Prison officials often point to these reward systems and the latitude provided by indeterminate sentences as major factors in controlling inmate behavior.

Prisoner Rights

For decades prisoners had virtually no rights. Courts followed what was known as a **hands-off doctrine,** a basic reluctance to intervene into prison matters. This doctrine denied prisoners access to the courts and provided prison administrators and correctional systems nearly free reign in governing those persons confined to prison. With prisoners considered as possessing few rights and with administrators managing facilities with very little accountability, it is not surprising that cruel punishments and shocking conditions of confinement occurred.

As it turned out, however, a hands-off doctrine was not to prevail forever. Due to changing attitudes, the civil rights movement, and other factors, courts began reversing the hands-off position and granted prisoners access to the courts. As a result, inmates in jails and state and federal prisons began filing cases, mainly in the federal courts, challenging nearly every facet of prison life. Through this litigation, prisoners won some important protection of rights and a variety of administrative practices and prison conditions of confinement were changed.

Most of the history of the nation's prisons has been marked by a hands-off doctrine. For one reason or another, courts did not hear prisoners' suits. Three reasons for this have been suggested:

1. the separation of powers doctrine
2. a low level of prison knowledge and expertise
3. a fear that court intervention would undermine administrators and their methods of discipline[26]

During the hands-off era, the status of prisoners was best described in the case of *Ruffin v. Commonwealth* in 1871, which stated that a prisoner ". . . as a consequence of his crime, not only forfeited his liberty, but all his personal rights except those which the law in its humanity accords to him. He is for the time being the **slave of the state.**"[27]

Other cases continued to reinforce a hands-off position for a long time. For example, *Banning v. Looney*[28] in 1954 stated that "courts are without power to supervise prison administration or to interfere with the ordinary prison rules or regulations." This followed a similar opinion delivered three years earlier in *Stroud v. Swope*.[29] Hands-off was enunciated once again as late as 1962 in *Sutton v. Settle*[30] which stated that ". . . courts have no power to supervise the management of disciplinary rules of such institutions."

However, around the midpoint of the twentieth century a few courts began considering prisoners' suits. The case of *Cooper v. Pate*[31] in 1964 is an example of a case that marked the decline of the hands-off doctrine. In this case, the court ruled that prisoners have a right of access to the courts. Subsequent cases clarified and enhanced this principle. As a result, thousands of cases were filed to protect offenders' rights during the 1970s, continuing into the 1980s. This trend has resulted in substantial change and had far-reaching impact for corrections. In one instance, *Newman v. Alabama*[32], the court even removed the entire state correctional system from those authorized to operate the system and placed it under the control of a special master. In seeking remedies, inmates began frequently utilizing, as a basis for their rights, three amendments in the Bill of Rights and a piece of legislation dating back to 1871.

Sources of Rights

Prisoners' rights have been won essentially through case law that has been established by court decisions rather than statutory law enacted by legislatures. Many of the successful challenges have been based upon the First, Eighth, and Fourteenth Amendments to the Constitution and Section 1983 of the Civil Rights Act of 1871, enacted just after the Civil War.

First Amendment Rights Congress shall make no law respecting an establishment of religion, or prohibiting the free exercise thereof; or abridging the freedom of speech, or the press; or the right of the people peaceably to assemble, and to petition the Government for a redress of grievances.

Eighth Amendment Rights Excessive bail shall not be required, nor excessive fines imposed, nor cruel and unusual punishments inflicted.

Fourteenth Amendment Rights SECTION 1. All persons born or naturalized in the United States, and subject to the jurisdiction thereof, are citizens of the United States and of the State wherein they reside. No State shall make or enforce any law which shall abridge the privileges or immunities of citizens of the United States; nor shall any State deprive any person of life, liberty, or property, without due process of law; nor deny to any person within its jurisdiction the equal protection of the laws.

Section 1983 of the Civil Rights Act of 1871
Every person, who under color of any statute, ordinance, regulation, custom or usage, of any State or Territory, subjects, or causes to be subject, any citizen of the United States or any other person with the jurisdiction thereof to the deprivation of any rights, privileges, or immunities secured by the constitution and laws shall be liable to the party injured in an action at law, suit in equity, or other proper proceeding for redress.[33]

Prisoners wishing to seek redress must file a **writ of habeas corpus** or a **writ of certiorari,** written documents which bring their case to court for review. With a writ of habeas corpus an inmate has a means to challenge the legality of their imprisonment. A writ of certiorari calls a case up for review from a lower court to a higher court. These writs are much used means by which inmates challenge conditions of their confinement.

Categories of Rights

Prisoners sought and won many rights in nearly every aspect of prison operation. Individual cases have included issues related to discipline, classification, reading material, medical care, and living conditions among others. Rights won by inmates can be organized into four categories:

1. Access to court
2. Protection from cruel and unusual punishment
3. Civil rights
4. Protection of rights in decisions when adverse consequences are possible[34]

Access to Court

The decline and reversal of the hands-off doctrine was in large measure achieved by a group of cases that provided prisoners **access to courts.** There is really no precise date when this occurred nor any one case by itself that marks this reversal, although some point to the importance of *Cooper v. Pate*[35] in 1964. This case ruled that prisoners could sue the warden for depriving them of constitutional rights under title 42 of the United States Code, section 1983 (42 U.S.C. 1983), based on protections granted by the Civil Rights Act of 1871. By this ruling, state prisoners could now challenge the constitutionality of their imprisonment in the federal courts. As is usually true with case law, a position evolves and emerges over some period of time in a building process from case to case. By the late 1960s and early 1970s, it was reasonably well established that prisoners had in fact gained access to the courts, and in that sense, the hands-off doctrine was essentially reversed.

The case of *Ex Parte Hull*[36] in 1941 marked one beginning for **court intervention.** It has been stated that:

> The right to access to the courts, established in the case of *Ex parte Hull,* is held by courts to be the most basic of all rights guaranteed to prisoners. As such, it is the right most tightly protected. Without this careful guarantee of the right of access to the courts, as well as the means by which to bring claims, no other legal or constitutional rights of prisoners can be consistently protected. This right has provided the principal vehicle which lifts the barrier between prisoners and the Bill of Rights. In the absence of such protection, complaints about any features of the conditions of incarceration and assertions of deprivation of other rights could be brought to court only at the discretion of prison administrators.[37]

However, *In re Green*[38] made clear that such access was not absolute and should not be abused. Filing repeated and frivolous claims by a prisoner could limit access for that inmate. Although implementation was slow in being realized, additional cases provided prisoners basic constitutional rights by which they might petition federal courts for grievances. *Brown v. Allen*[39] in 1953 and *Wildwording v. Swenson*[40] also held that state inmates could seek relief in federal courts under Section 1983 of the 1871 Civil Rights Act by filing a federal writ of habeas corpus. Under *Johnson v. Avery*,[41] decided in 1969, the court held that one prisoner could provide legal assistance to another prisoner in preparing a writ of habeas corpus. Previously, this practice of providing "jail-house lawyers" was restricted and had the effect of denying access to illiterate and poorly educated prisoners.

Younger v. Gilmore[42] recognized that access to court also required the availability of sufficient legal libraries so that inmates could prepare adequate petitions. *Bounds v. Smith*[43] called for prison officials to provide inmates with adequate libraries for preparation of legal actions or legal counsel.

By the early 1970s access to court was established. No better example exists than the ruling in *Crug v. Hauck*[44] which stated that "ready access to courts is one of, perhaps *the* most fundamental constitutional right."

No Cruel and Unusual Punishment

A wide variety of prison and jail conditions and practices have been challenged on the basis of protection from **cruel and unusual punishment** provided by the Eighth Amendment. Major rights have been won in this area, although this protection has generally been granted in cases involving only the very worst of practices. Courts have shown some reluctance to extend this constitutional right in some situations for fear of interfering with administration of the prison.

Prison cases involving issues of cruel and unusual punishment can be divided into two categories.[45] One category goes beyond incarceration and deals with inflicting a degree of punishment exceeding that issued by the court. A second category considers those everyday conditions of confinement in prison that might constitute cruel treatment.

First, solitary confinement and use of force are two practices that have generated numerous cases charging excessive punishment. For example, courts have not declared such segregation itself unconstitutional but have reviewed certain conditions of such confinement and noted excessive punishments as in *McCray v. Burrell*.[46] *Hutto v. Finney*[47] upheld a lower court finding that solitary confinement for more than thirty days constituted cruel and unusual punishment. Limits were also placed on the number of persons who could be placed in a single isolation cell. *Wright v. McMann*[48] declared it unconstitutional to deprive an inmate clothing necessary for warmth. In *Fulwood v. Clemmer*[49] it was held that the effects of solitary confinement on the mental condition of an inmate must be reviewed on an individual basis.

Another question in this area is, when does a use of force constitute punishment in violation of the Eighth Amendment? As a general guideline, *Landman v. Royster*[50] asserted that whether property or lives are in danger is the test for determining reasonableness in the use of force.

Both *Jackson v. Bishop*[51] and *Inmates of Attica Correctional Facility v. Rockefeller*[52] held that corporal punishment in the form of whipping to enforce prison discipline was unconstitutional.

The second category involving cruel and unusual punishment focuses on general conditions of daily living while confined to prison. Simply poor living conditions will not result in eighth amendment protection. According to *McLaughlin v. Royster*,[53] what is required are inmates being exposed to conditions so dangerous and uninhabitable as to be shocking. Such was ruled in the infamous *Holt v. Sarver*[54] case which declared the entire Arkansas Penitentiary System unconstitutional. It was noted that:

> The court found a complex pattern of inhumane conditions in Arkansas, including inadequate supervision of living quarters, which resulted in murders and homosexual assaults; a ''trusty'' system, which permitted favored inmates to dominate and abuse other inmates; filthy, rodent-infested solitary confinement cells; and an absence of recreational or training programs, which resulted in idleness and violence. Eventually the court ruled in favor of the plaintiffs and ordered the state to make extensive improvements in the conditions of its prisons.[55]

Many conditions of confinement were also ruled unconstitutional in *Ruiz v. Estelle*[56] against the Texas Department of Corrections. Changes were ordered in many areas, including health care, number of inmates per cell, fire and safety regulations, and the use of inmate guards, called building tenders, to guard other inmates.

The availability of medical care while confined in prison has also been a specific subject of much litigation. Although the minimum level of care required by law is low, some level of medical treatment must be provided. *Edwards v. Duncan*[57] held that total and intentional denial of all medical care is unconstitutional. *Estelle v. Gamble*[58] (a case from a prison for women) held that deliberate indifference to serious medical needs of prisoners violated the Eighth Amendment.

The use of force to maintain prison control is an additional issue that has involved the Eighth Amendment. An Oregon inmate was shot in the leg by a correctional officer who was attempting to quell a prison riot. In *Whitley v. Albers*[59] the court held this action did not violate freedom from cruel and unusual punishment, as long as force was used in a good faith effort to restore order. In *Hudson v. McMilliam*[60] the court ruled that excessive force, in this case a beating, did violate the cruel and unusual punishment clause of the Eighth Amendment even though no serious injuries resulted.

Civil Rights

Protection of **civil rights** by prisoners has also been sought, many times under the **equal protection clause** of the Fourteenth Amendment. One early example was *Jackson v. Godwin*.[61] The effect of the ruling was that Jackson, a black inmate, could subscribe to publications for blacks, just as whites subscribed to publications for whites. Failure of prison officials to permit subscriptions by Jackson violated Fourteenth Amendment rights.

Inmates have been granted limited protection from the censorship of their mail by prison officials. Both incoming and outgoing inmate mail can be inspected and censored, and at times, residents can be denied the right to delivery of mail. The court in *Procunier v. Martinez*[62] established the principle that censorship must be based upon substantial governmental interests of security, order, rehabilitation, or other related areas. If an inmate's mail is censored, the inmate must be notified, the source of the correspondence must have a reasonable opportunity to protest, and the decision to censor or deny delivery must be reviewed by a prison official other than the individual who initially screens the correspondence. When inmates receive legal mail from attorneys, courts, or judicial related offices, it must be opened and inspected in their presence, so as not to interfere with or raise questions about their access to the legal system.

Under provisions of both the Fourteenth and First Amendment, which provide for the free exercise of religion, prisoners have gained some rights of religious practice. Although *Brown v. Johnson*[63] restricts religious practices for reasonable justification, correctional managers must provide sufficient rationale if a religious practice is not permitted. Black Muslim's in a series of cases, notably *Banks v. Havener*[64] and *Knuckles v. Prasse,*[65] have been successful in getting the courts to recognize the Black Muslin faith as an established religion, thereby permitting its practice in prison.

In the 1972 case of *Cruz v. Beto,*[66] it was held that inmate Cruz, a Buddhist, should be granted the opportunity to practice his unconventional religious beliefs under rights granted by the First Amendment. This case clarified the right of all inmates to be afforded a reasonable opportunity to practice their religious beliefs.

In a related development, the Religious Freedom Restoration Act became law in 1994 and applied to all persons, including prison inmates. Under this law, governmental agencies may not forbid a person's free exercise of religion unless there is some compelling state interest. Future court cases will define the application of this law in a prison setting.

Protection When Adverse Consequences Possible

A fourth group of cases provides constitutional protection in cases where officials make decisions that may carry an adverse consequence for an inmate. A landmark case was in the area of prison discipline. In *Wolf v. McDonnell*[67] the Supreme Court ruled that institution officials did have the right to conduct disciplinary proceedings and that inmates were not entitled to the full constitutional rights afforded a free citizen, but that the incarcerated were not wholly stripped of constitutional protections. Utilizing the Fourteenth Amendment, the court ruled that inmates were entitled to limited due process rights in disciplinary proceedings where they were subject to loss of good time, confinement in solitary, or other significant losses. The court established some basic guidelines for disciplinary proceedings when significant loss was possible.

- A neutral hearing officer.
- Advanced written notice of the claimed violations and a written statement of the fact-findings as to the evidence relied upon and the reasons for the disciplinary action taken.

- An inmate facing disciplinary proceedings should be allowed to call witnesses and present documentary evidence in his own defense.
- Confrontation and cross-examinations are not required since they could present greater hazards to institutional interests. However, where an inmate is unable to collect and present evidence necessary for an adequate and comprehensive defense of the case, he can seek the aid of a fellow inmate.

Two years later it was decided in *Baxter v. Palmigiano*[68] that prisoners do *not* have a right to counsel at disciplinary hearings.

Numerous other protections have also not been granted; examples encompass a variety of topics. In *Jones v. North Carolina Prisoners' Labor Union, Inc.,*[69] it was held that a ban on the prisoners' labor union did not violate provisions of the Fifth Amendment. The effect of *Washington v. Harper*[70] was that inmates with severe mental disorders could be treated with antipsychotic drugs against their will, even without a court hearing. A ruling in *Hudson v. Palmer*[71] allows inmate cells to be searched without a warrant or probable cause, and in *Meachum v. Fano*[72] the court held that inmates could be transferred from a lower-security institution to a maximum security institution without a due process hearing.

Courts have extended rights to decision making in the area of revoking parole and probation. Such decisions determine whether one remains in the community or goes to prison, and therefore, they carry considerable importance. In such cases as *Morrisey v. Brewer*[73] and *Gagnon v. Scarpelli*[74] various due process procedures were required and are discussed in chapters 11 and 12.

Outcomes of Court Intervention

Enforcement of court ordered rights may be aided by judges' authority to involve **contempt powers** and to appoint special officers of the court to monitor and oversee efforts to comply with court orders. A variety of such officials have been used since the decline of the hands-off doctrine. They are

- Masters—mainly fact finders for the court
- Receivers—to hold, manage, or liquidate property
- ''Special Masters''—to perform a wide variety of duties, including develop and implement a compliance plan
- Monitors—to observe the implementation process and report to the court
- Ombudsmen—to hear inmate grievances, conduct hearings, and make recommendations to the court[75]

Without question, inmates' rights have been expanded by court intervention. Prisoners have gained access to the courts and won protection in a number of areas. Some dehumanizing practices such as whippings and ''hot boxes'' have been eliminated. Some civil rights have been protected and limited due process interjected into decision-making regarding inmates. *Guthrie v. Evans*[76] provides an exemplary case wherein the Georgia State Prison at Reidville was transformed from a poor physical facility, operating with many constitutional violations, into a constitutional and accredited institution.

Such changes, however, have not been without problems. Changes have been unwelcomed by some correctional officials. Implementing these rights has been considered a "nuisance" and an unwarranted intrusion. Due to obstruction by some prison officials, compliance has been delayed or prevented in some situations.

Other administrators have made efforts to bring about required changes. However, this is not easy. Change may be costly, and familiar routines of employees may be altered. Implementing court-ordered changes usually requires a presence of commitment, time, energy, and dollars to be successful.

The twentieth century saw the pendulum swing from a hands-off period to a time of considerable court activity. Table 7.2 indicates what a panel of experts has considered as the top ten prison and jail cases. As we enter the next century, there is evidence of a retreat from the peak of judicial activism and intervention into prison operations. In 1979, *Bell v. Wolfish*[77] held that a variety of practices for pre-trial detainees, including searches, and double bunking (placing two prisoners in one cell) were not unconstitutional. In 1981, *Rhodes v. Chapman*[78] held that double bunking did not constitute cruel and unusual punishment. The most far-reaching recent case in this regard is *Turner v. Safley*.[79] The case centered around whether restrictions on inmate-to-inmate correspondence were constitutional. In this case, the court said the restrictions were reasonable and then issued a four-part reasonableness guideline to be considered in developing prison rules and regulations. Consideration is to be given to the following:

1. Is there a rational link between a regulation and a legitimate prison interest?
2. Do alternative means to exercise rights exist?
3. What will be the impact on officers and other inmates?
4. Do alternative regulations exist?

Although interpretations will be determined by future cases, the effect of *Turner* is that it is now easier for prison management to claim a legitimate prison interest—for example, security, safety, or health when establishing a regulation.

Additional cases eroding judicial activism are *Wilson v. Seiter*[80] in 1991 and *Rufo v. Inmates of Suffolk County Jail*[81] in 1992. *Wilson* held that in order for inmates to recover damages, prison officials must have acted with deliberate indifference, thereby making it much more difficult. In *Rufo,* the court allowed that a previously agreed to consent decree could at a later time be changed.

In sum it can be said that various court decisions have changed the basic relationship between correctional officials and inmates. Inmates are no longer looked upon as a powerless and unrepresented group who must willingly accept any policy or action of correctional administration. Access to the courts by prisoners changed the very nature of correctional institutions and placed inmates more in a realm of a consumer of services than a "slave of the state." At the

Table 7.2

The* Correctional Law Reporter *List of the Top Ten Prison and Jail Cases*

The following list notes the key issue of each case, and includes selected comments from the survey respondents.

Case	Key issue and comments
1. **Wolff v. McDonnell,** 418 U.S. 539 (1974)	Inmate discipline. **Wolff** ". . . probably had more effect than any other court decision in introducing the idea of legality into prison life" (Boston).
2. **Bell v. Wolfish,** 441 U.S. 520 (1979)	The double celling and deference to administrators decision that many consider ". . . the high water mark in corrections litigation" (Lund) that "brought us almost full circle back to the 'hands off' approach . . ." (Bennett).
3. **Turner v. Safley,** 107 S.Ct. 2254 (1987)	Defines a balancing test for cases where inmate rights conflict with institutional interests. Three words: ". . . for obvious reasons" (Stark).
4. **Bounds v. Smith,** 430 U.S. 817 (1977)	Access to the courts; the "law library case." One result is that ". . . issues, like how many books, which ones, and location of the library, continue to this day" (Dale).
5. **Estelle v. Gamble,** 420 U.S. 97 (1976)	Established the "deliberate indifference to serious medical needs" standard of reviewing medical questions under the Eighth Amendment. The ". . . source of the infamous deliberate indifference standard which has spread into various other subject areas besides medical care" (Breed).
6. **Rhodes v. Chapman,** 101 S.Ct. 2392 (1981)	Linked with Bell on double celling, this decision requires objective evidence, not just expert opinion. Rhodes ". . . had an effect on legislators . . . because it allows (at least at first) expansion of capacity 'on the cheap' " (Midgely).
7. **Wilson v. Seiter,** 111 S.Ct. 2321 (1991)	Adds state of mind as an issue in conditions cases and severely restricts the totality of conditions approach, thus ". . . reopening the question whether good intentions can place inhumane conditions beyond the reach of the courts" (Boston).
8. **The Early Cases**	Decisions such as **Holt v. Sarver,** 309 F.Supp. 362 (E.D. Ark., 1970), the various opinions in the Alabama conditions cases, and various other major early decisions. "This is where it all started, where the court rejected the historical 'hands off' approach . . ." (Bennett).
9. **Meachum v. Fano,** 427 U.S. 215 (1976)	The inmate transfer case that began to articulate the "state created liberty interest" test of Due Process. **Meachum** "insulates from scrutiny . . . the arbitrary and oppressive treatment of prisoners as long as prison officials . . . employ the right verbal formulae in the policies and regulations" (Boston).
10. **Rufo v. Inmates of Suffolk County,** 112 S.Ct. 748 (1992)	Modification of consent decrees. **Rufo** has the ". . . potential to undo many of the prisoners' earlier substantive gains" (Robbins).

**As identified by polling a panel of experts in correctional law.*
Note: Name in parentheses identifies respondent.
Source: Reproduced with permission from: Civic Research Institute, Inc., Correctional Law Reporter *(Kingston, New Jersey: June, 1994, Vol. VI, No. 1), pp. 1–16.*

same time, however, critics have pointed out some problem areas.[82] Much litigation has been on a piecemeal basis, examining an issue in finite detail but failing to consider how this relates within the total prison environment or how one decision may relate to another. As a result, little is produced in the way of general and consistent guidelines for the future. Court decisions are often rendered in a very detailed and abstract manner without understanding in human terms ongoing daily realities of prison life. Less well-intentioned administrators can often find means and loopholes by which to circumvent or evade court-ordered remedies for prisoners.

It is suggested that:

> In the final analysis, the continuing plight of prisoners in America is attributable largely to the fact that there are the above-mentioned limitations, and, more, to the amount of relief that can be obtained by means of litigation and judicial action. Neither the general public nor state legislatures are much more concerned with the problems and rights of prison populations today than they were fifty years ago. Only when this changes will there be widespread and lasting improvement in the plight of prisoners in America.[83]

Summary

Prisons present a special environment in which one must live, although considerable variation in conditions exists between different institutions. Characterized by Goffman as a total institution, prison requires adaptations to living with other inmates, in relationship to guards, and to numerous restrictions and enforced activities.

Literature concerning prison life explains the role an inmate subculture can play as one adapts to living in such an environment. A deprivation model considers negative aspects of prisons as mainly responsible for the development of an adaptive subculture. However, an importation model considers the origin to be from experiences and influences brought along with inmates from outside. More contemporary reviews are critical of both of these approaches. This model sees multiple inmate codes structured along various racial, ethnic, and friendship groupings which are more common in the diverse populations found in contemporary prisons.

Prison life presents numerous day-to-day problems. Life tends to be very routine and regimented. Tensions and anxieties abound and exploitation is prevalent. Assaults and violence can be problems, especially in large maximum security prisons. Overcrowding is often common. Control is maintained through a disciplinary system. Depending upon severity, violation of rules may bring minor sanctions, such as a temporary loss of privilege, to solitary confinements for severe offenses.

Life in prison in the United States is designed to be a life of sexual deprivation. Within this context, homosexual activity occurs, but accuracy of information and understanding is hampered due to various fears, myths, and stereotypes. Sexual activity involving inmates is often assaultive but is sometimes consensual. Many inmates disapprove of homosexuality and report no such experience while in prison.

For decades prisoners had practically no rights and were considered ''slaves of the state.'' Courts, under a hands-off doctrine would not intervene into the operation of prisons. However, changes occurred around and just beyond the mid-point of the twentieth century. As a result of court intervention, prisoners gained access to the courts and were granted some protection from cruel and unusual punishment, gained some civil rights, and won some due process procedural rights. Although compliance and implementation of court orders have presented some problems, there is no question that court involvement extended rights and changed many practices found in prisons. Court intervention has slowed, but inmates can still access the courts and no longer are powerless ''slaves of the state.''

Discussion Topics

1. Is it useful to characterize prisons as total institutions? Why or why not?
2. Which view of an inmate subculture do you think more closely resembles what is true in today's prisons?
3. Should prison policies of sexual deprivation be maintained?
4. Why did a decline of the hands-off doctrine come about?
5. What do you think are the most important rights won by inmates?
6. How valid are the criticisms of prisoner rights litigation?

Additional Readings

Del Carmen, Rolando V., Susan E. Ritter and Betsy A. Witt. *Briefs of Leading Cases in Corrections*. Cincinnati, OH: Anderson Publishing Company, 1993.

Early, Pete. *The Hot House: Life Inside Leavenworth Prison*. New York, NY: Bantam Books, 1992.

Fishman, Laura T. *Women at the Wall; A Study of Prisoners' Wives Doing Time on the Outside*. Albany: State University of New York Press, 1990.

Knight, Barbara B. and Stephen T. Early, Jr. *Prisoners' Rights in America*. Chicago: Nelson-Hall Publishers, 1986.

Notes

1. Erving Goffman, *Asylums: Essays on the Social Situation of Mental Patients and Other Inmates* (Garden City, NJ: Doubleday, 1961), 6.
2. Miles D. Harer, *''Recidivism Among Federal Prison Releasees in 1987: A Preliminary Report''* (Washington, D.C.: Federal Bureau of Prisons, December 9, 1993), 41.
3. Lloyd W. McCorkle and Richard Korn, ''Resocialization Within Walls,'' *Annals of the American Academy of Political and Social Science* 293 (1954): 88–98; Gresham M. Sykes and Sheldon L. Messinger, ''The Inmate Social System,'' in *Theoretical Studies in Social Organization of the Prison,* ed. R. A. Cloward et al. (New York, NY: Social Science Research Council, 1960), 5–19.
4. Gresham M. Sykes, *The Society of Captives: A Study of a Maximum Security Prison* (Princeton: Princeton University Press, 1958).
5. Clarence Schrag, ''Some Foundations for a Theory of Corrections,'' in *The Prison: Studies in Institutional Organization and Change,* ed. Donald R. Cressey (New York: Holt, Rinehart and Winston, 1961); and John Irwin and Donald R. Cressey, ''Thieves, Convicts, and the Inmate Culture,'' Social Problems 10 (1962), 142–55.
6. Charles W. Thomas and David M. Peterson, *Prison Organization and Inmate Subcultures* (Indianapolis: Bobbs-Merrill, 1977), 51–52.
7. Neal Shover and Werner Einstadter, *Analyzing American Corrections* (Belmont, CA: Wadsworth Publishing Co., 1988), 84.
8. James B. Jacobs, *New Perspectives on Prisons and Imprisonment* (Ithaca, NY: Cornell University Press, 1983), 75.
9. John Irwin, ''The Changing Social Structure of the Men's Prison,'' in *Corrections and Punishment,* ed. David F. Greenberg (Beverly Hills, CA: Sage Publications, 1977), 32.
10. Jacobs, New Perspectives, 33.
11. Ibid.

12. Hans Toch, ''Revisionist View of Prison Reform,'' *Federal Probation* 45, no. 2 (June 1981): 3.
13. Sandra E. Gleason, ''Hustling in Prison,'' in Norman Johnston and Leonard D. Savitz, *Legal Process and Corrections* (New York: John Wiley & Sons, 1982), 214.
14. Thomas W. Foster, *The Prison Subculture in Microcosm: An Exploratory and Typological Analysis of Mushfaking* (Academy of Criminal Justice Sciences Conference, 11–14 March 1981, Philadelphia), 1.
15. Hans Toch, *Living Prison* (New York: The Free Press, 1977), 181.
16. Matthew Silberman, *A World of Violence* (Belmont, CA: Wadsworth Publishing Company, 1994).
17. Dan A. Fuller and Thomas Orsagh, ''Violence and Victimization Within A State Prison System,'' *Criminal Justice Review* 2, no. 2 (Fall 1977): 49.
18. Hans Toch, *Police, Prisons and the Problems of Violence* (Washington, D.C.: National Institute of Mental Health, 1977), 52.
19. Alice M. Popper, *Prison Homosexuality, Myth and Reality* (Lexington, MA: Lexington Books, 1981), 49.
20. Ibid., 178.
21. Ibid., 189.
22. Charles W. Thomas et al., ''Structural and Psychological Correlates of Prisonization,'' *Criminology* 16, no. 3 (November 1978): 389.
23. Donald P. Schneller, *The Prisoner's Family: A Study of the Effects of Imprisonment on the Families of Prisoners* (San Francisco: R. and E. Research Associates, 1976), 78.
24. S. Wheeler, ''Socialization in Correctional Communities,'' *American Sociological Review* 26 (1961), 363.
25. 418 U.S. 539, 1974.
26. Ronald L. Goldfarb and Linda R. Singer, ''Redressing Prisoners' Grievances,'' *George Washington Law Review* 39 (December, 1970): 175–320.
27. 62 Va. (21 Gratt) 790, 796 (1871).
28. 213 F. 2d 771 (10th Cir.) Cert. denied, 348 U.S. 859 (1954).
29. 187 F. 2d 850 (9th Cir.) 1951.
30. 302 F. 2d 286 (8th Cir.) 1962.
31. 378 U.S. 546 (1964).
32. 466 F. Supp. 628 (M.D.Ala.1979).
33. 42 U.S.C. Section 1983.
34. Goldfarb and Singer, ''Redressing Prisoners' Grievances,'' 175–320.
35. 378 U.S. 546 (1964).
36. 312 U.S. 546 (1941).
37. Barbara B. Knight and Stephen T. Early, Jr., *Prisoners' Rights in America* (Chicago, IL: Nelson-Hall, 1986), 15.
38. 669 F 2d 779 (1981).
39. 344 U.S. 443 (1953).
40. 404 U.S. 249 (1971).
41. 393 U.S. 483, 485 (1969).
42. 404 U.S. 15 (1971).
43. 430 U.S. 817 at 828 (1977).
44. 404 U.S. 59 (1971).
45. Geoffrey P. Alpert, *Legal Rights of Prisoners* (Lexington, MA: D.C. Heath, 1978), 111.
46. 516 F. 2d 357, 367 (4th Cir. 1975).
47. 437 U.S. 678 (1978).
48. 460 F. 2d 126, 129 (2d Cir. 1972).
49. 206 F. Supp. 370 (D.D.C. 1962).
50. 346 F. Supp. 297, 312 (E.D. Va. 1972).
51. 404 F. 2d 571 (8th Cir. 1968).
52. 453 F. 2d 12, 23 (2d Cir. 1971).

53. 346 F. Supp. 297, 311 (E.D. Va. 1972).
54. 442 F. 2d 308 (8th Cir. 1971).
55. Shover & Einstadter, *Analyzing American Corrections*, p. 158.
56. 503 F. Supp. 1265 (S.D. Texas 1980). Cert. denied, 1035 S. Ct. 1438.
57. 355 F. 2d 993, 994 (4th Cir. 1966).
58. 429 U.S. 97, 104 (1976).
59. 106 S. Ct. 1078 (1986).
60. 60 U.S.L.W. 4151 (1992).
61. 400 F. 2d 529 (5th Cir. 1968).
62. 416 U.S. 396 (1974).
63. 743 F. 2d 408 (6th Cir. 1985).
64. 234 F. Supp. 27 (E.D. Va. 1964).
65. 302 F. Supp. 1036 (ED Pa. 1969); affd. 435 F 2d 1255 (CA 3 1970).
66. 405 U.S. 319 (1972).
67. 418 U.S. 539 (1974).
68. 96 S. Ct. 1551 (425 U.S. 308 1976).
69. 433 U.S. 119 (1977).
70. 58 U.S.L.W. 4249 (1990).
71. 468 U.S. 517 (1984).
72. 427 U.S. 215 (1976).
73. 408 U.S. 471 (1972).
74. 411 U.S. 788 (1973).
75. "Comment," *The Wyatt Case: Implementation of a Judicial Decree Ordering Institutional Change*, 84 Yale Law Journal 1338 (1975).
76. Civil Action No. 73–3068, S.D. Ga. (1973). For a detailed account see Bradley S. Chilton and Susette M. Talarico, "Politics and Constitutional Interpretation in Prison Reform Litigation: The Case of *Guthrie v. Evans*," in *Courts, Corrections, and the Constitution*, ed. John J. DiIulio, Jr. (New York: Oxford University Press, 1990), 115–137.
77. 441 U.S. 520 (1979).
78. 452 U.S. 337 (1981).
79. 107 S. Ct. 2254 (1987).
80. 59 U.S.L.W. 4671 (1991).
81. 60 U.S.L.W. 4100 (1992).
82. Knight and Early, *Prisoners' Rights*, 314.
83. Ibid.

Chapter 8

Prison Programs and Management

Learning Objectives

After studying this chapter students should be able to:

1. Identify various goals that have been associated with prison industry programs.
2. Differentiate between sheltered and open markets.
3. Provide reasons and identify sources of opposition to prison industry programs.
4. Illustrate various models of contemporary prison industry programs.
5. Identify research indicating a beneficial effect between prison educational programs and post-release recidivism.
6. Conceptualize prison as a community of persons living and working together.
7. Describe the setting for managing contemporary prisons.
8. Identify and describe the five basic management functions.
9. Understand the nature of work performed by correctional officers.
10. Contrast styles of prison management.
11. Explain the impact of Maslow's theory of motivation on management.
12. Explain the impact of the Hawthorne effect on management.

Key Terms

Ashurst-Summers Act
Autocratic management style
Centralized management structure
Controlling outcomes
Convict lease
Corporate model
Correctional education programs
Correctional officers
Eclectic approach
Group counseling
Hawes-Cooper Act
Hawthorne effect
Hierarchy of needs
Individual counseling
Leading
Leisure time activities
Management functions
Managers
Mental health programs
Open markets
Open system
Organizational chart
Organizations
Organizing
Participative management style
Partnership model
Planning
Prison chaplains
Prison community
Prison industry
Private-sector model
Psychological service
Rehabilitation programs
Self-help groups
Sheltered markets
Situationally specific management style
State-use system
UNICOR
Vocational education
Warden

Overview

When society places a person behind walls and bars, it has an obligation—a moral obligation—to do whatever can reasonably be done to change that person before he or she goes back into the stream of society.[1]

It is not acceptable to merely lock up inmates and throw away the key. As the former Chief Justice Warren E. Burger suggests, society and correctional institutions must seek to change the behavior and attitudes of offenders. Work, education, training, and other programs in prisons provide the formal mechanisms to attempt this change.

Managing prisons in order to achieve custody and also to provide rehabilitative opportunities presents major challenges. One immediate concern is the battle against idle time. Typically, much of inmates' time has been utilized doing the work necessary to operate the prison on a daily basis and by laboring in various forms of prison industries. To varying degrees, work programs have been used to serve both custody and rehabilitative goals. In addition to work programs, a variety of rehabilitative programs having individual and group orientations are to be found in the nation's prisons.

Providing all the necessary management functions for programs, prisoners, and employees rests with wardens and their staff. Prison communities are much like any outside community in that a variety of people complete numerous tasks necessary for daily living. Accomplishing this in a safe, humane and constitutional manner when many people (i.e., prisoners) are sentenced to live there presents very special challenges to prison managers.

Prison Industry Programs

Work programs, which often meant hard labor, have been associated with prisons in the United States from the very beginning. In most instances work has been required of virtually all inmates. It is generally understood that prisoners do the physical labor associated with the day-to-day running of a prison. This includes an immense variety of tasks such as general building maintenance, food preparation and service, operating farms, and performing clerical work, among others. Such work is obviously very essential to the internal operations of the prison and constitutes a major program component for many inmates.

Work programs known as **prison industry** have a different purpose, which is to produce goods for the marketplace. Through the years, prison industry has been utilized to meet a variety of goals, including

1. To make a profit for the prison
2. To reduce idle time
3. To enforce prison discipline
4. To punish
5. To rehabilitate

An inmate at work in a prison industry program.
© Bob Coyle

At times these goals have overlapped, been in conflict, been given different priorities, and been mixed in various combinations.

One of the earliest debates that emerged in the development of prisons centered over work by inmates. The Auburn and Pennsylvania prison systems differed in their willingness to allow inmates to work in groups. History has recorded that the first prison factory was established in the Auburn, New York, correctional facility in 1824. As is evidenced by today's prisons, the concept of congregate labor won out over the Pennsylvania system of isolated industries. This early debate was won primarily on an economic basis; the economic advantages of congregate work became clear quite rapidly. The primary goal of prison industries and other forms of work during the first 100 years of their existence was to make prisons self-supporting.[2]

Inmate labor has been justified on the basis that individuals who are incarcerated need to establish positive attitudes toward work. A high percentage of inmates are unemployed at the time of the commission of their crime. Acceptance of the work ethic and the development of marketable job skills are both rationales for inmate labor. Prison work programs were also once considered as a means to reduce serious physical and emotional disabilities associated with long-term idleness.[3]

From the point of view of a correctional administrator, probably the strongest justification for inmate labor is the calming effect it can have on an institution. The notion that "idle time is the devil's plaything" has real application to a correctional institution. Providing inmates with constructive and meaningful activities while incarcerated has long been the goal of most correctional administrators. One factor that generally emerges in the aftermath of major prison disturbances has been that there was insufficient activity to occupy the residents' time. Prison labor can and certainly has been used as a control device for many years. Inmate labor has also been justified as a portion of punishment; in fact, some states still commit individuals to "hard labor." "The work ethic was seen as the essential ingredient in the reformation of America's criminals."[4]

Although prison labor was a generally accepted practice, it was provided with a legal justification in the post-Civil War Thirteenth Amendment to the United States Constitution: "Neither slavery nor involuntary servitude, *except as a punishment for crime* whereof the party shall have been duly convicted, shall exist within the United States . . ." During the post-Civil War period, revised convict lease systems maintaining many aspects of slavery for blacks and poor whites were begun in the southern states.[5] The notion that prisons should not be a burden upon governmental coffers and that inmates should be required to work was widely accepted throughout the 1800s.

Sheltered versus Open Markets

Two basic systems for marketing prison goods developed. **Open markets** had prison-made goods in direct competition with private-sector products. There were several open market arrangements that were developed. Under one system known as contract labor, private business would set up manufacturing equipment inside prisons and then purchase inmate labor from the state to make products to be sold on the open market. In some cases the contractor agreed to purchase the manufactured goods for an agreed price. In contrast to these contractual arrangements, some prison industries were operated totally by the state, which freely sold products on the open market.

Under **convict lease** arrangements, care and custody of inmates were turned over to private entrepreneurs who then worked these convicts in a variety of industries outside the prison. Convict labor was utilized in agriculture, mining, railroads, lumber, construction, and other industries. J. Thorsten Sellin has provided remarkably detailed documentation of the brutality, sickness, and corruption that accompanied this system.[6] It was not until the early 1900s that these arrangements were declared illegal.

Sheltered markets refer to the use of prison labor for the public good but not in direct competition with private industry. There have long been public works systems that utilize inmates. These public works activities have ranged from the notorious southern chain gangs, which replaced the lease system in many states, to utilization of inmates to work in parks and recreation centers. Another form of sheltered market labor is the so-called **state-use system.** This state-use system was developed in the 1800s in response to rising opposition to the open market approach. Under the state-use arrangement, inmate-produced products can only be used by or in conjunction with the operation of governmental activities. Products such as school desks, license plates, and forms for governmental agencies fall within this category.

Opposition to Prison Industries

As the United States developed as an industrial power, so did opposition to the open market prison industries. This opposition came from three sources: private businesses that felt they were hurt by competition with prison-made goods; workers who made the goods; and citizens working to change the harsh conditions of prisoners.

Economic hard times of the 1920s and 1930s, coupled with the strong public sentiment against open marketing of prison-made goods, brought mandated change to prison industries. The **Hawes-Cooper Act** of 1934 and **Ashurst-Summers Act** of 1935 coupled with later state and federal legislation effectively ended open marketing of prison-made goods.

Hawes-Cooper declared that prison-made goods transported from one state to another were subject to the laws of the importing state. The law made it effectively possible for any state to prohibit prison-made goods from being produced or sold within its boundaries. The Ashurst-Summers Act made it a violation of federal law to transport in interstate commerce prison-made products for private use. By 1944, thirty states had adopted restrictive legislation prohibiting the operation of private-sector prison industries.[7] The change in prison production was dramatic; in 1905, 75 percent of prison-made goods entered the open market, but by 1940 less than 12 percent were in competition with private industry.[8]

Model Programs

As contemporary prisons struggle with rapidly increasing populations, inmates serving more time, and rising costs, there has been an expansion of prison industries to help deal with these problems. Depending upon the degree of involvement by the private sector, four prison industry models have been identified. They are (1) the private-sector model, (2) the partnership model, (3) the corporate model, and (4) the state-use model.[9] Although there has been some easing of restrictions, a state-use model remains the most prevalent form. Sales of goods produced are restricted to sheltered government markets. Examples may be found in a number of states. Control and administration in this model resides mainly with the public sector, since there is little, if any, private involvement.

By contrast, a **private-sector model** seeks to maximize involvement of the private sector. Until recent revitalization, this model was unused for decades. Its demise was brought about by severe abuses and restrictive legislation, such as the Hawes-Cooper Act, during the Depression of the 1930s.

A **partnership model,** as the name implies, represents a joint venture between the public and private sectors. Goods are generally sold in the traditional sheltered state-use market, but private enterprise has involvement in the production. Prison industry may purchase recognized materials from a private supplier or may be involved in the assembly of privately manufactured goods. Technical assistance may also be provided by the private sector.

Application of this model is growing. Partnership ventures seem to reduce fears in the private sector concerning perceptions of unfair competition from prison industries. With this model control is shared, a potential for increased markets exists, and start-up costs can be reduced.

Emphasis in a **corporate model** is placed on managing a prison industry on established businesslike principles found among successful private corporations. The industry operates as an independent business enterprise with connection to the prison for security purposes. Organizational structure includes a chief executive officer and a policy board. Goods are sold in the traditional sheltered government-use market. Experimental models have been utilized in Florida's PRIDE corporation as well as corporations in California and Georgia. However, the primary example for this model is Federal Prison Industries, Inc., which operated by the trade name **UNICOR.**

UNICOR

UNICOR was established by an act of Congress and executive order in 1934. It operates as a wholly owned government corporation and provides employment and training for prisoners in federal correctional facilities. Operations include such activities as

- providing products to federal agencies at fair market prices;
- ensuring optimum inmate employment;
- developing products that minimize competition with private-sector industry and labor;
- providing inmates opportunity to earn funds;
- providing inmates "real work" training opportunities; and
- reducing inmate idle time.[10]

The extent and diversity of UNICOR industrial operations is shown in table 8.1. About 30 percent of all federal inmates are employed by UNICOR.[11] Sales from a variety of products total hundreds of millions of dollars annually with major sales in the areas of electronics products and metal/wood products. Dozens of plants throughout the United States produce their products for many federal agencies.

Future of Prison Industries

The future of prison industries remains unclear. Some signs point toward expansion while others point toward restriction. Once again, as in the past, industry groups have promoted legislation and other restrictions aimed at curbing prison industry activities. Some specific restrictions are provided by the following passage:

> Examples of these restrictions include the enactment of interstate highway sign legislation in 1987 which, with the intense lobbying support of the Traffic Safety and Sign Manufacturers Association, states that the future production of highway signs by correctional industries cannot exceed the dollar volume produced in 1986.
>
> Also the Federal legislation authorizing UNICOR to borrow requires establishment of a formal process of private sector notification and involvement in its new product and product expansion plans, in order to minimize potential adverse impact on private competition. More recently, the Business and Institutional Furniture Manufacturers Association (BIFMA) voted in new guidelines to exclude government agencies from its membership, effectively eliminating correctional industries.[12]

On the other hand, there are some examples of cooperation and partnership resulting in shared benefits for both private industry and prison industries.[13] In New York a partnership arrangement involves private industry in the areas of furniture and textile products, including Burlington Industries. The private sector gains greater access to a government market, and correctional industries gain a new product with reduced start-up costs.

Another example of mutual benefit has been worked out between UNICOR and the Association of Federal Drapery Contractors.

> One telling case in point about the mutual benefits to be gained through cooperation between correctional industries and the private sector is the situation that existed between Federal Prison Industries and the Association of Federal Drapery Contractors. What began as an adversarial relationship in 1984

Table 8.1

UNICOR—Industrial Operations

Industrial Operations		
Divisions	***Metal/Wood Products***	***Textile & Leather Products***
Location and Number of Plants	Allenwood, PA *23 Plants Ashland, KY Duluth, MN El Reno, OK La Tuna, TX Leavenworth, KS Lewisburg, PA Lompoc, CA Milan, MI Morgantown, WV Otisville, NY Petersburg, VA Pleasanton, CA Seagoville, TX Talladega, AL Tallahassee, FL Terminal Island, CA Texarkana, TX	Alderson, WV *20 Plants Atlanta, GA Big Spring, TX Butner, NC Englewood, CO Leavenworth, KS Lexington, KY Maxwell AFB, AL Miami, FL Oakdale, LA Otisville, NY Pleasanton, CA Ray Brook, NY Safford, AZ San Diego, CA Sandstone, MN Seagoville, TX Terre Haute, IN Tucson, AZ
Products and Services Provided	Executive furniture, general office furniture, modular systems furniture, ergonomic seating, dormitory and quarters furniture, household furniture, metal bunk beds, ID tags, lockers, storage cabinets, metal shelving, pallet racks, brushes, office accessories, furniture refinishing	Cotton and wood textiles, canvas goods, synthetic textile products, mattresses, apparel, weather parachutes, brooms, bed linens, gloves, draperies, bedspreads, knit products, boxer drawers and pajamas, military uniforms, undergarments, laundry services
Major Federal Customers	Department of Defense General Services Administration U.S. Postal Service Veterans Administration Immigration & Naturalization Service Federal Aviation Administration Bureau of Prisons Social Security Administration	Department of Defense General Services Administration U.S. Postal Service Veterans Administration Bureau of Prisons
Inmates Employed	4,718 Not including support positions	3,032 Not including support positions
Sales	$124,260,843	$80,048,765

Data/Graphics	Electronics, Plastics, & Optics	Divisions
Alderson, WV *15 Plants Ft. Worth, TX Leavenworth, KS Lexington, KY Lompoc, CA Marianna, FL Otisville, NY Petersburg, VA Pleasanton, CA Ray Brook, NY Sandstone, MN	Bastrop, TX *14 Plants Big Spring, TX Boron, CA Butner, NC Danbury, CT Englewood, CO Lexington, KY Lompoc, CA Loretto, PA Memphis, TN Oxford, WI Petersburg, VA Phoenix, AZ Rochester, MN	*Total Number of Plants* 72 Plants*
Data encoding services (key-to-tape, word processing, telecommunications transmission). Printing services of all types from duplication to 4-color process. Micrographics and textile printing. All types of signs (traffic, safety, and recreational symbols) and decals (silk-screened, die-cut) manufactured on aluminum, plastic, wood, and fiberglass, architectural signage, and metal photo-engraved plaques.	Wiring devices of all types, electrical cable assemblies, electronic wiring harnesses, light kits, printed circuits, electronic systems support, and the remanufacture of vehicular electronics systems. Thermo plastics, fiberglass and Kevlar reinforced products, prescription and non-prescription eyewear, safety eyewear, and radio mounts.	*Products and Services Provided*
Department of Agriculture Department of Commerce Department of Defense Department of Interior Department of Justice Department of Transportation Administrative Office of the U.S. Courts General Services Administration U.S. Postal Service Federal Retirement Thrift Investment Board	Department of Defense General Services Administration Veterans Administration Bureau of Indian Affairs Bureau of Prisons	*Major Federal Customers*
814 Not including support positions	3,651 Not including support positions	Total Number of Inmates 13,330**
$22,494,949	$109,059,539	$335,864,096***

**Some locations with multiple plants.*
***This includes an additional 1,115 inmates employed in support positions.*
****Does not include transfer sales of $13.2 million.*
Source: UNICOR, Annual Report, *1988 (Washington, D.C.: U.S. Department of Justice, 1989), pp. 16–17.*

when UNICOR announced it would expand its drapery manufacturing industry turned into a solid working relationship after much compromise. In brief, a series of meetings led to opening channels of communication, whereby UNICOR agreed to utilize private drapery contractors for the raw materials, installation, and hardware portion of the government contracts it receives. Moreover, both UNICOR and the Federal Drapery Contractors now have a better understanding of each other's concerns. As a result, the Drapery Contractors have been an advocate of correctional industries, speaking out in support of UNICOR on several occasions.[14]

It is clear that agreements can be reached that benefit both parties involved. It is reasonable to assume that taxpayers and inmates also benefit. However, arriving at such joint ventures takes commitment and hard work. Opportunities need to be identified and relationships developed where both the concerns of private industry as well as merits of successful joint ventures can be addressed. The extent to which such communication is accomplished will probably determine the future of prison industry expansion.

Rehabilitation Programs

It must be remembered that in a prison setting security rather than rehabilitation receives priority. Classification determines a level of custody and prison assignment appropriate for an inmate. Delivery of **rehabilitation programs** is often compromised and handicapped by the totality of negative and intrusive features which accompany life in prison. In most instances, treatment programs operate mainly as extensions of security functions and as a means for inmates to reduce idle time. Thus rehabilitation programs in prison usually place more emphasis on the "program" aspect than on the "rehabilitation" aspect.

Rehabilitation has also labored under an unproven assumption that "nothing works". When such an assumption is made, meaningful rehabilitation efforts subside, replaced by simple notions of incapacitation, crime control, and "get tough" approaches. Prison populations increase and inmates simply mark time. Rehabilitation technology needs additional study and research. What has not worked many times is the willingness to commit to the goals of rehabilitation and to effectively deliver treatments.

Right to Treatment

Courts have generally held that rehabilitative treatment is not a constitutional right for adult felons. In *Padgett v. Stein,*[15] for example, no constitutional obligation was found for correctional systems to provide rehabilitative opportunities, although this could be considered together with other factors when looking at the total conditions of confinement. In some instances there may be court-ordered participation in treatment, in which case a prisoner does not have a right to refuse. Such treatment might be ordered for sex offenders or alcoholics. In general, rehabilitative treatment programs are established as a matter of public policy rather than being mandated by the Constitution.[16] Prison managers have been left free to develop and administer such programs within the context of constitutional requirements and without cruelty. We now review basic programs that operate in one form or another within most of the nation's prisons.

Education Programs

Education and prisons have been tied together since the opening of the Walnut Street Jail in Philadelphia, which provided instruction in reading, writing, and math. Many early educational efforts were conducted by the clergy who held Sabbath schools using the Bible as their text. In 1847, New York became the first state to mandate that correctional education be available in all institutions. The development and growth of the reformatory system further advanced the connection between education and institutions. Throughout the twentieth century, education programs for inmates have increased in size and importance. Major boosts to prison education came via the introduction of federal support through such pieces of legislation as the Adult Education Act of 1966 and the Basic Education Opportunity Grant program initiated in 1972.

There are thousands of correctional educators working in the nation's prisons. These teachers are serving students who have had little success in schools and are lacking in both basic academic and vocational skills. Many inmates function two to three grades below the level completed in school and most have had no vocational training. Many are illiterate, unable to read such items as newspapers and traffic signs.

Typically **correctional education programs** include adult basic education (ABE), General Educational Development (GED), adult continuing education (ACE), and post-secondary education (PSE). Educational activities generally represent the largest and most visible area of prison programming to benefit inmates.

Vocational education training programs are available in most prisons in the United States. These programs have as their goal:

> . . . the development of job related skills through a combination of on-the-job training and classroom experience within the institution. Some of these programs may include the more specific goal of the acquisition of a trade or technical certification.[17]

Training is provided in many fields including welding, carpentry, data processing, office equipment repair and food service.

Vocational training, along with social education, has focused on job readiness. The concern in these areas is life skills. Many inmates lack the basic social life skills to function effectively as law-abiding citizens. These inmates may need help in writing checks, applying for jobs, budgeting their money, and other everyday activities. Vocational and social educators recognize that if inmates are to reenter society and abstain from criminal activity, they must be employable and have the basic tools necessary to function as responsible citizens.

Recent in-depth research reported by Miles D. Hares reveals the beneficial effects of participation by inmates in educational programs while in prison, and a subsequent reduction of post-release recidivism. In general, post-release recidivism was less within all types of educational programs (that is, ABE, GED, ACE and PSE). In addition, this relationship between educational participation and recidivism was found to exist regardless of prior criminal activity, preprison educational attainment level, preprison employment and post-release employment.[18] Such findings indicate that education is crime prevention across a broad spectrum. It is therefore disheartening that survey results indicate many prison educational programs have been reduced.[19]

Mental Health Programs

There is increasing evidence that the need for **mental health programs** within correctional institutions is on the rise. Changing standards of basic care established through professional standards, court intervention, and shifts in public expectations have all provided pressure on correctional facilities to expand mental health programs. Another factor affecting the need for additional mental health services has been the marked drop in the number of individuals cared for in mental hospitals in the United States. Because of court action and changing standards, the populations of mental hospitals have significantly declined. All too often these former mental hospital patients have found their way into our nation's prisons. Mental health programs are sometimes directed at inmates with special needs such as substance abuse and sex offenses.

Psychological Service

Most correctional institutions have either a full-time or part-time **psychological service** team that works within the facility. A program staff is generally composed of licensed psychologists and trained individuals working under the direction of a licensed psychologist.

The institutional psychologist's responsibilities include the administration and interpretation of standardized tests. Many correctional institutions and parole authorities require a psychological profile of inmates under their care. The generation and reporting of these psychological profiles consume a great deal of the staff's time. The psychological staff also becomes involved in crisis intervention, group and individual counseling, and referrals.

Institutional staffs frequently have consultant arrangements with psychiatrists who work within the facility on a part-time or as-needed basis. Psychological programs within prisons are plagued by many of the same problems that face other staff. These problems include large caseloads, requirements of extensive documentation (which leads to volumes of paperwork), and an inability to compete economically with private practice salaries.

Counseling

With the introduction of trained professionals in the mental health field into correctional facilities, it became possible to begin to provide ongoing counseling and crisis intervention. A variety of approaches have been taken, utilizing both group and individual counseling. **Individual counseling** between an inmate and counselor does take place in the correctional institution. These individual sessions are often part of a larger treatment program and may include any one of a variety of counseling philosophies, such as rational emotive therapy or a Rogerian approach. These individual sessions, particularly if they require a large number of visits, are often available to only a limited number of inmates. The high ratio of inmates to counseling personnel in most correctional institutions makes it difficult to conduct an extensive individual counseling program.

Group counseling is planned activity in which three or more people are present for the purpose of solving personal and social problems. Group counseling sessions that have one or two counselors working with five to ten inmates at a time are commonly used in correctional institutions. Group sessions are relied upon partly because of the savings in both time and personnel that they afford. Group counseling is also very practical because there is an availability of clients with similar types of problems that lend themselves to a group approach. Group counseling differs from group therapy which also takes place in some correctional institutions. The primary differences between the two techniques can be found in the level of treatment and training of the personnel involved. Group counseling generally focuses on personal and social issues and is conducted by paraprofessional, or nonlicensed, individuals. Group therapy is generally concerned with deep-seated psychological problems and should be directed by a highly trained psychologist or other mental health professionals.

In all counseling relationships, there is a prerequisite need for trust, openness, honesty, and a sincere desire to seek change among the participants, if the process is to prove successful. The nature of the correctional setting provides some obstacles to meeting these basic conditions for counseling. Particularly in the group setting, there can be real barriers to trust between the inmate and the counselor. The counselor is not only an individual trained and ready to work with the inmate but, in most cases, is also an employee of the correctional system. The inmate may well be hesitant to reveal his or her true feelings to a correctional employee charged with the responsibility for completing a psychological or social report to a parole board or other releasing agency.

Eclectic Approach

Often the most effective programs are those that assume an **eclectic approach.** By incorporating various components of different techniques into a unified program, the correctional staff has the greatest opportunity to effect change. One program cited as a model for its use of an eclectic approach is the Sex Offender Program of the Connecticut Department of Corrections. The program's goal is the diagnosis and treatment of personality dysfunctions of individuals committed to the Connecticut Department of Corrections for rape and other sexual offenses. Following a clinical assessment of each offender, a treatment plan is developed which could include chemotherapy, psychotherapy, group or individual counseling, courses in human development, sex education, and behavior modification. The program staff recognizes that not all cases can be successfully treated and includes incapacitation as one of their treatment modalities.[20]

In the Connecticut program and many others, it is recognized that there is not one simple or sure means to cope with the mental health problems of offenders. By utilizing a variety of approaches, the correctional system and its staff have the best opportunity to assist inmates in changing their patterns of behavior.

Prison Chaplains

The first services to be offered in prisons were religious. The involvement of religious groups in providing for the spiritual needs of inmates dates back to the first correctional institution. **Prison chaplains** were the first educators, social workers, and counselors in prisons. As prisons added more helping professionals, the chaplains were able to focus their attention solely on religious matters.

The prison chaplain, who is generally an employee of the correctional institution, oversees the operation of the prison chapel. Generally the chapels are interdenominational, and the chaplain must serve the needs of individuals of many faiths. The chaplains are often very active in bringing community religious groups into the correctional facilities to work with the inmate population.

The prison chaplain remains not only a spiritual leader in the prison but also a counselor and an advisor. It is often the chaplain who is called upon to inform an inmate of the loss of a loved one or other personal tragedy. Many chaplains take an active role in their institutions in advocating the rights of prisoners and the maintenance of humane conditions. Religious services are an important component of programs available in most institutions.

Leisure Time Programs

Inmates have a need not only to learn to read, gain employment skills, and gain emotional stability but also to learn **leisure time activities.** There are a variety of recreational programs operated in most correctional institutions. These programs not only help to occupy the inmates' time but also attempt to have an impact on their self-image, resocialization, and ability to express themselves.

Most correctional facilities have personnel assigned to provide recreational programs. These recreational specialists arrange programs such as movies, sporting events, touring groups, and arts and crafts. Art, theater, music, and other cultural activities have long been a part of some prison recreation programs. Craft industries in correctional institutions provide both a recreational activity and a means for inmates to earn extra money through the sale of their handiwork. Some prisons have even provided inmates with summer garden plots. All of these recreational activities recognize that inmates need to learn how to better utilize leisure time if they are to succeed upon release and also that planned activities are an effective means of making the experience of incarceration more bearable for many inmates.

Self-Help Groups

Many correctional facilities encourage inmates to participate in various **self-help groups.** These groups include Seventh Step, Alcoholics Anonymous, Dale Carnegie Courses, and the Jaycees. The self-help groups are generally operated outside the normal program areas and are often directed by the inmates with a staff advisor to act as a liaison. These programs provide the inmates with self-help activities, additional links to the community, and an opportunity to participate in a noninstitutionally sponsored activity of their own choosing.

Management of Prisons

Chief responsibility for management of prisons resides with wardens or superintendents and their staff. The management process in a prison community is unique because it involves managing people (prisoners) who do not want to be there. They have been forced to live there as a consequence of their law violation and subsequent sentence. Within this context security must be maintained, work and rehabilitation programs operated, buildings maintained, and constitutional rights upheld, a situation that presents formidable challenges. Prisons are in many respects like communities on the outside. Most of the same daily needs and services related to governing, feeding, and working necessary for living in the outside community must also be provided in a **prison community.**

Materials in this section will be presented around a series of five questions designed to foster insight and understanding into the complexities involved in management of prisons. These questions concern:

1. What constitutes prison organization and managers?
2. What is the setting for prison management?
3. What are management functions?
4. Who is managed in a prison community?
5. How do styles of prison management differ?

Managers

Managers are the people engaged in the process of management at different levels of an organization. Their activities may involve matters internal or external to the organization. Included would be such tasks as planning, attending meetings, supervising, and communicating with other people. Managers may be classified in three basic ways:[21]

By Title	*By Position*	*By Level*
Executive	Top management	Third-level management
Managers	Middle management	Second-level management
Supervisors	First-line management	First-level management

Within a prison the title of **warden** or superintendent is usually given to the top management position, while the title of correctional officer (CO) is usually given to the first level. The roles and tasks of these two positions vary greatly, but both are managers. A correctional officer is concerned with day-to-day activities at the first-line level internal to the organization. The warden is little involved in performing first-level tasks but is responsible for overall operations both internal and external to the prison.

Organizations

Organizations may be thought of as a group of persons pursuing a special purpose or goal. In our society, organizations of all types abound. Manufacturers, schools, the military, government, and prisons are all organized in some manner to facilitate meeting particular objectives. Organizations provide structure that defines roles and establishes authority by which persons in these groups come together, associate, and interact with one another, attempting to achieve their goals. Organizations thus attempt to coordinate the work activities of its people.

Organizations are often depicted by means of a formal table of organization or **organizational chart.** Such charts detail formal positions and relationships within organizations. Positive titles, levels of authority, span of control, chain of command, and grouping of similar tasks are components commonly contained in organizational charts. Such charts can be useful since they indicate what departments have been established, each employee's job title, and location within an organization. Such charts also detail a chain of command, locating responsibilities and showing who is accountable to whom.[22]

It is equally important to remember however, that some very important aspects of organizations never show on a table of organization. The whole network of informal structure and relationships does not appear, although such relationships can be very important management considerations. Organizational charts do not indicate job descriptions of actual daily duties and responsibilities, patterns of communication, informal levels of authority and power, or how closely workers are controlled.[23]

Setting for Prison Management

During much of our history, prisons were isolated with only minimal exchanges with the outside environment. Courts adhered to a hands-off doctrine and prison administrators were little accountable for how prisons were managed.

The setting for managing contemporary prisons, however is quite different. Today, prison administrators find they operate in a subsystem placed within a larger and broader system. Such an orientation views prison organization as an **open system.** There are numerous input and output exchanges with other public and governmental units, three examples of which follow.

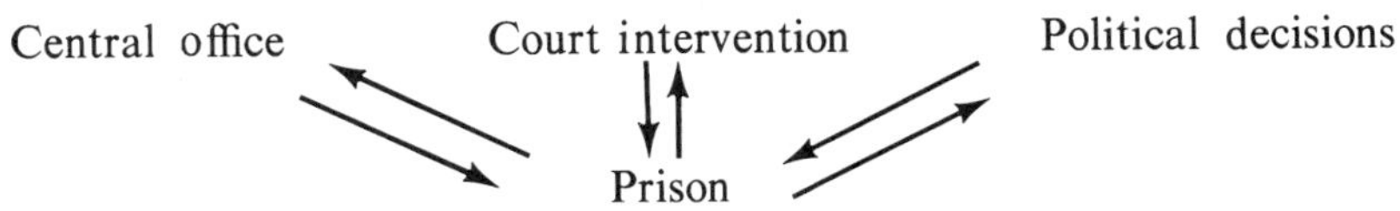

One major external influence on prison operations was the decline of the hands-off doctrine. As a result, persons under sentence gained access to the courts. Discussed throughout this textbook are many legal cases that have contributed to the development of correctional law. Rulings in such cases have management implications that have affected a wide range of practices, including medical care, disciplinary procedures, religious practices, and living conditions. Conceptualizing prisons as an open system can be an advantage in understanding, analyzing, and planning for eventualities brought forth by judicial decision.

A closely related aspect is that contemporary wardens function as part of a **centralized management structure.** For decades wardens were authoritarian, governing prisons with virtually unchecked power and little outside influence or accountability. Management of corrections and prisons today is centralized in statewide departments or bureaus of corrections. Wardens and superintendents are placed in an organizational bureaucracy along with dozens of other managers and executives, as illustrated by figure 8.1. Some outcomes for wardens of this centralized management have been a loss of power, an increased accountability to other officials, a closer monitoring by governors, and a need to develop skills and attitudes to work with others in the organization.

Considering prisons as open systems allows for an analysis of the interactive patterns with the environment. Some approaches to organization and management have been criticized for being too "closed" and "static" in outlook. Organizational charts and controls have been emphasized at the neglect of considering organizations as existing as a portion of a larger environment. Considering the efficacy of an open system approach, Duffee's opinions are that:

> The use of an open-system model does not change the processes by which managers understand the organization, but it does not seem more effective than other models for classifying the elements of organization and predicting basic organizational behavior.[24]

Figure 8.1 Bureau of Corrections organizational chart

Source: Adapted and projected from Bureau of Justice Statistics, Prisoners in 1981 *(Washington, D.C.: Government Printing Office, 1982), p. 41.*

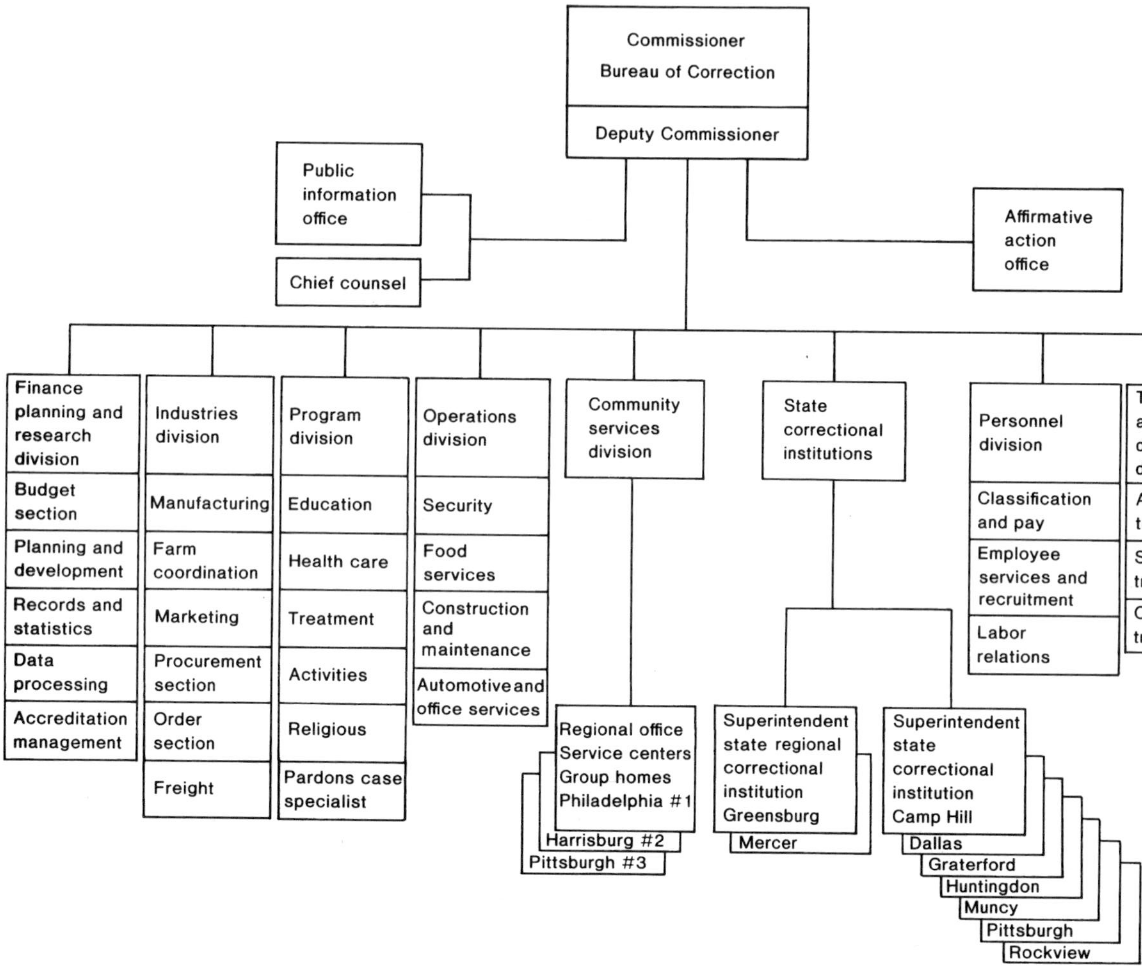

Management Functions

Management involves a wide range of activities often categorized into five basic groups which constitute a totality of **management functions.** Specific activities within each function are summarized as follows:

1. **Planning.** Establishing goals and objectives; developing a ''road map'' of procedures; projecting to some future occurrence.
2. **Organizing.** Establishing a structure to meet function; developing roles and accountability; measuring and awarding performance; establishing departments, establishing lines of communication and coordinating work of employees—because each needs to know what their jobs are to have a sense of accomplishment.
3. **Staffing.** Matching people to jobs based on the organizational process; bringing in and developing the human resources of the organization; the ''living'' part of management: involves recruitment/selection/retention; and training/career development/rewards.
4. **Leading.** Providing motivation and leadership for employees to get the job done; treating employees fairly and maintaining morale; meeting employees' need for recognition, and suggestions for improvement through communication by managers; providing opportunity for growth; key person is individual supervisor; building loyalty to organization.
5. **Controlling Outcomes.** Monitoring activities to determine if organization is meeting goals; compare performances to standards; utilizing feedback information to compare actual to planned performances and making long-term corrections as necessary; involves budgeting, cost analysis, possible experimentation, and employee acceptance.[25]

There is no end point in the process of management. These five managerial functions comprise a circular, interrelated set of processes. *The degree of effective and efficient management attained is determined by how well these managerial functions are accomplished.* A wide range in the degree of effectively managed organizations is to be found. Some prisons would rank as more effectively managed than others. Many have been found to be managed with a low degree of effectiveness.

Managing People in a Prison Community

Unlike most organizations that are managed to turn out some product, prison management involves managing people. Essentially, there are two groups of people that comprise a prison community: employees and inmates. Prison management involves to a great extent managing relationships among employees, between employees and inmates, and between inmates. Prisons are most unique in that a majority of the people (i.e. the inmates) are forced to live there according to terms of their conviction and sentence; they do not want to be there. Such a situation obviously presents awesome challenges, to manage such a community of people that is secure, safe, and livable. The people on the front line around the clock dealing with such challenges are the correctional officers. Other than inmates, this is the largest group of people requiring management attention within the prison community.

Correctional Officers

It is through the work of **correctional officers** (COs) that management controls and operates the prison community. In many ways they are extensions of the warden in carrying out orders and rules. These officers comprise about 75 percent of all prison employees and are the single largest group of correctional workers in the nation. Their work is absolutely essential to the operation of prisons. Although the nature of work performed by COs varies based on the individual assignment and the particular institution in which they are employed, there are duty assignments that can be identified. Lucien X. Lombardo identified seven general categories of work assignments held by correctional officers:

- Cell block officers—oversee housing units
- Work detail supervisors—enforce work assignments
- Industrial shop and school officer—maintain order and security
- Yard officers—patrol; watch for trouble
- Administration building assignments—building security; receive public; little contact with inmates
- Wall posts—perimeter security; little contact with inmates
- Relief officers—work whenever and wherever needed.[26]

Most experienced correctional officers have worked in a number, if not all, of the various identified work assignments. In some correctional institutions, it is standard operating procedure to rotate officers to new duty assignments. These reassignments serve to reduce boredom with familiar tasks, prevent the establishment of routinized patterns of behavior that might increase an officer's vulnerability, and finally, to reduce the opportunity for officers to engage in illicit activities. Correctional officers are, in a sense, also prisoners and are frequently dealt with in a similar fashion by administrators. The reassignment of correctional officers to new job stations frequently resembles the inmate classification system, which has been criticized for basing its decision more upon prison than individual needs. Correctional officers are often viewed as pawns in the institutional chess game to control inmates.

The correctional officer, like other correctional workers, must often function with conflicting goals. The primary responsibility of the CO is maintaining order and preventing escapes. There are, however, numerous secondary goals ranging from meeting various production quotas, to assisting troubled inmates, to dealing with the general public. For the correctional officer who works in the wall post, the goals are clear and relatively easy to meet. The officer in the tower must prevent escapes and monitor activity near and within the institutional compound. For most correctional officers, however, the task is far more complex.

Barriers to employing women as correctional officers have been declining in recent years.

© D. MacDonald/Unicorn Stock Photos

The correctional officer staff is divided over twenty-four hours, while many social services and administrative personnel work a traditional eight-hour shift. In many institutions, one officer may be responsible for fifty or more inmates at a time. This high ratio can lead some officers to merely go through the paces of the job, thus creating a safe distance between themselves and the inmates. The physical construction of the correctional institution can also influence the correctional officer's work environment. Some institutions, by their very construction, can be both threatening and depressing. Many newer institutions, however, provide a safer environment in which to work.

Another factor affecting the correctional officer's job is the organizational climate. The officer working in a large maximum security institution faces more challenges and problems than does another officer in a small minimum security facility. The amount of actual or perceived administrative support along with the formal training of the officer also affects the individual's perception of the job. It is not possible to make a blanket statement about the organizational climate and its effect on the CO's job, but it can accurately be determined that each correctional institution has an identifiable atmosphere that influences those who work within it.

Women as Correctional Officers

For many years the only opportunities for women to work as correctional officers (or matrons as they were called) was in female correctional institutions. "In many settings there is still a 'prison matron' philosophy. Namely, that females are basically capable of working at only line positions in all female institutions.''[27]

In recent years many of the barriers to employing women as correctional officers in all-male institutions have been dropped. In a number of prisons, women regularly serve as line correctional officers. Women were traditionally excluded from CO positions in order to protect inmate privacy, for the safety of the women, and because of fears that women would be unable to control inmates. Despite some court action that has upheld the right to exclude women from contact positions in certain types of correctional facilities,[28] there has been an increasing number of females employed as line correctional officers throughout the country. It would appear that many of the traditional assumptions and fears about women as correctional officers were not based upon facts, but rather myths and misconceptions.

The addition of female correctional officers to all-male institutions may have positive outcomes in terms of normalizing the prison experience. ''Seeing women in the prison somewhat in proportion to the numbers they represent in the community can lend an air of naturalness and reality to the prison community.''[29] The positive effects the female correctional officers can have upon an institution may help to alter the public image of both institutions and guards.

Styles of Prison Management

Managing prisons according to sound governing and management principles has been neglected. Too often prison managers have not incorporated valid management ideas into their operations. John I. DiIulio, Jr. has explained ''if most prisons have failed it is because they have been ill-managed, undermanaged, or not managed at all.''[30] However, now more than ever, due to such factors as increasing public scrutiny and rising demands for accountability, there is a need to bring sound management ideas to strengthen managerial effectiveness. Recognition of such a need has existed for some time and was stated by Seth Allan Bloomberg who said that:

> Corrections can no longer afford to exist in a vacuum, set apart from all other knowledge; it should and must utilize an interdisciplinary approach, drawing on the expertise of related fields if it is to accomplish its various goals. The isolation of corrections has been diminished as a result of the erosion of the judicial ''hands-off'' doctrine, but this change has been forced on corrections by the judiciary; and, at best, it can represent but a piecemeal attack on the system. Insofar as management techniques are concerned, corrections remains as isolated as ever. The time has come to incorporate management theory into correctional practice.[31]

It was previously noted that the concept of organization provides a structure that brings people together for a particular purpose. The choice of management style has a profound effect on the nature of the organizational climate and the environment in which people work. Wardens and top management

Reprinted with special permission of King Features Syndicate.

personnel play a key role in this regard since they are in leadership positions and the style of management they select determines in large part the nature of the prison environment for staff and inmates.

Types of management styles may be conceptualized on a continuum ranging from an autocratic to a participative style. The climate prevailing in an organization is substantially a consequence of the management style favored and practiced by a small group of top managers. If these managers are autocrats, the system below them will reflect this fact. If, on the other hand, they are participative, and consciously share with those under them the making of decisions and the rewards of organizational accomplishment, the organization likely will become more democratic and participative.

Inasmuch as the managerial style of key decisionmakers determines in a major way the climate of an organization, it is appropriate to consider management style and organizational climate simultaneously.

Autocratic Style

Historically, most prisons have operated according to an **autocratic management style.** Wardens have held absolute power wielded in an arbitrary fashion. In such a prison, emphasis is placed on giving orders, obeying rules, and unswerving dedication and loyalty to the warden. An autocratic warden operates with personal decision-making authority according to a paramilitary model of management. Employees have no input in the management process. Inmates often have been subject to harsh living conditions and severe punishment. Relationships among employees tend to be very impersonal and subject to the warden's absolute power. Employees tend to have negative feelings toward management.

Many wardens have operated according to an autocratic style. Probably the best known has been Joseph Ragen, warden of the Stateville, Illinois Penitentiary from 1936 to 1961. His rein of unchecked power and human abuse has been well documented by Jacobs.[32]

An autocratic approach concentrates on formal management and organizational structure. Emphasis is placed on the division of labor and on controlling workers according to outlined levels of authority. Such a structure emphasizes vertical levels of increasing authority by which lower level workers are subordinated to higher level supervisors. Such organizations have been characterized by such terms as *rigid, static,* and *inflexible* for changing needs. Organizations are thought of as machines with little or no consideration given to the human side or the environment in which it exists.

Douglas McGregor has developed a set of basic assumptions about human nature that flow from an autocratic perspective. Specifically, the average human being is seen as having an inherent dislike of work; most people must be controlled and threatened with punishment to get them to put forth adequate effort toward the achievement of organizational objectives; and the average human being prefers to be directed, wishes to avoid responsibility, has relatively little ambition, and wants security above all.[33]

In another description, Rensis Likert's research suggested that the nature of this style is such that leadership has no confidence or trust in subordinates; fear and punishment characterize motivational forces; most communication flows in a downward direction; most decisions are made at the top level of the organization; goals are established through issuing orders; and review and control are highly concentrated in top management.[34] Despite broad criticism, this approach continues to be found in the management of prisons.

Participative Style

By contrast, a **participative management style** makes a quite different set of assumptions about human nature. This approach concentrates on developing a set of assumptions according to a human resources perspective. The expenditure of physical and mental effort in work is seen as being as natural as play or rest. External control and the threat of punishment are not the only means for bringing about effort toward organizational objectives. People will exercise self-direction and self-control in the service of objectives to which they are committed. Commitment to objectives is a function of the rewards associated with their achievement. The average human being learns, under proper conditions, not only to accept but to seek responsibility and has the capacity to exercise a relatively high degree of imagination, ingenuity, and creativity in the solution of organizational problems. Finally, under the conditions of modern life, the intellectual potentialities of the average human being are only partially utilized.[35] Among major influences contributing heavily to the development of participative management were Maslow's behavior approach and the surprising results obtained from the Hawthorne studies.

Maslow's Hierarchy of Needs

Behavior approaches to management and organization have included principles and theories of human motivation. Although complex, motivational factors move people to action. Behaviors are prompted by inner personal urges or needs which in turn lead to rewards. Therefore, behavioral management approaches have sought to determine human needs and then structure work situations to reward these needs.

Abraham Maslow's **hierarchy of needs,** a theory of motivation, has been a significant contributing force to a behavioral approach. Maslow contended that it is necessary to consider the situation or environment in which a person finds himself. His approach suggests there are five basic human needs ranked in order of importance, beginning with basic needs and continuing through safety, belonging, esteem, and self-actualization needs.

As lower level needs on the ladder are satisfied, succeedingly higher level needs emerge, thus serving as a motivational source. Basic needs are physiological needs. People are first motivated to seek air, food, water, and physical comfort. When physiological needs are relatively satisfied, there emerge needs to feel safe and secure in one's environment. People seek to feel stability, freedom from fear and chaos, protection, and familiar surroundings rather than the unknown.

Next, a basic need exists for one to feel part of family and friendship groups, the need to both give and receive affection. Unmet needs are likely to bring feelings of loneliness and rejection. Peoples' ego needs center around desiring a positive evaluation of themselves. Human beings have a need for self-respect, self-confidence, and achievement. Also existing is the desire for recognition, importance, and appreciation by others. Unlike lower level needs, esteem needs are usually not relatively satisfied. People generally seek increased recognition, sense of importance, and esteem.

The ultimate need is seeking the complete realization of one's potential—being all that one can be. This need is highly individualized and may find expression in a variety of forms. As with esteem needs, self-actualization needs are seldom met.[36]

Maslow's work helped establish the importance of the dimension of motivation. As others have elaborated on this basic idea, it is generally accepted that motivation is a key component in a person's urge to find job satisfaction. Behavioral approaches to management have attempted to develop organizational structures and management styles to meet these needs.

Hawthorne Studies

During the late 1920s and early 1930s, a series of experimental studies were conducted among telephone equipment assemblers at the Hawthorne works of the Western Electric Company in Chicago, Illinois. These studies examined employee attitudes, their motivations, and the influences exerted by informal worker groupings.

Experiments were begun to determine the relationship between the amount of illumination and production output. Researchers expected to find an increase in production when lighting was improved and a decrease when it was lessened. One experiment divided workers into two groups of equal size having about the same work experience. Specifically:

> One group, called the "test group," was to work under variable illumination intensities; the other group, called the "control group," was to work under an intensity of illumination as nearly constant as possible. The groups were located in different buildings in order to reduce the influence of any spirit of competition. The test group worked under three different intensities of light, 24, 46, and 70 foot candles, while the control group worked under a more or less constant level of 16 to 28 foot candles.[37]

But the results did not correspond with what was expected. Quoting from reports of the study, it was noted that:

> This test resulted in very appreciable production increases in both groups and of almost identical magnitude. The difference in efficiency of the two groups was so small as to be less than the probable error of the values. Consequently, we were again unable to determine what definite part of the improvement in performance should be ascribed to improved illumination.[38]

As a result it became apparent to investigators that psychological factors, "human" in nature, were involved. Researchers then embarked on an extensive series of additional experiments and interviews to discover attitudes and feelings workers held toward their jobs. In some instances production increased even when advantages such as rest periods were taken away, because the experimental situation created a positive atmosphere wherein close personal attention was given to the workers, findings commonly referred to as the **"Hawthorne effect."** These were very insightful findings. Subsequent research and development flowing from the concept of the Hawthorne effort resulted in a management style identified as participative.

Managers fostering a participative climate will be group oriented and perceive a managerial role as involving the integration of the work group and its development into an effective team. Toward this end, a participative manager believes in maintaining an informal, friendly relationship with employees. Besides sharing information with them, managers solicit and respect employees' opinions and input about the work situation.

Autocratic versus Participative Style

The issue concerning whether an autocratic or participative style fosters a higher degree of effective and efficient prison management is far from resolved. Both approaches have supporters and detractors although a base of solid research is lacking for either position. According to Holt,[39] power should be shared. He suggests that inmates in the prison community should participate with effective power in management and policy decisions. Research efforts are sparse but Grusky[40] and Wilson[41] concluded that inmate cooperation increased under participative management practices. Also, communication with staff increased

		Participative	
		Yes	No
Democratic	Yes	A Have say Have vote	C Have no say Have vote
	No	B Have say Have no vote	D Have no say Have no vote

hostility toward staff decreased, and disciplinary problems were less likely than under autocratic practices.

On the other hand, many wardens and scholars do not share such a view. For example, DiIulio has concluded that "the experience of most prison systems seem to count against participative prison management,"[42] and elsewhere states that "there is not a single example of a system of inmate self-government—formal or informal—in a higher-custody prison that has resulted in a safer, cleaner, more productive facility."[43] He draws heavily upon the failed experiment of inmate self-government within the Walla Walla Penitentiary at Washington State to support this contention. Others have suggested that this failure resulted from insufficient joint efforts by all involved to infuse the attempt with sufficient democratic values.[44] However, it may be extremely valuable to note that one very successful executive manager, who wholeheartedly practiced participative management, bluntly states that "participative management is not democratic. Having a say differs from having a vote."[45] Such an idea might be expanded to develop the analytical framework shown above.

The implications of such a conceptualization are profound when considering prison management. Although inmates cannot have a vote on their release, for example, perhaps having a say about situations that affect them could improve conditions for all who live and work there. For example, a report by Miles D. Harer states that:

> Staff and inmate "climate" surveys serve several management and normalizing purposes. They provide management with insight into staff and inmate perceptions about the effectiveness of policies, operations, and programs. They help management identify ineffective practices (in terms of normalization, custody, and security), underlying causes, and corrective measures that are needed to correct these problems. Survey responses also help identify staff and inmate needs and concerns that are not being addressed and, therefore, can indicate new policies, operations, or programs that are needed.[46]

It is clear that there is much more to be discussed, researched, and learned concerning effective management styles for prisons. It could be that it is a question of selecting neither an autocratic nor a participative style to fit all conditions, but rather, developing a more **situationally specific management style.** Rather than trying to find the one best way to organize and manage under all conditions, a situationally specific approach considers a specific set of conditions at hand and examines the functioning of organizations (i.e., prisons) in relation to needs of specific members within the organization. Such an approach begins with the premise that no one approach is superior under all circumstances. It has been noted that "the beginning of administrative wisdom is the awareness that there is no one optimum type of management system."[47]

These words would appear to be particularly well suited for prison management because of numerous variables and situational factors; for example, a specific situational condition of prison is the level of custody. It is plausible to consider that an autocratic style (but not inhumane) might more effectively manage inmates in a maximum security prison but that a participative approach might be more effective under conditions of minimum security. Similarly, the type of offender, length and type of sentence, and opportunity to earn good time are specific conditions that could suggest a situational approach to management.

An additional example focuses on correctional officers and other prison employees. It is generally understood that present-day workers will accept challenges and that perhaps more than anything else they want to be involved and to make a contribution. Traditionally, autocratic leadership has resulted in staff and inmates being managed by the same approach. Some results of this have included arbitrary hiring and dismissal practices, poor communications, poor morale, and lack of recognition for achievements. In some organizations officers have felt unappreciated and in some instances have sought (and gained) union representation. By contrast, a plausible result of a situationally specific style of management in a maximum security prison might include a more participative approach involving officers and a less participative approach toward inmates.

John J. DiIulio, Jr., suggested that "there is nothing inherent in the nature of prisons or their clientele that makes better prisons impossible." Furthermore, "poor prison conditions are produced by observable and, it appears, remediable defects in the way that prisons are organized and managed."[48] Improvements would likely result insofar as prison managers adopt roles as change agents who proactively advocate and implement innovative and more effective approaches. The challenge is for corrections officials to do the necessary homework and solid research, and for the American public and lawmakers to ensure that whatever styles are utilized, America's prisons are at the very least safe, secure, humane, livable, and within constitutional compliance for all who live and work there.

Summary

Various forms of work programs have always been associated with prisons in the United States. This has included hard labor, work necessary to operate the prison, as well as the production of prison-made goods. Prison industries have, at times, been used both to serve custody and rehabilitative goals as well as make a profit for the prison.

From time to time prison industries have received restrictions from outside industry groups and workers. On the other hand, there are examples of cooperation and partnership that have resulted in shared benefits. A primary model of prison industry is Federal Prison Industries, Inc. which operates by the trade name UNICOR.

Delivering rehabilitation programs is difficult in prisons since security receives priority. In many cases such programs operate as extensions of security functions to reduce idle time. Numerous categories of programs exist such as educational, counseling, vocational, and self-help.

Managing people and programs in a prison community presents special challenges. Prison managers today are more visible and more accountable than during previous times, due in part to the development of a centralized corrections management structure. Within this framework, wardens perform five basic management functions: planning, organizing, staffing, leading, and controlling outcomes. Effective management is determined by how well these functions are accomplished.

Prison management involves relationships with two basic groups of people comprising a prison community: inmates and employees. Correctional officers may be seen as extensions of the warden especially in relation to security functions. Officers perform a variety of work assignments in this capacity. This job has often been the topic of research and analysis involving such matters as job stress, view of inmates, gender, and role conflict. Securing inmates is a special concern of management, especially in such areas as controlling escapes, contraband, and riots.

It is recognized that a management style employed by top administrators determines in a major way the climate and nature of relationship between people of an organization. Prisons have typically been governed by an autocratic style characterized by giving orders, motivation by fear, and no upward flow or feedback of information.

By contrast a participative style influenced by Maslow's work and the Hawthorne studies, seeks input from employees, fosters a team approach to problems, and rewards achievements. At issue is the extent to which and under what situationally specific conditions this style could be adapted to prison management. Solid additional research is needed to provide management styles that are most likely to result in safe, humane, and constitutional prisons.

Discussion Topics

1. Can a partnership prison industry model be expanded to more prisons?
2. Should rehabilitation be attempted in a prison setting?
3. Is it advantageous to view prison from an open systems perspective?
4. Should educational opportunities be provided to prison inmates? Why or why not?
5. Why is it we say that there is no end point in the five management functions?
6. If you were warden of a prison housing male inmates, would you hire any female correctional officers? Why or why not?

7. Under what conditions would you prefer an autocratic or participative style of management? Why?

Additional Readings

DePree, Max. *Leadership is an Art.* New York: Doubleday, 1989.

DiIulio, John J., Jr. *Governing Prisons.* New York: The Free Press, 1987.

Hares, Miles D. "Recidivism Among Federal Prison Releasees in 1987: A Preliminary Report." Washington, D.C.: Federal Bureau of Prisons, December 9, 1993.

Lombardo, Lucien X. *Guards Imprisoned.* 2d ed. Cincinnati: Anderson Publishing Company, 1989.

Sellin, J. Thorston. *Slavery and the Penal System.* New York: Elsevier Scientific Publishing, 1976.

Zimmer, Lynn E. *Women Guarding Men.* Chicago, IL: The University of Chicago Press, 1986.

Notes

1. Remarks of Warren E. Burger, Chief Justice of the United States at the University of Nebraska, Lincoln, NE, 16 December 1981.
2. Grant R. Grisson and Conan N. Louis, "The Evolution of Prison Industries," *Corrections Today* (Nov./Dec. 1981): 42.
3. Jack Schaller, "Work and Imprisonment: An Overview of the Changing Role of Prison Labor in American Prisons," *The Prison Journal* 72, no. 2 (Autumn–Winter 1982): 3.
4. Ibid.
5. J. Thorsten Sellin, *Slavery and The Penal System* (New York: Elsevier, 1976), 145.
6. Ibid., 145–162.
7. Barbara Auerbach, "New Prison Industries Legislation: The Private Sector Re-enters the Field," *The Prison Journal* 72, no. 2 (Autumn–Winter 1982): 26.
8. Grisson and Louis, "The Evolution of Prison Industries," 43.
9. Robert C. Grieser, "Model Approaches: Examining Prison Industry That Works," *Corrections Today* (August 1988): 176.
10. UNICOR, *Annual Report, 1988* (Washington, D.C.: U.S. Department of Justice, 1989), 4.
11. Ibid., 2.
12. Robert C. Grieser, "Do Correctional Industries Adversely Impact the Private Sector?" *Federal Probation* 53 (March 1989): 19.
13. Ibid., 24.
14. Ibid.
15. 406 F Supp. 287 (MDPO 1976).
16. *Preston v. Ford,* 378 F Supp. 729 (EDKY 1974).
17. Raymond Bell et al., *Correctional Education Programs for Inmates* (Washington, D.C.: U.S. Department of Justice, 1979), 5–6.
18. Miles D. Hares, "Recidivism Among Federal Prison Releasees in 1987: A Preliminary Report" (Washington, D.C.: Federal Bureau of Prisons, 9 December 1993), 50–54.
19. Jamie Lillis, "Prison Education Programs Reduced," *Corrections Compendium* XIX, no. 3 (March 1994): 1–11.
20. A. Nicholas Groth, *Men Who Rape* (New York: Plenum Press, 1980), 214–219.
21. Gary Dessler, *Management Fundamentals: A Framework* (Reston: Reston Publishing Company, Inc., A Prentice-Hall Company, 1977), 2.
22. Ibid., 107.
23. Ibid.
24. David Duffee, *Correctional Management: Change and Control in Correctional Organizations* (Prentice-Hall: Englewood Cliffs, 1980), 59.
25. Dessler, *Management Fundamentals,* 4.
26. Lucien X. Lombardo, *Guards Imprisoned* 2d ed. Cincinnati: Anderson Publishing Co., 1989), 51–57.

27. Robert J. Wicks, *Guards! Society's Professional Prisoners* (Houston: Gulf Publishing, 1980), 3.
28. Dothard v. Rawlinson, 433 U.S. 321, 1977.
29. Wicks, *Guards!* 96–97.
30. John J. DiIulio, Jr., *Governing Prisons* (New York: The Free Press, 1987), 7.
31. Seth Allan Bloomberg, ''Participatory Management: Toward a Science of Correctional Management,'' *Criminology* 15 (August 1977): 154.
32. James B. Jacobs, *Stateville* (Chicago: University of Chicago Press, 1977).
33. Douglas McGregor, *The Human Side of Enterprise* (New York: McGraw-Hill, 1960), 33–34.
34. Rensis Likert, *The Human Organization: Its Management and Value* (New York: McGraw-Hill, 1967), 4–10.
35. McGregor, *The Human Side of Enterprise,* 47–48.
36. Abraham H. Maslow, *Motivation and Personality* (New York: Harper and Row, 1954), 35–47.
37. F. J. Roethlisberger and William J. Dickson, *Management and the Worker* (Harvard University Press: Cambridge, 1966), 16.
38. Ibid.
39. Norman Holt, ''Prison Management in the Next Decade,'' *The Prison Journal* 57 (Autumn–Winter, 1977): 24.
40. O. Grusky, ''Organizational Goals and the Behavior of Informal Leaders,'' *American Review of Sociology* 65 (1959): 59–76.
41. T. Wilson, ''Patterns of Management and Adaptations to Organizational Roles: A Study of Prison Inmates,'' *American Journal of Sociology* 74 (1968): 146–57.
42. DiIulio, *Governing Prisons,* 39.
43. Ibid., 38.
44. Charles Stastny and Gabrielle Trynauer, *Who Rules the Joint? The Changing Political Culture of Maximum-Security Prisons in America* (Lexington, MA: Lexington Books, 1982), 4.
45. Max DePree, *Leadership is an Art* (New York, N.Y.: Doubleday, 1989), 22–23.
46. Miles D. Harer, ''Recidivism Among Federal Prison Releasees in 1987: A Preliminary Report,'' 91.
47. Tom Burns and G. M. Stalker, *The Management of Innovation* (London: Travistock Publications, 1966), 125.
48. DiIulio, *Governing Prisons,* 235.

Chapter 9

Women in Prison

Learning Objectives

After studying this chapter students should be able to:

1. Identify and describe historical isolation processes resulting in women prisoners receiving unequal treatment.
2. Provide an understanding of the number and location of women's prisons throughout the United States.
3. Illustrate incarceration trends for women prisoners.
4. Identify major differences between male and female inmate subcultures.
5. Describe the pain, problems, and challenges of incarcerated mothers who are separated from their children.
6. Provide examples of program equality between male and female prisoners.
7. Illustrate legal challenges to program inequality and provide some alternatives for more equitable treatment alternatives.
8. Define what constitutes a coed prison and provide an understanding of the number and location of such facilities in the United States.

Key Terms

Coed prisons
Community-based corrections
Cool
Deprivation model
Equitable treatment alternatives
Female inmate subcultures
Importation model
Isolation process
Life
Mother/child programs
Play families
Pooling prison resources
Program inequality
Retaining mother/child ties
Separation of children
Square
Total institution

Introduction

Until the 1970s, research and study of female criminality and imprisonment was largely neglected. Most prison research focused on male prisons with little attention given to women's prisons. This was due mainly to the relatively smaller number of women incarcerated and to biased attitudes which viewed women's incarceration as somehow less important. In recent years women's prisons have been the subject of an increasing number of studies and reports. Some traditional assumptions have been challenged and while prison life for males and females is similar in some respects, the history and experiences of women prisoners are significantly different, especially in such areas as equitable programs and services, the nature of the inmate subculture, and the consequences affecting children and families. Because of these fundamental differences, this topic can be more effectively presented in a separate chapter. Such differences are examined after first providing a background and brief historical review.

Historical Review

Until the latter 1800s most women and children were confined together with men in prisons, poorhouses, and jails. Conditions were very poor, resulting in sexual abuse and exploitation. There was no classification of inmates. Both sexes and all types of offenders were housed together. As the penitentiary movement emerged with its emphasis on silence and separation of inmates, women gradually began to be housed together in a separate area of the prison. However, they were not placed in separate sleeping cells as the men were.

It has been written that:

> The lack of accommodations for female inmates made isolation and silence impossible for them, and productive labor was not considered an important part of their routine. The neglect of female prisoners, however, was rarely benevolent. Rather, a pattern of overcrowding, harsh treatment, and sexual abuse recurred throughout prison histories.[1]

Nicole Rafter has identified three stages in this **isolation process** by which women prisoners received unequal treatment.[2] In the first stage, women were confined in large rooms or perhaps individual cells and were considered part of the general population. During the second phase, they were placed in separate areas within or attached to the main section such as an annex, upper floor, or attic kitchen. In the third phase, they were further isolated to a separate building on or near the main prison.

During this time there was also a common perception that a female offender was much worse than a male offender. She was considered to be the most depraved of women, morally corrupt, and hopeless. These and other factors came together to produce the dreadful conditions, isolation, and neglect of women prisoners throughout most of the 1800s. As a consequence:

> Gender-based perception—that female prisoners were the source of sexual mischief; that they could not earn as much as men; that they had gone beyond the pale of redemption—combined with the problem of smaller numbers to create a situation in which women's needs were slighted. The same factors operated to ensure that, once the double standard of care developed in informal practice, it continued and, in some institutions, intensified as the years went by.[3]

Furthermore, it was noted that:

> Probably lonelier and certainly more vulnerable to sexual exploitation, easier to ignore because so few in number, and viewed with distaste by prison officials, women in custodial units were treated as the dregs of the state prisoner population.[4]

Reform Efforts

Not surprisingly, reform efforts were led by women themselves.[5] The source, direction, and inspiration for American reformers came largely from the words and writings of Elizabeth Fry (1780–1845) in England. Fry was a Quaker with a strong belief in the spiritual equality of the sexes. After visits to Newgate prison in London beginning in 1813 and viewing the miserable conditions of women, she worked the rest of her life for positive change. She labored to offer spiritual guidance and to alleviate physical suffering; she argued that female prisoners could be reformed.

Additional observations and sources of change came from women matrons who worked with women prisoners and from groups of women, often Quakers, who periodically visited women's prisons. Efforts by these activists often brought benefits to individual prisoners and helped identify that a real source of problems was housing women in institutions designed for men. Thus, a fourth stage in the isolation and separation process became apparent—removing women to prisons of their own.

This fourth phase did not begin in earnest until the reformatory movement of 1870. However, foreshadowing what was to come, the first women's prison in the United States, which was designated by legislation action, was the establishment of the Mount Pleasant Female Prison at Ossining in New York in 1835.[6]

The potential and promise of the reformatory movement for women prisoners went largely unfulfilled. There was considerable regional variation in such matters as harshness of custody and choice of programs beyond those stereotypically considered "women's work." In some instances, building a reformatory resulted in widening the net, placing women here who would otherwise have been placed in community-based corrections. The long-standing tradition of providing different and unequal treatment for female prisoners continues to be a problem.

By 1940, twenty-three states had separate women's prisons. Today, although small in number as compared to prisons for men, the American Correctional Association directory lists prisons for women in nearly every state and at the federal level, as shown in table 9.1.

Population Profile

The number of female prisoners has been increasing during recent decades as shown in figure 9.1. The current female prison population in state and federal institutions is estimated to be 57,000 which is a 325 percent increase from the 13,420 female inmates in 1980.[7]

As shown in table 9.2, research indicates a more rapid rise in the incarceration rate for females (that is, the number of inmates per 100,000 residents) than for males. From 1980 through 1992 the incarceration rate for males increased from 287 to 636 per 100,000, an increase of 122 percent. During this same period of time, the rate for females increased from 12 to 35 per 100,000, an increase of 192 percent. The percent of all inmates who are female has remained nearly constant, increasing from 4.1 percent in 1980 to 5.7 percent by the end of 1992.

Similar to their male counterparts, many women offenders are undereducated, unemployed, and poor. A disproportionally high number are from racial and ethnic minority groups. However, the abuse patterns of females differ from males. These characteristics are shown in table 9.3. Blacks and Hispanics comprise 60 percent of the total population. Only 17 percent are married and 45 percent never married. Over 60 percent did not graduate from high school, although 15 percent had at least some college. Only about one-third were employed full-time prior to their arrest.

Many female inmates report physical and sexual abuse prior to incarceration and at a much higher overall rate than male prisoners. As contained in table 9.4, female prisoners, as contrasted with males, were about three times more likely to report any prior abuse and at least six times more likely to report sexual abuse or abuse since age eighteen. Additionally, as shown in table 9.5, over three-fourths of women inmates who had a history of abuse reported being sexually abused. Over 56 percent of these abused female inmates indicated their abuse had involved a rape, and another 13 percent reported an attempted rape.

The offense pattern of females also differs from males as indicated in table 9.6. Although females are found in all crime categories, overall, they are less likely than are men to commit violent offenses, but are more likely to commit property and drug offenses. Especially notable is the higher proportion of property offenses such as larceny, forgery, and fraud. In recent years there has been a large increase in the percentage of women in prison for drug offenses. For example, in 1986, 12 percent were incarcerated for drug offenses and just five years later this figure jumped to nearly 33 percent.[8]

Unlike most males, many women sentenced to prison have sole responsibility for babies and small children, as discussed later in this chapter. Many women may be lead into trouble because of the ongoing criminal involvement

Table 9.1

Prisons for Women in the United States

State	Location
Alabama	
Julia Tutwiler Prison for Women	Wetumpka
Alaska	
Meadow Creek Correctional Center	Eagle River
Arizona	
So. Arizona Community Corrections Center	Tucson
Arkansas	
Tucker Unit for Women	Tucker
California	
California Institution for Women	Corona
Central California Women's Facility	Chowchilla
Northern California Women's Facility	Stockton
Colorado	
Women's Correctional Facility	Canon City
Women's Correctional Facility	Pueblo
Connecticut	
Correctional Institution	Niantic
Delaware	
Women's Correctional Institution	Claymont
Florida	
Jefferson Correctional Institution	Monticello
Florida Correctional Institution	Lowell
Broward Correctional Institution	Pembroke Pines
LEAF	Pinellas Park
Georgia	
Metro Women's Correctional Institution	Atlanta
Baldwin Correctional Institution-Death Row	Hardwick
Metro Transitional Center	Atlanta
Washington Correctional Institution	Davisboro
Women's Diversion Center	Atlanta
Idaho	
Pocatello Women's Correctional Center	Pocatello
Illinois	
Dwight Correctional Center	Dwight
Cook County/Division 4	Chicago
Kankakee MSU	Manteno
Indiana	
Indiana Women's Prison	Indianapolis
Women's Intake Center	Indianapolis
Rockville Training Center	Rockville
Women's Work Release Center	Indianapolis
Iowa	
Correctional Institution for Women	Mitchellville
Women's Residential Center	Des Moines
Kentucky	
Correctional Institution for Women	Pewee Valley
Dismas House	Owensboro
Waddy Group Home	Waddy

Table 9.1 (continued)

State	Location
Louisiana	
Correctional Institute for Women	St. Gabriel
Maryland	
Correctional Institution for Women	Jessup
Massachusetts	
Correctional Institution	Framingham
Neil Houston House	Roxbury
Michigan	
Crane Women's Facility	Coldwater
Scott Regional Correctional Facility	Plymouth
Minnesota	
Correctional Facility	Shakopee
Missouri	
Rankin County Community Work Center	Pearl
Chillicothe Correctional Center	Chillicothe
Renz Correctional Center	Cedar City
Montana	
Women's Community Correctional Facility	Billings
Women's Correctional Center	Billings
Nebraska	
Nebraska Center for Women	York
Nevada	
Women's Correctional Center	Carson City
Conservation Camp	Silver Springs
New Hampshire	
State Prison-Women's Facility	Goffstown
New Jersey	
Edna Mahan Correctional Facility	Clinton
New Mexico	
Women's Correctional Facility	Grants
New York	
Correctional Facility	Albion
Correctional Facility	Bedford Hills
Taconic Correctional Facility	Bedford Hills
Rose M. Singer Center for Women	East Elmhurst
Lansing Center Satellite Home	Elmira
North Carolina	
Correctional Center for Women	Black Mountain
Fountain Correctional Center for Women	Rocky Mount
Correctional Institution for Women	Raleigh
Correctional Center for Women	Raleigh
Residential Facility for Women	Wilmington
Ohio	
Reformatory for Women	Marysville
Franklin Pre-Release Center	Columbus
Northeast Pre-Release Center	Cleveland
Oklahoma	
Mabel Bassett Correctional Center	Oklahoma City
Dr. Eddie W. Warrior Correctional Center	Taft

(continued on next page)

Table 9.1 (continued)

State	Location
Oregon	
Women's Correctional Center	Salem
Pennsylvania	
State Correctional Institution	Cambridge Springs
Rhode Island	
Women's Division	Cranston
South Carolina	
Leath Correctional Institution	Greenwood
Women's Correctional Center	Columbia
Tennessee	
Inman Group Home	Tullahoma
Texas	
Gatesville Unit	Gatesville
Mountain View Unit	Gatesville
Willoughby House	Fort Worth
Hilltop Unit	Gatesville
Hobby Unit	Marlin
Utah	
Women's Correctional Facility	Draper
Virginia	
Correctional Center for Women	Goochland
Washington	
Corrections Center for Women	Gig Harbor
Wisconsin	
Taycheedah Correctional Institution	Fond du Lac
Female Youth Development Center	Union Grove
Wyoming	
Wyoming Women's Center	Lusk
Federal Bureau of Prisons	
Federal Prison Camp	Alderson, WV
Federal Medical Center	Lexington, KY
Federal Medical Center-Carswell	Ft. Worth

Source: From Juvenile and Adult Correctional Departments, Institutions, Agencies and Paroling Authorities, *1995 Edition. Copyright © 1995 American Correctional Association, Laurel MD. Reprinted by permission.*

of a boyfriend or husband, rather than on their own initiative. This might be particularly true with regard to drug offenses and being an accomplice to a crime. The rate of return to prison is less for women than men.[9]

Prison Life for Women

In many respects the experience of living in prison for women is similar to that of men, as described in chapter 7. By their very nature, women's prisons also fit Goffman's description of a **total institution.** Depending mainly on the designated level of security, some fit the description more clearly than others. Daily

Figure 9.1 United States female prison population—state and federal institutions

Source: Adapted and projected from Bureau of Justice Statistics, Prisoners in 1992 *(Washington, D.C.: Government Printing Office, 1993), p. 4.*

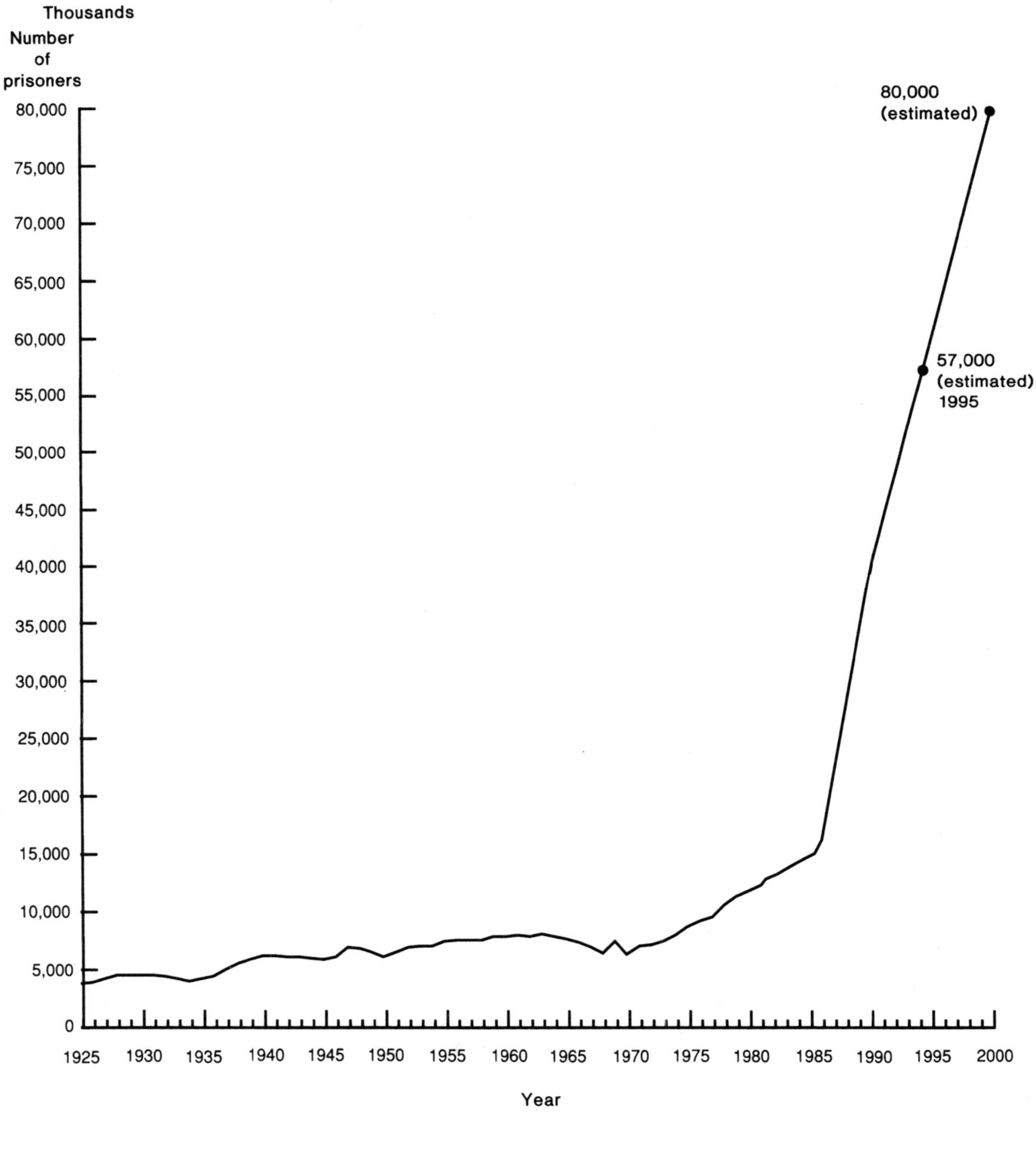

Table 9.2

Number of Inmates and Incarceration Rate by Sex and Percent Female

	Number of Inmates		*Incarceration Rate*		
	Female	*Male*	*Female*	*Male*	*Percent Female Inmates*
1980	13,420	316,401	12	287	4.1
1992	50,409	833,184	35	636	5.7

Sources: Adapted from Bureau of Justice Statistics, Women in Prison *(Washington, D.C.: Government Printing Office, March 1991), p. 7, and Bureau of Justice Statistics,* Prisoners in 1992 *(Washington, D.C.: Government Printing Office, May 1993), p. 4.*

life is regimented and routine and subject to numerous rules. Currently overcrowded conditions and confining quarters often prevail. Largely cut off from the outside world, inmates must adjust to living day after day in such an environment among other prisoners and guards. Sexual deprivation is intended. Adapting to such a life is thought to account for the development and existence of a prison inmate subculture.

Inmate Subculture: Women's Prison

Studies of **female inmate subcultures** have not been as extensive as in prisons for men. A foundation for research and analysis during the past several decades was provided by two major works: Ward and Kassebaum in 1965 and Giallombardo in 1966.[10] A major focus in these works was directed toward the social relationships women prisoners formed. Particular importance was attached to homosexual relationships which existed. Such relationships were considered as the only adaptation made by women to the deprivation and stress of prison life. Overlooked, were other forms of adjustments that may have existed.

Women in prison were seen as relating to each other on the basis of traditional female roles. Such roles are validated by attracting men and providing nurturing and support. When imprisoned, women were cut off from close and supportive relationships experienced on the outside. As an adaption to this loss, female inmates often develop **play families** and kinship networks, not found in male prisons. Such families often were quite elaborate with roles developed for father, mother, daughter, sister, and so forth. In Giallombardo's words:

> The prison homosexual marriage alliance and the larger informal family groupings provide structure wherein the female inmate's needs may find fulfillment and expression during the period of incarceration. Kinship and marriage ties make it possible for the inmates to ascribe and achieve social statuses and personalities in the prison other than that of inmate which are consistent with the cultural expectations of the female role in American society.[11]

Such conclusions were based on assumptions that recent critics contend were biased;[12] that is, an assumption that women see themselves as important only

Table 9.3

Characteristics of Female State Prison Inmates

Characteristic	*Female Inmates Percent*	
Race and ethnicity		
White non-Hispanic	36.2	
Black non-Hispanic	46.0	
Hispanic	14.2	
Other races	3.6	
	100.0	
Age		
17 or younger	.1	
18–24	16.3	
25–34	50.4	
35–44	25.5	
45–54	6.1	
55 or older	1.7	
	100.0	
Marital status		
Married	17.3	
Widowed	5.9	
Divorced	19.1	
Separated	12.5	
Never married	45.1	
	100.0	
Education		
8th grade or less	16.0	
Some high school	45.8	
High school graduate	22.7	
Some college or more	15.5	
	100.0	
Prearrest employment		
Employed	46.7	
Full-time		35.7
Part-time		11.0
Not employed	53.3	
Looking		19.2
Not looking		34.1
	100.0	100.0

Source: Adapted from Bureau of Justice Statistics, Women in Prison *(Washington, D.C.: Government Printing Office, March 1994), p. 2.*

relative to men. Their self-image is validated by their sexuality and their success in attracting men. Their principal concern is homemaking with little propensity to compete in the job market. Such assumptions lead to a failure to see factors other than homosexuality as a basis for relationships. In later research, for example, Heffernan[13] described a variety of relationships based on factors other than homosexuality, such as close friendship and mutual interests. Overt homosexuality was present but was not the only way women prisoners related to

Table 9.4

Prior Physical or Sexual Abuse of State Prison Inmates			
	Percent of State Prison Inmates		
	Total	Female	Male
Ever physically or sexually abused before current incarceration:			
No	86.1%	56.8%	87.8%
Yes	13.9	43.2	12.2
Before age 18	11.9%	31.7%	10.7%
After age 18	4.2	24.5	3.0
Physically abused	11.3	33.5	10.0
Sexually abused	6.8	33.9	5.3
Number of inmates	700,475	38,109	662,367
Relationship of abuser to inmate:			
Intimate	11.2%	49.8%	3.0%
Spouse/ex-spouse	6.1	30.5	1.0
Boyfriend/girlfriend	6.6	27.5	2.2
Relative	68.1	56.1	70.6
Parent/guardian	53.7	37.7	57.1
Other relative	22.5	26.5	21.7
Friend/acquaintance	22.8	20.1	23.4
Someone else	21.2	19.6	21.6
Refusal to answer	1.1	1.3	1.1

Source: Adapted from Bureau of Justice Statistics, Women in Prison *(Washington, D.C.: Government Printing Office, March 1994), p. 5.*

Table 9.5

Type of Abuse Reported by Female Inmates	
Type of Abuse	Percent of Female Inmates Who Were Abused
Physical only	21.2%
Sexual only	21.9
Both	56.9
Type of sexual abuse	
Total	78.8%
Completed rape	55.8
Attempted rape	13.0
Other sexual abuse	10.0
Refusal	.8

Source: Adapted from Bureau of Justice Statistics, Women in Prison *(Washington, D.C.: Government Printing Office, March 1994), p. 6.*

Table 9.6

Most Serious Offense of State Prison Inmates, by Sex		
	Percent of Prison Inmates	
	1991	
Most Serious Offense	Female	Male
All offenses	100.0%	100.0%
Violent offenses	32.2%	47.4%
Murder	11.7	10.5
Negligent manslaughter	3.4	1.7
Kidnapping	.4	1.2
Rape	.4	3.7
Other sexual assault	1.3	6.2
Robbery	7.3	15.2
Assault	6.2	8.3
Other violent	1.1	.5
Property offenses	28.7%	24.6%
Burglary	4.5	12.9
Larceny/theft	11.1	4.5
Motor vehicle theft	.7	2.3
Arson	1.0	.7
Fraud	10.2	2.4
Stolen property	1.0	1.4
Other property	.1	.5
Drug offenses	32.8%	20.7%
Possession	11.8	7.3
Trafficking	19.8	13.0
Other/unspecified	1.3	.4
Public-order offenses	5.7%	7.0%
Weapons	.5	1.9
Other public-order	5.1	5.1
Other offenses	.6%	.4%
Number of inmates	38,462	665,719

Source: Adapted from Bureau of Justice Statistics, Women in Prison *(Washington, D.C.: Government Printing Office, March 1994), p. 3.*

each other. Based on her research, Heffernan developed three categories of adaptation by women to the inmate world:

1. "square";
2. "life";
3. "cool".

These terms, which also exist on the outside, are argot labels used by inmates to describe various inmate groupings in prison. A **square** holds conventional norms and values and is not involved in a life of crime. She probably was not in prison before, follows prison rules, and does what is expected of her. **Life** refers to involvement in a complex of deviant behaviors. The term reflects a lifestyle that includes prostitution, drugs, and petty thievery. Many of these people had done prior prison time. A **cool** attempts to remain in control and manipulate the situation to her advantage. Each of these three inmate groups has a different perspective and assumes differing roles in adjusting to prison life.

These adaptations were based largely on the outside environments of the women and provided different goals, behavior codes, and means of support while imprisoned. As with prison for men, there has been some disagreement on whether adjustment by women to prison life results from an **importation** or **deprivation model.** Traditionally, research among women prisoners has supported an importation viewpoint; that is, factors brought with the women from the outside appear to be more important in determining adjustment to prison life. It remains to be seen whether changing roles for women in the outside environment will eventually bring changes in the inmate subcultures found in prison for women.

In summary, researchers and scholars have suggested that some of the major differences between male and female inmate subcultures include the following:

- Extended ''play'' family relationships in female but not in male prisons
- More consensual homosexuality in female prisons
- Less economic exploitation and underground economy in prisons for women
- Women prisoners less committed to convict code of conduct
- More support for importation model in prisons for women
- Women's prisons less committed to risk taking, such as riots and attacks on staff

Women Prisoners with Children

A majority of women in prison—over 70 percent—have children who were living with them prior to their incarceration.[14] This means that, due to incarceration, thousands of children are separated from their mothers during their formative years. Recent survey data, shown in table 9.7, indicates this figure to be 71.7 percent, compared to 52.9 for male prisoners. These data also show that:

- Nearly three-fourths of the women have two or more children
- Nearly 85 percent had legal custody before entering prison
- Three-fourths of the children now live with someone other than their mother/father (50 percent with grandparents)
- 85 percent plan to live with their children after release from prison

The **separation of children** from their mothers presents enormous problems and challenges. As has been described:

> While separation from children is also painful for men in prison, women in prison are more likely to have been the primary caretaker and, in many

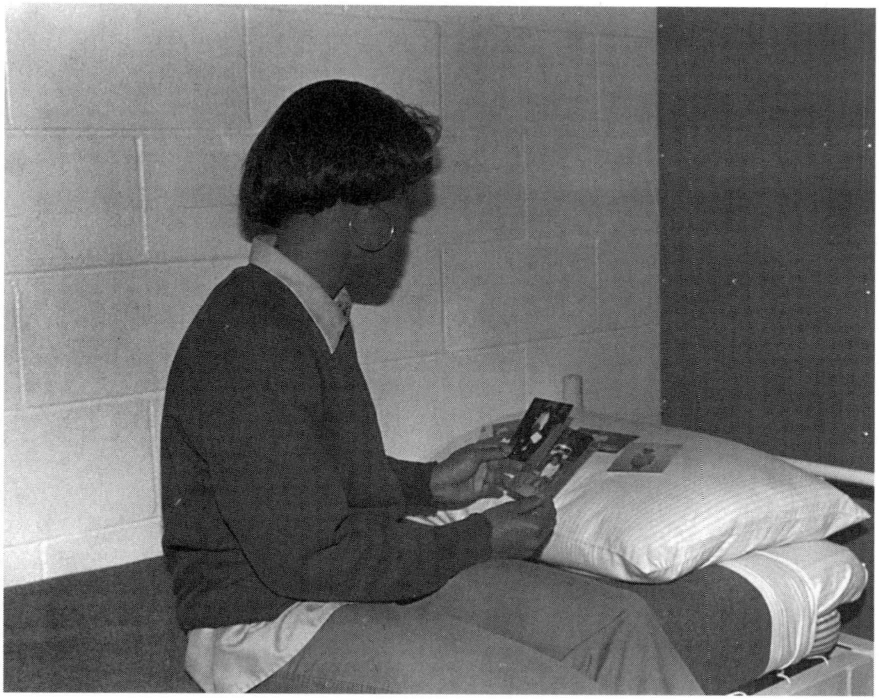

It is often painful for mothers in prison to be separated from their children.

> instances, the sole parent before incarceration. The pain of separation is thus likely to be a profound hurt that affects not only the incarcerated mother but her children as well.[15]

One real fear of many of these convicted mothers is that they will permanently lose custody of their children. Legal approaches vary, and in some jurisdictions a mother automatically loses custody once she becomes a convicted felon, while in some states she may retain custody. Standards and guidelines for determining custody in such situations frequently are vague and open the way for misuse. The best interests of the children involved are often not considered. In some cases, mothers voluntarily give up custody, feeling this would best serve the child. About one-half of imprisoned mothers retained custody of their children.[16]

Added to such problems as being undereducated, unemployed, and lacking in medical care, these mothers often bear considerable shame and guilt. A period of separation may reinforce feelings of inadequacy and foster fears and anxieties about what type of care their child is receiving. If mothers are able to be involved in placement decisions prior to incarceration they are likely to feel more satisfied. Research indicates that most children (eight in ten) of incarcerated mothers are cared for by some other family members and relatives such as grandparents, aunts, uncles, and brothers or sisters. About one-third live with the biological father. Occasionally, children are placed with friends, put in a foster home, or are on their own. Significantly more black than white

Table 9.7

Children of State Prison Inmates by Sex of Inmates

Characteristic	Percent of State Prison Inmates: Female Percent		Male Percent	
Have children				
No	21.9		36.1	
Yes	78.1		63.9	
Under age 18		66.6		56.1
Adult only		11.4		7.6
	100.0		100.0	
Number of children under age 16				
1	37.3		43.2	
2	29.9		28.9	
3	18.1		15.2	
4	8.5		6.8	
5 or more	6.1		5.9	
	100.0		100.0	
Lived with child(ren) under 18 before entering prison				
No	28.3		47.1	
Yes	71.7		52.9	
	100.0		100.0	
Had legal custody of child(ren) under 18 before entering prison				
No	15.1		53.1	
Yes	84.9		46.9	
	100.0		100.0	
Where child(ren) under 18 live(s) now				
Child's mother/father	25.4		89.7	
Grandparents	50.6		9.9	
Other relative	20.3		2.9	
Friends	4.1		.4	
Foster home	8.6		1.7	
Agency/institution	2.1		.5	
Alone	2.0		1.1	
	100.0*		100.0*	
Plan to live with child(ren) under 18 after release from prison				
No	10.9		36.7	
Yes	85.2		51.9	
Don't know	3.9		11.3	
	100.0		100.0	

**Percents add to more than 100% because inmates with more than one child may have provided multiple responses.*

Source: Adapted from Bureau of Justice Statistics, Women in Prison *(Washington, D.C.: Government Printing Office, March 1994), p. 6., and* Woman in Prison *(March 1991), p. 6.*

children live with the woman's parents, and more white than black children live with the child's father or nonrelative.[17]

Most mothers want to:

1. honestly explain to their children the reason for the separation;
2. want to stay in contact; and
3. plan to reunite when released from prison.

Mothers feel that honesty provides a basis for respect and improves chances for a longer term relationship. Most stay in touch by letters and telephone. Visits are not possible at some institutions and where permitted are usually brief, infrequent, and require transportation.

Retaining Mother/Child Ties

For decades the needs of incarcerated mothers and their children were largely ignored. Only during recent years have significant efforts taken place to provide opportunities beyond traditional visitations for **retaining mother/child ties.** Such **mother/child programs** are not found at every facility but are scattered about and may be divided into three categories:

1. Nurseries for newborns
2. Extended visiting programs
3. Parenting programs[18]

The importance of developing ties between mother and newborns is generally recognized. However, this opportunity is usually not provided for mothers who deliver babies while incarcerated. An infant is placed elsewhere for care and "the mother may not even be allowed to touch the baby again until her release, if ever."[19] One rare exception has been the nursery at the Women's Correctional Institution at Bedford Hills, New York, that operates on the top floor within the prison. Mothers may live for up to a year in the nursery spending their time caring for their child.

Numerous facilities have developed some form of program to provide extended time for visitation and more contact between mothers in prison and their children. Quite often, extended daytime visitation programs have been developed. Such arrangements allow for longer periods of time together and away from a typical visiting room environment. A separate area is often set aside for this purpose and might be designated as a parenting or children's center. Structured activities, meals, and free time are customary.

Some prisons have instituted programs allowing overnight visits by children, usually on weekends for up to forty-eight hours. Such programs frequently carry names and acronyms such as PACT (Parent and Child Together) in South Carolina and MOLD (Mother Offspring Life Development) in Nebraska and Kentucky. Such programs may operate as a privilege, be based on a screening criteria, and may be withdrawn for disciplinary reasons. These programs provide for significant time to be spent together with mothers assuming responsibility for children at all times during the visit. Such programs may also incorporate instructional sessions for mothers to learn and to improve communication and parenting skills, which they have opportunity to "try out" during the visits.

Despite problems and extra effort necessary to operate such programs, it is reasonably clear that they are feasible and can be successful. Since the future of any nation rests with its children, it seems wise to support and assist incarcerated parents (including father) with childrearing and retaining ties with their children. Baunach concluded her research by saying "given this need, reality-oriented programs designed to develop a positive sense of self, adequate parenting skills, and to encourage responsible decision-making among inmate mothers are not luxuries, but necessities."[20]

Prison Programs for Incarcerated Women

When compared to their male counterparts, women prisoners have not had equal access to programs and services. Disparities exist in type of facilities, number and type of job-training programs available, prison industry jobs, and other services. In some instances women have begun to seek more equal treatment through the courts.

Such **program inequality** has arisen in part due to the much smaller number of female prisoners. The relatively much larger group of male prisoners has made possible a greater number of prisons located throughout the nation, providing access to more industrial operations for learning job skills and access to a variety of other programs and services. In addition, sexist attitudes viewed only certain types of jobs as appropriate for women, and therefore, no need was seen for the same type of training and job skills as men. As a result, correctional systems, legislators, and others have not been aggressive in providing equal program access and opportunity for women prisoners.

Some Specific Inequalities

Research by the United States General Accounting Office on this problem reveals a number of specific inequalities that accrue to women in prison. Many of these relate to the numbers, type, and location of prisoners. It was found that

- Men may be placed in an institution more appropriate to the type of security their individual cases require.
- As their need for higher security levels diminishes, men may transfer to less secure institutions, thereby having more personal freedom.
- As their release dates near, men may be placed in a facility nearer their home community so they can reestablish family ties and find jobs.

- Many men may participate in work release programs because their institutions are near community resources.
- Men may get the opportunity to transfer between institutions for programs, training, or services.
- Men's institutions more often house industrial operations or vocational training programs.

In contrast, women generally have little opportunity to transfer between institutions because they are usually housed in one or two central institutions within a state or in one of a small number of federal facilities. Because of the small number of women's facilities, the following situations exist:

- Women may be placed in an institution housing inmates with a range of security levels. Consequently, women who are low security risks may have less personal freedom than their male counterparts.
- Women may not have the opportunity to transfer to a less secure institution as they become safer risks.
- Women may often be incarcerated long distances from their home and community. Moreover, they may not have the opportunity to be incarcerated in their home community when they are near release.
- Women may have little opportunity to participate in outside work release programs, since many women's prisons are located in rural settings far from community resources.
- Women may not be able to transfer between institutions to get programs, training, or services.
- Women's institutions often do not include major industrial operations or vocational programs.[21]

Furthermore, it was noted that men generally receive training in skilled trades or go on to work release programs, while women's institutions often limit the vocational programs to traditional, low-paying female occupations. Men often work in industrial operations, frequently for pay, while work release opportunities in female prisons are often limited because of institutional locations and the lack of segregated housing for those on work release. And, finally, men often have access to full-scale hospital and mental health facilities, which are often located within the prison system itself or at a nearby location, while women frequently have to be transported to distant community facilities for the same treatment.[22]

Court Challenges to Work Program Inequalities

Beginning during the 1970s inequalities affecting women prisoners began to be challenged in court. Many suits were based on the equal protection clause of the Fourteenth Amendment. Several sex discrimination cases were filed alleging unequal access and lack of parity in work programs for men and women prisoners. Table 9.8 provides examples of some cases that led the way.

Table 9.8

Court Cases that Challenged Work Program Inequalities Affecting Women	
1974	*Barefield v. Leach* No. 10282 (D.N.M. 1974). This was one of the earlier cases in which the state was ordered to achieve reasonable parity for male and female prisoners. The court found that the state had not provided program parity in vocational programming, wage-paying work within the institution, and facilities for projects.
1977	*Grosso v. Lally* No. 4–74–447 (D.Md. 1977). In a consent decree, it was agreed that opportunities, programs, and conditions for female prisoners would be "no less favorable, either quantitatively or qualitatively," than for male prisoners. Women were to be allowed participation in educational and drug programs, vocational programs, and community corrections, and to receive equivalent wage rates.
1979	*Glover v. Johnson* 478 F. Supp 1075 (E.D.Mich. 1979). Female inmates throughout the state were found to have fewer and inferior educational and vocational programs than did males. A "parity of treatment" test is to be applied. Although prison officials are not required to provide identical programs, they are to provide women with programs substantially equivalent in substance and based on the needs and interests of female inmates. The plaintiffs noted in this case that males had access to twenty-two vocational courses leading to marketable skills, but females had access to only five courses that did *not* lead to marketable skills.
1982	*Canterino v. Wilson* 546 F. Supp 174 (W.D.Ky. 1982). Based to a great extent on *Glover,* the court reaffirmed programs for women equivalent in substance to programs for men, even though they may not be exact duplicates in form. Included in this instance was the expansion of some existing vocational programs and the addition of others.

In such cases, courts have decided in favor of women prisoners and have indicated that an adequate defense for continuing unequal practices cannot be based on:

1. the administrative convenience;
2. the expense of providing equal situations; nor
3. the small number of women prisoners.[23]

Despite some gains, changes have been piecemeal. Such changes have not swept across the nation. Most women prisoners remain in situations of inequality. This would seem to be an area ripe for additional suits and judicial intervention. Only time will tell the extent to which continued challenges and possible additional changes will occur.

Alternatives for More Equitable Treatment

More **equitable treatment alternatives** may be grouped into three categories:

1. Sharing institutions with men
2. Pooling local, state, and federal prison resources
3. More extensive use of community corrections[24]

The use of **community-based corrections** for women is uneven across the United States and seems to vary with a jurisdiction's philosophy of corrections. In some instances, women are incarcerated when less restrictive and less costly community-based corrections would be appropriate. Little study and research have been done to learn about behaviors and outcomes of women placed on probation and parole. Greater use of community-based corrections for women could provide the necessary supervision and more equitable access to resources and services as compared to a female prison setting. Advantages to such an approach noted by one state included the following:

- Individuals did not lose their employment.
- Individuals can earn wages, support dependents, pay fines, and make restitution.
- Minor offenders do not come in contact with more sophisticated criminals.
- Offenders can serve their sentences without major disruptions to their lives and the lives of their children.[25]

Joint efforts by federal, state and local governments in **pooling prison resources** could also create a more equitable environment for female prisoners.[26] Each level of government has its own facility for female inmates, sometimes in the same location. The number of inmates housed in each facility in small programs are limited, and often there is only one institution for women so there is no choice of placement as there is for men. Pooling and consolidating resources could improve equity by:

- providing a choice of institutions;
- improving classification for females according to age, custody level, type of crime and length of sentence; and
- increasing opportunities for programs and services.[27]

However, despite some efforts, such resource pooling has seldom been accomplished, and with current overcrowding it seems highly likely that its potential for improving equity will remain a lost opportunity.

A third approach is shared facilities, or the so-called **"coed" prisons.** After decades of housing the sexes together, a movement toward single-sex prisons began in 1870 and ended in 1971 when the Federal Co-Correctional Institution (FCI) at Fort Worth, Texas was opened as a sexually integrated prison.[28] In general terms, males and females live apart but share programs and space under one administration.[29]

With some exceptions, many decisions to establish coed prisons have resulted mainly from cost considerations and administrative convenience rather than a policy aimed toward providing equity for female prisoners.[30] For example, establishing a coed facility could help solve problems of overcrowding at a male prison and underpopulation at a female prison.

In some shared facilities, women may be isolated "out of the way" and fare no better than if they were in a single-sex prison. This would not be a coed prison but rather a prison within a prison. In order to be truly coed, Weisheit and Mahan suggest that:

> The ideal ratio of female to male prisoners in a sexually integrated prison is 50–50. When the difference in the ratio of females to males is greater than 40–60 or 60–40 the coed policy loses its effectiveness.
>
> Those held together must be classified as similar. Drastic differences in ages and backgrounds between males and females interferes with the effectiveness of sexual integration.
>
> The administration of a coed prison must include females as well as males in decision-making positions. Women are usually at a disadvantage in a prison run strictly by men regardless of the policy.
>
> A coed prison must provide opportunities for real interaction between males and females with equal status. For primary, helping relationships to develop between men and women they must associate in the settings of everyday life: meals, work, learning, play, etc. However, overt and offensive sexuality must be discouraged, sanctions for homosexuality must be equal to those for heterosexuality, inmate-staff sexual involvement must be discouraged and pregnancies during incarceration must be avoided.[31]

The extent to which these conditions prevail determines the extent to which a shared or co-correctional prison could accurately be considered as coed.

Coed prisons are certainly feasible as was demonstrated within Federal Correctional Institutions at Lexington, Kentucky, and Fort Worth, Texas, for example. While not a panacea, research indicates some advantages to coed prisons, which include the following:

- More extensive and nontraditional training programs more available to females.
- Women prisoners can be located closer to their homes by increasing the number of placement locations.
- A more normalized environment improves inmates' dress, language, and grooming habits.
- Fewer fights and assaults result in a safer environment for both staff and inmates.
- Transition to the community upon release is improved.[32]

All three of these alternatives could offer some improvement in equitable treatment for women prisoners. There is little evidence, however, that the public and legislators are inclined to implement such arrangements. Commenting on this situation, Rafter has concluded that:

> There are few signs that the prison system or the public has enough commitment to equal rights to put these solutions into action. It is more likely that in the future, as in the past, incarcerated women will continue to be punished not only for their crimes but also for their minority statuses in a system planned for men.[33]

Summary

Until recently the study of crime and imprisonment of women was neglected. Research that has been done indicates that incarcerated women have different experiences than men in such areas as equitable programs, nature of the inmate subculture, and concerns involving children and families.

Women prisoners were at first confined together with men—there was no classification of inmates, and abuse of women was commonplace. In successive stages, women began to be segregated and isolated, eventually to single-sex prisons where treatment and programs traditionally were not on a par with men. Various individuals such as Elizabeth Fry and others worked to improve the plight of women prisoners during the 1800s.

Presently, separate facilities exist just for female prisoners in most states. Women prisoners constitute about 5 percent of the nation's total prison population, although since 1930 the rate of increase has been greater than among men. Many women prisoners are poor, undereducated, and unemployed. Many were abused as children and now have children of their own. Many women appear to become involved in crime because of romantic involvement with males who are committing criminal acts.

Early studies of the inmate subculture focused almost exclusively on homosexuality. While such activities are present, not all female inmates are involved and other forms of adaptation to prison life have been identified.

A major concern and pain for many women in prison is the separation from their children. There is great variance among jurisdictions as to whether problems and needs in this are recognized and solutions sought. Many mothers want to retain ties with their children, and a few programs have been established that help them to do so.

Programs for incarcerated women have been marked by inequality. Access to meaningful job programs has been very limited. Some court cases have brought scattered changes, but many women prisoners remain in situations of inequality. Several alternatives to help bring about more equitable treatment exist but have not been significantly implemented by legislators and corrections officials.

Discussion Topics

1. Do female prisoners fare better in coed facilities or single-sex prisons?
2. Why do you think earlier studies of the female inmate subculture concentrated almost exclusively on homosexuality?
3. To what extent do you think inequalities for women prisoners will be reduced during the twenty-first century?
4. Incarceration should not be the only factor in deciding whether a mother gives up custody of her children. Do you agree or disagree? Why?
5. If you were a prison superintendent, what programs would you establish for mothers to retain ties with their children?

Additional Readings

Baunach, Phyllis Jo. *Mothers in Prison*. New Brunswick, NJ: Transaction Books, 1985.

Giallombardo, Rose. *Society of Women: A Study of Womens' Prisons*. New York, NY: Wiley and Sons, 1966.

Pollock-Byrne, Joycelyn M. *Women, Prison, and Crime*. Pacific Grove, CA: Brooks-Cole Publishing Company, 1990.

Rafter, Nicole Hahn. *Partial Justice*. Boston, MA: Northeastern University Press, 1985.

Notes

1. Estelle B. Freedman, *Their Sisters' Keepers* (Ann Arbor, MI: The University of Michigan Press, 1981), 15.
2. Nicole Hahn Rafter, *Partial Justice: Women in State Prisons* (Boston, MA: Northeastern University Press, 1985), 9–10.
3. Ibid., 13.
4. Ibid., 21.
5. Ibid., 13–16; and Freedman, *Their Sisters' Keepers*, 22–45.
6. Rafter, *Partial Justice*, 16.
7. Bureau of Justice Statistics, *Women in Prison* (Washington, D.C.: Government Printing Office, March, 1991), 46–47.
8. Bureau of Justice Statistics, *Women in Prison* (Washington, D.C.: Government Printing Office, March, 1994), 3.
9. Ann Goething and Roy Michael Howsen, ''Women in Prison: A Profile,'' *The Prison Journal* 53 (Autumn-Winter 1983): 35.
10. David A. Ward and Gene G. Kassebaum, *Women's Prison* (Chicago, IL: Aldine, 1965); and Rose Giallombardo, *Society of Women* (New York: Wiley and Sons, 1966).
11. Giallombardo, *Society of Women*, 185–186.
12. Ralph Weisheit and Sue Mahan, *Women, Crime, and Criminal Justice* (Cincinnati, OH: Anderson Publishing Company, 1988), 72.
13. Heffernan, Esther, *Making It in Prison* (New York: Wiley and Sons, 1972).
14. Bureau of Justice Statistics, *Women in Prison* (Washington, D.C.: Government Printing Office, March, 1994), 6.
15. Susan K. Datesman and Gloria L. Cales, '''I'm Still the Same Mommy': Maintaining the Mother/Child Relationship in Prison,'' *The Prison Journal* 53 (Autumn-Winter, 1983): 142.
16. Phyllis Jo Baunach, *Mothers in Prison* (New Brunswick, NJ: Transaction Books, 1985), 122.
17. Ibid., 29.
18. Baunach, *Mothers in Prison*, 75. Also see Virginia V. Neto and LaNelle Marie Bainer, ''Mother and Wife Locked Up: A Day With the Family,'' *The Prison Journal* 63 (Autumn-Winter, 1983), 124–141.
19. Weisheit and Mahan, *Women, Crime, and Criminal Justice*, 77.
20. Baunach, *Mothers in Prison*, 130.
21. United States General Accounting Office, *Women in Prison: Inequitable Treatment Requires Action* (Washington, D.C.: Government Printing Office, 10 December 1980), 14–15.
22. Ibid., 17.
23. Ibid., 8.
24. Rafter, *Partial Justice*, 186–187.
25. General Accounting Office, *Women in Prison*, 29–30.
26. Ibid., 30.
27. Ibid.
28. Weisheit and Mahan, *Women, Crime and Criminal Justice*, 77.
29. Rafter, *Partial Justice*, 186.
30. Ibid., 187.
31. Weisheit and Mahan, *Women, Crime and Criminal Justice*, 79–80.
32. General Accounting Office, *Women in Prison*, 27–28.
33. Rafter, *Partial Justice*, 188.

Unit 3

Alternatives to Incarceration

A majority of persons under correctional supervision are being managed without incarceration. Chapters in this unit examine a wide range of intermediate punishments and supervision alternatives carried out in a community setting.

Chapter 10 Community-Based Corrections
Chapter 11 Probation
Chapter 12 Parole

Chapter 10

Community-Based Corrections

Learning Objectives

After studying this chapter students should be able to:

1. Define community-based corrections.
2. Understand the concept of reintegration.
3. Identify reasons for the development of community-based corrections.
4. Explain the concept and provide examples of intermediate sanctions.
5. Explain important considerations to follow when planning intermediate sanctions.
6. Describe the importance of and strategies for seeking community acceptance.
7. Provide different examples of community corrections.
8. Provide a description of electronic monitoring.
9. Describe a typical corrections boot camp operation.
10. List items likely to improve the success of corrections boot camps.

Key Terms

Alternatives to incarceration
Basic probation
Boot camp
Citizens' advisory board
Community acceptance
Community-based corrections
Community involvement
Cost effectiveness
Deinstitutionalization
Detention centers
Diversion centers
Drug and alcohol testing
Electronic monitoring
Intensive probation
Intermediate sanctions
Public protection
Reintegration
Restitution/community service
Shock incarceration
Surveillance and control

Introduction

The idea of attempting corrections on a concentrated and broad scale involving the local community is reasonably new and also controversial. Although a scattering of halfway houses and other programs have existed since the mid-1800s, **community-based corrections** emerged during the latter part of the twentieth century. This trend was given particular emphasis by the 1967 Presidential Task Force on Corrections.

Probation and parole (analyzed in detail in separate chapters of this textbook) have traditionally been thought of as comprising corrections at the community level. They continue to form the backbone of community-based correctional efforts. However, in recent years a number of additional services and supplemental programs constituting community-based corrections have been developed. These activities are organized, administered, and funded in a variety of ways. Such efforts increasingly occur within a framework of an emerging concept identified as **intermediate sanctions.** The idea of this concept is to have in place a wide range or a graduated continuum of sentencing options rather than a simple dichotomy between basic probation and a substantial prison term, as will be explained in a separate section in this chapter.

Obviously, a central theme is providing services within a setting of **community involvement.** This may be accomplished without a person experiencing any incarceration, or as a supplement to or following a period of incarceration. Thus, any and all activities *involving the community* in efforts to reintegrate offenders can appropriately be defined as community-based corrections. Notice that we speak of activities in the community to reintegrate offenders. Just because a facility is located in a local community does not automatically mean that programs involving the community are operating to promote reintegration. If a local facility provides only a custody function, then it would be inappropriate to refer to this as a community-based effort, although this is frequently the case. Local jails provide an example. In recent years, facilities providing mainly security and punishment functions have increasingly been called community-based but provide for little involvement of the community, according to a philosophy of reintegration as described in the next section.

Reintegration

Providing correctional opportunities in a community-based setting is based to a large extent upon the idea of **reintegration.** Reintegration considers the difficulties an offender has already had while living in the community and is likely to have when released from an atmosphere of control and supervision found in an institution as compared to the relative freedom of living in the community. Reintegration involves day-to-day living experiences and participation in community institutions. Such participation includes family involvement, attending school, locating or having work, and engaging in recreational activities. Reintegration is based on the premise that if a person is able to become involved in

A community corrections resident working in a manufacturing warehouse.

major social institutions and activities within a community, then the chances of law-abiding behavior are increased.

When discussing ideas of reintegration, the Presidential Task Force on Corrections stated:

> This point of view does not deny the importance of increasing individual capacity, but it does make clear that correctional techniques are nearsighted when they fail to take into account and make needed changes in an offender's social and cultural milieu. Successful adjustment on his part will also usually require conditions within the community that will encourage his reintegration into nondelinquent activities and institutions.[1]

The Task Force also emphasized:

> The task of corrections therefore includes building or rebuilding solid ties between offender and community, integrating or reintegrating the offender into community life-restoring family ties, obtaining employment and education, securing in the larger sense a place for the offender in the routine functioning of society. This requires not only efforts directed toward changing the individual offender, which has been almost in the exclusive focus of rehabilitation, but also mobilization and change of the community and its institutions. And these efforts must be undertaken without giving up the important control and deterrent role of corrections, particularly as applied to dangerous offenders.[2]

Reasons for the Development of Community-Based Corrections

Dissatisfaction with Institutions

Developing prisons was originally viewed as a reform movement in large measure to replace corporal punishment and capital punishment. It was hoped that period of confinement coupled with strict supervision would promote law-abiding behavior when persons were released from this experience and returned to live in their local communities. However, history indicated that law-abiding behavior in many instances was not promoted. Many people return to prison, sometimes two or three times or more, for continued law violations. In addition, prisons are expensive to build and have come to be characterized by:

- overcrowding;
- inadequate budgets;
- being riot prone;
- being unsafe for inmates and staff;
- extreme idleness; and
- lack of meaningful programs.[3]

Movements toward community-based corrections have occurred largely as a result of this disenchantment with prisons and institutionalization. In fact, this has lead in some instances to a process known as **deinstitutionalization.** This is a process whereby large traditional institutions have been closed in favor of smaller community-based facilities. In some cases the degree of custody has been lessened while those identified as potentially dangerous are still confined. This process has occurred in various states and at the federal level.

Humanitarianism

Given the nature of most prisons, as just outlined, many would agree that humanitarian goals can better be met in a community setting than in a prison setting. All things being equal, this in itself is a plus factor for community-based corrections. Dehumanizing experiences have done little to correct, and only add to, resentments and frustrations—conditions not conducive to law-abiding behavior.

The National Advisory Commission on Corrections suggests:

> The humanitarian aspect of community-based corrections is obvious. To subject anyone to custodial coercion is to place him in physical jeopardy, to narrow drastically his access to sources of personal satisfaction, and to reduce his self-esteem. That all these unfavorable consequences are the outcome of his own criminal actions does not change their reality. To the extent that the offender can be relieved of the burden of custody, a humanitarian objective is realized. The proposition that no one should be subjected to custodial control unnecessarily is a humanitarian assertion.[4]

It must be kept in mind, however, that protection of the public must not be compromised. For this reason, effective screening and classification of offenders are of utmost importance.

Cost Effectiveness

Everyone—individual citizens, governmental leaders, correctional administrators—wants to receive the most from money spent. However, determining **cost effectiveness** for corrections can be difficult. Basic assumptions must be made regarding differing methods of figuring costs, quality of service offered, and other variables that effect conclusions. Commenting on these difficulties of cost-benefit analysis, Eskridge, Seiter, and Carlson approach this matter by suggesting:

> If it is presupposed that each correctional alternative to halfway houses, for example, is equally effective, it is necessary only to outline the important cost considerations. It is generally concluded that prison is more expensive than halfway houses (figuring both construction and per diem costs), that halfway houses are more expensive than parole or probation, and that these are all more expensive than no supervision. It is therefore, important to note the actual program alternative for any offender, rather than to make comparisons within generalities. . . .[5]

In some instances, community-based programs have been operated at lower cost per participant than comparable programs in an institutional setting. If offenders are able to earn wages and pay taxes as part of a community-based program, costs will usually be significantly lower. However, if community programs are very intensive, such as providing transportation, child care, and job training fees, costs might be similar to incarceration costs, which provide very little in the way of intensive programs. Therefore, comparisons involve more than simple dollar costs. We must analyze further to determine the nature and intensity of the program being delivered. It has been suggested that given the general ineffectiveness of traditional incarceration, greater emphasis should be given to community-based efforts because this offers some hope for a more effective approach.[6]

More Adequate Justice Administration

As previously noted, the setting for corrections across the United States is fragmented by federal, state, and local levels of government. Similarly, the justice system shares this same fragmentation, plus a further division on the basis of three major functions: policing, courts, and corrections. This fragmentation can lead to what is sometimes referred to as a ''nonsystem'' with regard to the administration of justice. As a result, there is often little sharing of information or continuity of response to an offender under the supervision of the justice system. Community-based corrections offers an opportunity for greater cooperation among policing, courts, and corrections at the local level. Most policing and court functions are at the local level. Community-based corrections increase correctional activities at the local level and thus provide increased possibilities for more coordinated and suitable management of justice.

Intermediate Sanctions

During the 1980s and 90s the term intermediate sanctions has been pushed to the forefront. It is typically used in a generic sense to describe a variety of options, usually thought of as something more than regular probation and less than a longer prison term. The high costs of prison space and prison overcrowding have been two major factors causing officials to seek a range of options that could be appropriate for a variety of lower-risk offenders. The idea is to have available not just probation or prison term as a sentencing option, but to have some intermediate alternatives between these two extremes. Some examples include:

- **Day fines** This type of fine is used in Europe and attempts to adjust the dollar amount of the fine to the ability of the offender to pay, based on employment and income. Some range is usually included based on the severity of the offense. Thus, the more severe the offense and the greater the ability to pay, the higher the fine.
- **Restitution/community service** The attempt here is to have an offender provide some "pay-back" or replacement for damage done. This may be in a monetary form or, as is quite common, in the form of labor performed as a public service.
- **Electronic monitoring** This is a technologically operated system capable of tracking offenders. With such a system, offenders can be confined to a given area; confinement to their residence constitutes house arrest.
- **Intensive probation** This form of supervision involves closer monitoring than basic probation. Strict conditions, frequent visits by probation officers, and greater participation in programs are features of intensive probation.
- **Boot camps** This is a military style option, presently very popular with politicians and the public. Boot camps are described in some detail in the next section.

Most states and the federal government either have or are developing and/or expanding intermediate sanctions. Significant funding is being provided for these efforts. Although the idea of intermediate sanctions "sounds good" and is popular, developing such options should be carefully planned. As documented by McGarry and Carter, this involves an array of important considerations. In summary fashion, this includes:

- **Establish a policy team** It is important to involve key criminal justice players—prosecutors, defense attorneys, judges, and probation officers.
- **Developing goals** Attention must be given to clarifying what is to be accomplished and what resources are available.

Table 10.1

A Day in Boot Camp

Upon arrival at the boot camp prison, male inmates have their heads shaved (females may be permitted short haircuts) and are informed of the strict program rules. At all times they are required to address staff as "Sir" or "Ma'am," must request permission to speak, and must refer to themselves as "this inmate." Punishments for even minor rule violations are summary and certain, frequently involving physical exercise such as push-ups or running in place. A major rule violation can result in dismissal from the program.

In a typical boot camp program for adult offenders, the 10- to 16-hour day begins with pre-dawn reveille. Inmates dress quickly and march to an exercise yard where they participate in an hour or two of physical training and drill. Following this they march to breakfast in a dining hall where they must stand at attention while waiting in line and move in a military manner when the line advances. Inmates are required to stand behind their chairs until commanded to sit and must eat without conversation. After breakfast they march to work sites where they participate in hard physical labor that frequently involves community service such as picking up litter in State parks or along highways. When the 6- to 8-hour work day is over, offenders return to the compound where they participate in more exercise and drill. Dinner is followed by evening programs that include counseling, life skills training, academic education, or drug education and treatment.

As their performance and time in the program warrants, shock incarceration inmates gradually earn more privileges and responsibility. A special hat or uniform may be the outward display of their new status. Those who successfully finish the program usually attend an elaborate graduation ceremony with visitors and family invited to attend. Awards are often presented to acknowledge progress made during the program, and the inmates may perform the drill routines they have practiced throughout their time in the boot camp.

Source: Doris Layton MacKenzie, "Boot Camp Prisons in 1993" (Washington, D.C.: National Institute of Justice Journal, *November, 1993), pp. 22–23.*

- **Creating a sentencing policy** Specific means of intermediate sanctioning must be selected as appropriate for stated goals.
- **Developing effective monitoring** Valid and reliable data must be collected to evaluate the extent to which goals are being met.[7]

Failure to approach the development of intermediate sanctions without an integrated and comprehensive plan invites ineffective results.

Boot Camps

Boot camp programs are a form of intermediate sanction that is being increasingly used and expanded. Boot camps typically require a short prison or jail term, usually called shock incarceration. The program approach is based on a military style, long utilized for basic training in military boot camps. This involves such things as unquestioned obedience to authority, hard physical labor and exercise, strict rules with few privileges, and mandatory program participation. Table 10.1 describes a typical day in boot camp.

It has been noted that "there are currently well over 7,000 beds devoted to boot camp programs. On average, offenders spend 107 days in the programs. Thus, more than 23,000 offenders could potentially complete programs in a 1-year period."[8]

Although physical activity is a main part of most boot camps, some increase has been seen in providing programs. Typically, a drug treatment component is being included as well as some form of education program. An evaluation of five such programs in Illinois, New York, Texas, Florida, and South Carolina indicates several specific items to improve success. These are:

1. Offenders should be carefully screened, permitting into the program only those with severe drug/alcohol abuse.
2. An in-prison phase of treatment should be for a minimum of three months.
3. Drug treatment personnel should be well trained and not required to follow a military model in therapeutic sessions.
4. Intensive community supervision should be provided upon release of the offender from shock incarceration to continue treatment and assist with employment and other community adjustment needs.
5. Valid and reliable scientific research should be designed to continually evaluate the treatment program.[9]

Boot camps have strong media appeal and have attracted growing interest. However, research has indicated a potential for some negative outcomes. Some boot camp environments have been characterized by inconsistent standards, contrived stress, and leadership styles likely to lower self-esteem. Indications are that such a setting may increase the potential for aggression by both staff and offenders and encourage an abuse of power.[10] Therefore, it seems prudent to suggest that "whether the point is to provide rehabilitation, to deter, or to direct people from prison, alternatives other than boot camp should be given careful consideration."[11]

Offender Involvement

Involvement by offenders in a community-based program may occur at numerous points in the criminal justice process. This can be described in four major segments.

1. *Prior to court conviction* In some instances, persons suspected of law violation and whose cases are in a pre-trial phase may be placed in a community-based program even before a formal court decision of guilt or innocence, sometimes as a condition of bail bond.

2. *After court conviction without incarceration* Many times it is decided that a person declared guilty of an offense can safely and effectively be dealt with in the community without a period of confinement in a prison or jail. Traditionally, probation has fulfilled this function. In recent years additional alternatives have been developed. This approach is

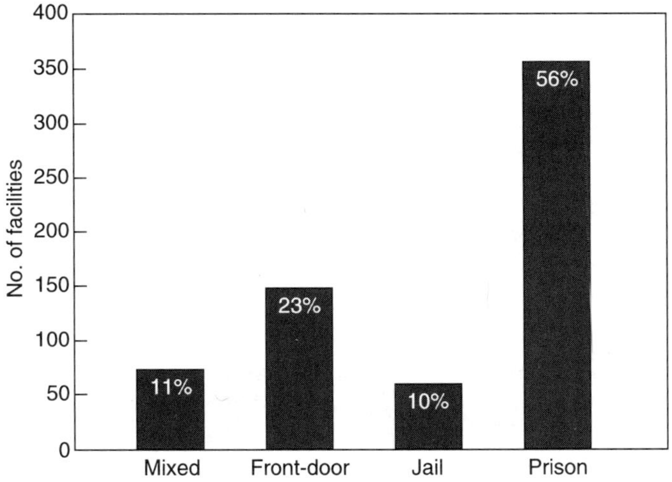

Figure 10.1 Role in criminal justice system—primary source of referrals

Source: Adapted from National Institute of Corrections, Residential Community Corrections Facilities: Current Practice and Policy Issues *(Washington, D.C.: August, 1992), p. 3.*

often used for first-time offenders whose crime is less serious. Most intermediate sanctions fall in this category. Examples include fines, restitution, electronic monitoring, and submitting for alcohol/drug treatment.

3. *During incarceration* In some instances incarcerated offenders have an opportunity to avail themselves of some community resources. This usually involves participation in study and work programs within the community. Some institutions provide programs under the heading of graduated release or prerelease programs which draw on community resources. The aim of such efforts is to ease the transition from living in an institution to living as a law-abiding citizen in the community.
4. *After release from incarceration* A majority of incarcerated offenders are released on conditional release. In addition to the usual parole supervision, supplementary programs have been organized and implemented in many jurisdictions. In some instances those persons released without parole may also participate in available programs.

A recent survey documents great diversity in the manner in which offenders are placed in residential community correctional facilities. Contained in this survey report is the following (figure 10.1):

> A dimension of great interest to policymakers is the method by which offenders are placed or channeled into residential community corrections facilities. The survey data indicate that the point of access to such programs from the criminal justice system is also quite diverse. Access can come at pre-trial, pre-sentence, direct sentence, probation, pre-release from prison or jail, or after conditional release from prison and violation of conditions. This means that access to such programs is influenced by numerous and diverse decisionmakers including sentencing judges, probation officers, community corrections boards, jail

administrators, state departments of corrections, paroling authorities, and offenders themselves. Figure 10.1 illustrates the distribution of programs represented in the survey data according to their primary source of referrals. "Front door" (147 facilities) indicates that the majority of referrals to the program came from probation or the court. "Jail" (62 facilities) indicates that the majority of referrals to the program came from a local institution. "Prison" indicates that the majority of referrals to the program came from a state (272 facilities) or federal (54 facilities) institution or a majority of referrals came from a parole board (38 facilities). "Mixed" (74 facilities) indicates that the facility's referral sources varied so widely that no single type of referral source is primary.[12]

Tables 10.2, 10.3 and 10.4 provide examples of some of these facilities. While no two programs are exactly alike, these examples would be typical of others within the same category.

Strategies for Development

A major challenge of establishing effective community-based corrections and intermediate sanctions is to involve citizens in a positive way. This is not an easy task, but it is essential for the ultimate establishment of programs. Since positive correctional efforts have not been visible to local citizens, they tend not to understand the nature and purpose of such programs and hence often mount opposition to organizing corrections within a community setting. There are, however, approaches that may be effective in gaining **community acceptance** and involvement, such as applying marketing strategies that have been successfully used by the business community.[13]

It has been demonstrated that an informed public will support such efforts.[14] This task is formidable, but as a Delaware Governor once stated:

> I believe the public will not only permit but will support intermediate sanctions. A case in point: When the Edna McConnell Clark Foundation asked hundreds of Alabama residents how they would sentence 20 convicted offenders, virtually all thought prison appropriate. After some explanation of costs and alternatives, the same people "resentenced" most of these cases to intermediate sanctions. This demonstrates that an educated public will support alternative sanctions.[15]

Initial time and effort can be usefully spent to educate the public of the nature and purpose of community-based corrections. First, it needs to be demonstrated that **public protection** remains as a primary goal; that is, only those persons deemed to be of minimum or no risk are permitted to participate. Second, it can be emphasized that depending on the particular program, control and supervision are not abandoned. It is important for the public and policymakers to understand that eliminating incarceration does not eliminate control.

Table 10.2

Cedar Rapids Community Corrections Center

The Cedar Rapids Community Corrections Center is an example of a residential community corrections facility run by a public agency and primarily serving what we have termed a "front-door" offender population, resulting from direct court commitments.

Population

The Cedar Rapids Community Corrections Center is a small, 44-bed residential facility that houses both men and women offenders. The typical offender is sentenced and placed at the Center as a condition of probation, or, less frequently, by order of the court as an alternative to a county jail sentence. On occasion, federal offenders are housed there as they are nearing completion of their prison terms, as a direct community confinement sanction, or on conditional pre-trial release. Most are non-violent offenders who have drug and alcohol problems and have had an uneven work history. Placement at the Center is most likely the third step for an offender, after a suspended sentence and a straight probation. Some are placed as a result of a probation violation. Offenders typically stay four to five months at the Center.

Facility

The Center is housed in a two-story structure, built specifically for the program. It is located in a commercial area of Cedar Rapids, Iowa, a city of just over 110,000 located in the east central part of the state. It has mostly two- and four-person rooms, a few single rooms, a dining and recreation area on the ground floor, and a lounge and offices on the second floor in addition to sleeping rooms. The front entrance opens onto a control desk where sign-ins, sign-outs, and various screenings are conducted. Staff offices and a small visitor's area are adjacent to the control desk. There is a very small area outside for recreation.

Services Offered; Daily Routine

The major focus of the Center is on stabilization of offenders and on finding and keeping employment. Services provided in-house by the Center include counseling, GED, drug and alcohol screening, and the general discipline offered by the Center's "level" system of graduated privileges earned through good behavior and maintenance of certain responsibilities and chores. The staff includes two full-time counselors (each of whom is assigned to individual residents) and a job developer. Other staff are administrative, clerical, security, and support.

Drug and alcohol services are provided by other agencies in the community. The location and nature of services depend to some extent upon ability to pay and whether offenders have insurance. There are services, however, for those who cannot pay.

Residents are expected to be dressed and ready for the day by 8 A.M. Meals are prepared by a staff cook, but cleanup and general maintenance of the facility are handled by residents as part of their responsibilities. Residents are to be engaged either in job search, employment, training, or programming of some kind at all times. Each resident meets with his or her individual counselor on a formal basis at least once every two weeks. Informally, they see their counselors several times weekly. The formal meetings are structured around the Case Management Classification (CMC) system with negotiated case plans as the vehicle by which offenders lay out goals and objectives and measure their progress during their time at the Center.

(continued on next page)

Table 10.2 (continued)

Security

The Center is a nonsecure facility. Offenders come and go through an unlocked door, although they are required to report their whereabouts at all times and their freedom to leave the facility is strictly limited, especially during their early weeks at the Center. Visiting among rooms is prohibited, and of course items such as weapons, drugs, and alcohol are strictly forbidden. Possession of such is grounds for dismissal from the program.

Organization and Funding

The state of Iowa passed legislation as early as 1973 creating locally controlled community corrections programs. The Cedar Rapids Community Corrections Center is an agency of Iowa's Sixth Judicial District Community Corrections Board. This Board is a public agency that hires staff and oversees services that are required by state legislation and funded through the state's regular budget process. As such, the Center's staff reports directly to the local Community Corrections Director and receives funding through the District. The State Department of Corrections oversees the Center through its regulatory and audit functions.

The budget of the Center is roughly $400,000 per year with an average daily cost per resident of $48.57. As part of the Center's program, clients are required to pay $8.50 per day toward the costs of their participation.

Source: Adapted from National Institute of Corrections, Residential Community Corrections Facilities: Current Practice and Policy Issues *(Washington, D.C.: August, 1992), pp. 34–35.*

Surveillance and control remain basic to sound community corrections. In fact, some programs are rigorous in this regard, maintaining a strict monitoring by **drug and alcohol testing** and checking for attendance at scheduled activities. Additionally, it can be stressed that an attempt is being made to increase opportunities for offenders to become not only law-abiding but also tax-paying citizens.

The formation of a **citizens' advisory board** may also be an effective strategy.[16] Stress has been placed on the importance of allowing such a group to be a ''working'' board; that is, allowing the members to become involved in the planning and decision making. This serves the purpose of involving local citizens and provides an opportunity for them to observe and learn about objectives of community-based corrections. There is no precise formula for determining membership on such a board, but selection of persons representing the overall spectrum of the community is desirable. For example, depending upon the makeup of the community, usually included would be persons from such groups as small business, minorities, industry, police, civic organizations, judges, politicians, and community service agencies.

It is of utmost importance to involve the courts, including defense and prosecution attorneys, and police in planning and implementing correctional programs at the community level. Without their support and cooperation, there is little chance of establishing effective community-based corrections. Too often the administration of justice has operated as a nonsystem with little coordination

Table 10.3

Volunteers of America, Sacramento

The Volunteers of America program in Sacramento, California, is an example of a residential community corrections facility that serves a mixed population of "front-door" and "back-door" offenders.

Population

The Volunteers of America (VOA) in Sacramento has been under renewable contracts with the Federal Bureau of Prisons since the program's inception. Four types of offenders reside at the Center, the first being those placed in the facility as a condition of their bond ("pre-trial cases"). Others are transferred to the VOA from a more secure federal institution to complete the remainder of their sentence ("pre-release cases"). The third group of offenders are those referred due to conditional release supervision requirements of the federal courts, U.S. Parole Commission, or U.S. Probation Department in lieu of prison. The fourth group are those sentenced to a Community Corrections Center directly by the court. The program is voluntary, and each referral is screened for appropriateness for this community-based option. The program does not generally accept referrals who are sentenced for, or have a history of, violent offenses, sex offenses, arson, use and/or possession of weapons, or a lengthy history of petty thefts.

The "typical" offender, as defined by program staff, is male and between the ages of 38 and 42. He generally has some higher education, is married, and remains involved in a relationship with his spouse. "Public law" offenders are typically first offenders with serious charges; pre-release offenders generally have a significant criminal history. The majority of the offenders are serving sentences for drug-related offenses or white collar crimes. While the average length of stay is approximately 120 days, actual program stays range anywhere from two days to one year.

Facility

The Volunteers of America (VOA) program of Sacramento is located in the heart of downtown Sacramento, occupying a structure that was formerly a garden apartment complex, in a quiet residential neighborhood. Located close to the local government and the state capitol building, the facility offers easy access to the community.

The VOA is a 62-bed facility, with 16 spaces allotted for women offenders. Apartments have been converted to housing areas by placing two beds in each bedroom, as well as in the apartments' living rooms. The units' kitchens have been disabled, but provide space for a work table and chairs. Although the VOA has inhabited the structure since 1978, only one minor structural change has been made to the complex, removing a wall, allowing for offices to adjoin a staff monitoring office and the residents' designated dining space. In some of the apartments, partitions have been used to divide living rooms into separate bedrooms. Residents are not permitted to enter one another's apartments. Comfortable community lounges are provided and equipped with televisions. The apartment units form a rectangle, surrounding an outdoor seating area.

Services Provided; Daily Routine

The Sacramento VOA is operated by a total of 14 full-time staff. Residents' activities are monitored around the clock by seven monitors and one supervisor. Substance abuse counseling services are provided by two in-house correctional counselors. Each resident is provided a counselor who meets with the resident weekly to establish a contract of activities which will facilitate a successful transition into the community. The program also offers the assistance of a job developer to aid in securing and maintaining employment. The remaining staff provide support services to the operation.

(continued on next page)

Services Provided: Daily Routine

Residents are referred to resources within the community for medical and educational needs. In-house counseling services are provided, but are limited to release readiness and substance abuse issues. Those with a need for more in-depth service are also referred to community resources. Residents are responsible for arising early and preparing for a day of work or job seeking. All residents have a schedule that they follow which has been approved by staff in advance, and may include leaving the facility for work or job seeking, counseling activities, or an hour of approved recreation in a neighborhood park. Residents dine together in a designated dining area and meals are provided to the facility by contract.

All residents have a daily curfew, although this varies depending upon the legal status of the offender. Curfews and home visitation privileges are mandated by the Federal Bureau of Prisons for pre-release and direct court commitment offenders. Pre-release offenders are able to earn their way through phases gaining additional privileges based on their successful progression through the program. The public law and direct court commitments have more stringent requirements governing their absences from the facility and their community privileges.

Security

The Sacramento VOA is an open facility, with structure provided to residents through the enforcement of rules and regulations. Residents enter the facility with the presumption of trustworthiness, and are monitored through random searches of their property and self, as well as visits and telephone calls to their activities. Drug and alcohol testing is accomplished routinely, and the VOA has long maintained a position of "zero tolerance" for the use of drugs in the program. Although an offender who tests positive for the use of alcohol is managed through in-house sanctions, those who test positive for drug use are automatically removed from the program.

Organization and Funding

The Volunteers of America was established by the Salvation Army and is nearing its 100th anniversary. It provides services where public agencies have left a void. The Sacramento branch of VOA is managed by a Board of Directors and it offers programs to the community in 18 distinct areas. One of these missions is to provide residential correctional services to offenders in the community.

The Sacramento Volunteers of America program is funded in total by the contract that they maintain with the Federal Bureau of Prisons. By using a formula of annual operational costs divided by the number of contract beds available, the VOA establishes a per diem rate that is charged to the Bureau of Prisons. The current contracted rate is $42.62 per day per bed filled. Residents reimburse the federal government by paying 25% of their gross earnings for their room and board throughout their stay in the facility. As a Federal Bureau of Prisons contract facility, the Sacramento Volunteers of America program must undergo an annual audit of its procedures.

Source: Adapted from National Institute of Corrections, Residential Community Corrections Facilities: Current Practice and Policy Issues *(Washington, D.C.: August, 1992), pp. 44–45.*

Table 10.4

Blue Ridge Work/Pre-Release Center

The Blue Ridge Work/Pre-Release Center is an illustration of a residential community corrections facility that provides what we have termed "back-door" services to offenders serving or completing terms of incarceration. This long-established program is also illustrative of a significant number of "back-door" programs begun during the 1970's, more closely approximating a correctional facility in its relatively large size and institutional nature, than some of the smaller, residence-like facilities found in this universe of programs.

Population

The Blue Ridge Work/Pre-Release Center is operated by the South Carolina Department of Corrections as a facility for male inmates who have been authorized to work in the community and as a transitional facility for inmates within 30 days of their release from the Department. It is a large facility, housing 141 inmates in the work-release program and 67 in pre-release status.

Those in the work-release program spend an average of four months and those on pre-release status spend an average of 30 to 45 days at the Blue Ridge Center. With the exception of offenders convicted of criminal sexual conduct, any offender who meets the agency's requirements is eligible for placement at the Center (although legislation prohibits those convicted of violent offenses from participating in work release in the county of conviction).

Facility

The facility was constructed 60 years ago as a tuberculosis sanitorium. It is located on a 20-acre tract of land in a mixed commercial and residential area of Greenville, South Carolina, a city of close to 300,000 in the northwest part of the state. Since the Department of Corrections took over the building in 1970, various modifications have been made, including the enclosing of extensive porches around the facility to create more dormitory space for residents. The facility is generally divided into two areas, one for work-release inmates, and one for those on pre-release status. Eating areas, classrooms, and dormitory areas are separate for the two populations.

Services Offered; Daily Routine

Services for work-release inmates revolve primarily around finding and securing employment in the community. One of three job placement counselors on staff is assigned to each arriving work-release inmate. Staff identifies employment opportunities, schedules interviews, and must approve acceptance of any employment. Once a job has been secured, the Center provides transportation to and from work. The inmates are required to turn over all earnings to the facility for deposit in an inmate account and are required to contribute 25% of their gross wages to defray the costs of room, board, and transportation. The Center regulates what the inmates may remove from their accounts and requires inmates to pay their own medical expenses, to contribute to the support of dependents, and to save.

Services for pre-release inmates revolve around preparations for re-entry into the community. Most are provided in a classroom setting through presentations and classes offered by community agencies, Center staff, and volunteers. Alcoholics Anonymous and Narcotics Anonymous are also offered, as are presentations by members of the Alston Wilkes Society, a South Carolina nonprofit group providing direct services to offenders in institutions and in the community.

(continued on next page)

Table 10.4 (continued)

Depending upon their status, inmates at the Blue Ridge Center spend each day at work or looking for work (if they are work-release inmates) or in classroom preparation for release (if they are pre-release inmates). Meals are prepared by Center staff but general cleaning and maintenance are accomplished through inmate job assignment. In addition to those inmates already mentioned, about two dozen long-term inmates reside at the Center to handle the majority of maintenance at the Center. Work-release inmates are eligible for community passes and furloughs after demonstrating their responsible performance at work and in the community.

Security

Security is handled primarily through control of offenders' movements in and out of the facility by way of a control desk. Work-release inmates are provided transportation to and from work in the community. Formal counts are held at each staff shift change. Those working outside the facility punch time cards to track their departures and arrival at the facility. Exterior doors at the pre-release part of the facility are locked and inmates are not allowed in the yard after dark. The portion of the facility dedicated to work-release is a bit more open. All offenders on the work-release program are screened for drug use at intake and randomly thereafter.

Organization and Funding

The Blue Ridge Work/Pre-Release Center is a facility of the South Carolina Department of Corrections. It is one of the community centers of the Department's Community Services Division. As such, it is funded through the Department and the state's routine budget process.

Source: Adapted from National Institute of Corrections, Residential Community Corrections Facilities: Current Practice and Policy Issues *(Washington, D.C.: August, 1992), pp. 40–41.*

and cooperation among police, courts, and corrections. Community-based correctional efforts offer an opportunity to develop more coordination among these three components in working toward a more effective system of justice.

Approaches to community-based corrections are varied. Many different programs exist across the United States. In the following sections selected examples are presented to illustrate some of the approaches taken. This discussion begins with a well-known example which operates in Georgia.

Georgia's Alternatives to Incarceration[17]

Like most states during the past several decades, Georgia was faced with problems of a rapidly growing offender population, limited resources, and prison overcrowding. Relying heavily on the role probation has always played in community-based corrections, officials undertook to develop solutions. These efforts resulted in the expansion of basic population and the development and implementation of a variety of additional **alternatives to incarceration.** The results have helped to reserve expensive prison offenders while at the same time satisfy the public's demand for both punishment and rehabilitation. The

Table 10.5

Comparative One-Year Costs in Georgia of Community-Based Options with a Prison Option

Option	*Yearly Cost*	*Community Cost as a Percent of Prison Cost*
Prison	$13,450	—
Basic Probation	$ 273	2%
Intensive Probation Supervision	$ 1,713	13%
Diversion Centers	$ 3,543	26%
Detention Centers	$ 4,023	30%
Shock Incarceration	$ 3,523	26%

Source: Adapted from "Probations Role in a Balanced Approach to Corrections." Georgia Department of Corrections. Undated (Circa 1988). Unpaged insert.

cost-effectiveness of Georgia's community-based options is shown in table 10.5. These options provide punishment, public protection, and rehabilitation opportunities at less cost than full-time prison incarceration.

There has been continued growth and expansion of each of these alternatives since they were begun. Georgia's innovative approaches to diversion and alternatives to incarceration have been nationally recognized. The Georgia Department of Corrections has received a Ford Foundation Award in the category of "Innovations in State and Local Government" for its "Alternatives to Incarceration Program." If not the best, this is certainly among the best and most balanced corrections approaches in operation that satisfy the public's need for punishment while at the same time providing community-based alternatives to meet a variety of offender needs. This program is now viewed not only as just a solution to prison overcrowding but as a beneficial program in its own right.

This program provides six options to incarceration in traditional state prison. These are identified as (1) basic probation, (2) community service, (3) intensive probation, (4) diversion centers, (5) detention centers, and (6) shock incarceration.

Basic Probation

Even though additional alternatives have been developed, **basic probation** remains very important in maintaining a balanced system. A large number of cases are handled with this option. Felons comprise about 60 percent of basic probationers.

In order to manage more effectively an increasing number of probationers, a needs/risk assessment is completed for each offender. Offenders are then classified and managed according to the (1) amount of surveillance needed, (2) degree of treatment required, and (3) frequency of supervision contacts necessary. The reported successful completion rate is around 87 percent.[18]

Community Service

This sentencing option has gained support and acceptance by the courts and communities as being productive and effective. Typically, community service is utilized: (1) as an additional condition of probation, (2) as a disciplinary step for some probation violators, (3) in lieu of fines and/or restitution and (4) as an alternative to jail. This approach saves taxpayers' money with its goal to help offenders while compensating the victim or community.

Some 1,500 court-approved agencies throughout Georgia utilize community service probationers. There is great variety in type of community service work done, ranging from such tasks as maintenance to marine biology. In this particular program, participation by local communities saves a considerable amount of money. Statewide, probationers perform over a million hours of work annually.

Intensive Probation Supervision (IPS)

Intensive probation is a "tough" form of supervision that features a high degree of structure and rigidly monitored supervision. It is considered one of the more effective and recognized alternatives. The program is comprised of three phases of supervision, with decreasing restrictions and requirements. Depending upon individual needs, offenders spend six to twelve months under IPS. They are transferred to basic probation upon successful completion of the IPS Program.

A home confinement component of this program is found in some judicial circuits within the state. This program component confines probationers to their homes except for approved work and treatment or emergencies.

The program utilized a team supervision approach with caseloads in the range of twenty-five to fifty-five. A member of the team has contact ranging from daily to once per week depending upon probationer's needs and program phase. Probationers must generally:

- be employed or a full-time student;
- complete community service;
- submit to random drug and alcohol testing;
- participate in treatment;
- abide by restrictive curfews and travel limits; and
- pay fees and other court costs.

IPS targets nonviolent felony offenders who present no evidence during screening of an unacceptable risk to community safety. Since these offenders would otherwise be incarcerated at the state prisons, a greatly reduced cost is realized. The program is able to divert a significant number of such offenders where they have low recidivism rates, maintain employment, and meet financial obligations.

Diversion Centers

Programs at these **diversion centers** provide a blend of punishment and rehabilitation. Twenty-four hour control and surveillance of offenders' activities is provided by correctional staff personnel. Coupled with the requirement that residents maintain gainful employment these alternatives make a middle-range sentencing option available for many offenders who could otherwise require long-term incarceration.

Diversion center programs are primarily aimed for nonviolent felony offenders. A screening process ensures selection of offenders who are in suitable physical and mental health and capable of fulfilling the responsibilities of full-time employment. Residents must participate in community service work and are monitored for alcohol and drug abuse. The length of stay averages four to five months, at which time those successfully completing are transferred to regular probation supervision.

Such programs play a major role in the management of offenders while being cost-effective. Money earned through employment enables offenders to pay their own room and board at the center, pay restitution to victims, continue financial support for dependents, and pay any other court-ordered obligations. Although there is still a public cost in building and maintaining these centers, the offenders' required contributions help offset these costs.

Detention Centers

This alternative is aimed at managing the special population of probation violators. Without this type of option, those whose probations were revoked would be confined in a high-cost, high-security state prison. **Detention centers** provide lower cost, minimum security facilities for lower risk offenders committed to state custody. As judges increase the percent of offenders in alternative probation programs, there will be some increase in the number of cases entering state prison through probation revocation. Detention centers provide an effective, lower cost way to control these offenders.

Confinement is for a period of 60 to 120 days. Detainees provide a variety of nonpaid labor. Rehabilitation and counseling programs are provided during the evening hours by probation staff.

Shock Incarceration

An alternative of **shock incarceration** places offenders in prison for a short period of time in hopes that the harsh reality of prison life will act as a deterrent from committing any future crimes. The target population is young, male offenders seventeen to twenty-five years of age with a felony conviction but without any previous incarceration in an adult prison.

Shock units are located at state correctional institutions housing a general inmate population, but shock incarceration offenders are located separately and

Table 10.6

How Electronic Monitoring Equipment Works

Electronic monitoring equipment receives information about monitored offenders and transmits the information over the telephone lines to a computer at the monitoring agency. There are two basic types: continuously signaling devices that constantly monitor the presence of an offender at a particular location, and programmed contact devices that contact the offender periodically to verify his or her presence.

Continuously signaling devices

A continuously signaling device has three major parts: a transmitter, a receiver-dialer, and a central computer.

The transmitter, which is attached to the offender, sends out a continuous signal. The receiver-dialer, which is located in the offender's home and is attached to the telephone, detects the signals sent by the transmitter. It reports to the central computer when it stops receiving the signal and again when the signal begins.

A central computer at the monitoring agency accepts reports from the receiver-dialer over the telephone lines, compares them with the offender's curfew schedule, and alerts correctional officials about any unauthorized absences. The computer also stores information about each offender's routine entries and exits so that a report can be prepared.

Programmed contact devices

These devices use a computer programmed to telephone the offender during the monitored hours, either randomly or at specified times. The computer prepares a report on the results of the call.

Most but not all programs attempt to verify that the offender is indeed the person responding to the computer's call. Programmed contact devices can do this in several ways. One is to use voice verification technology. Another is to require the offender to wear a wristwatch device programmed to provide a unique number that appears when a special button on the watch device is pressed into a touchtone telephone in response to the computer's call.

A third system requires a black plastic module to be strapped to the offender's arm. When the computer calls, the module is inserted into a verifier box connected to the telephone. A fourth system uses visual verification at the telephone site.

Source: Adapted from Annesley K. Schmidt, "Electronic Monitoring of Offenders Increases" (Washington, D.C.: National Institute of Justice, January/February, 1989), p. 4.

are considered in a probation status. The program is patterned after military boot camp with intense physical training, hard work, and little or no free time. The initial confinement is for a ninety-day period. Upon successful release from the unit, offenders are placed on regular probation supervision.

Electronic Monitoring[19]

In recent years electronic monitoring has been used to supervise offenders in the community. Although this technology is innovative and still in the early stages, it has generated much interest. As described in table 10.6, electronic devices enable officials to monitor and verify the whereabouts of offenders, such as being at home during the required hours. Most electronic surveillance programs are linked with traditional probation and parole supervision in the community, although some have involved pretrial releasees and others whose cases are on appeal.

Much of the appeal of electronic monitoring is linked to caseloads and expenditures. Probation officers realize a reduction in time required by traditional supervisions, while at the same time, being able to identify those who are a good risk for release and those who need continued supervision. In some programs offenders rather than taxpayers pay the costs of monitoring, which acts as a fine and an economic plus for the community. In some other instances costs to taxpayers have increased.

Most states have some form of electronic monitoring in place, usually involving a limited number of offenders with wide variations in program structure and administration. Two states, Florida and Michigan, have lead the way in placing offenders in this type of program, although the administration is quite different.

In Michigan, the state Department of Corrections monitors most offenders, and local courts, sheriffs, or private agencies monitor the rest. In contrast, the Florida Department of Corrections monitors only a little over half the participating offenders. Another quarter are monitored by city or county agencies, including sheriff's offices, local departments of corrections, and police departments. Most of the rest are monitored by one of several private agencies that offer monitoring services, and a very small number are monitored by a federal demonstration project.

Florida is a microcosm of the country as a whole in that monitoring activities take place in all areas—large metropolitan areas, medium-sized cities, small towns, and rural areas—by all levels of government. The government may provide the service with its own staff or contact for it. These public agencies represent all elements of the criminal justice system, including police departments, sheriffs, courts, correctional systems, and probation and parole agencies.

Initial Problems

Some programs have had difficulty gaining acceptance within their agencies. In some jurisdictions, judges have refused to utilize electronic monitoring.

Offenders had to learn to handle the equipment properly and understand what was expected of them. Their families also had to adapt to limiting their use of the telephone so the computer calls could be received.

Other problems were related to the equipment itself. In some jurisdictions, there was a "shakedown" period when operators learned to use the equipment correctly, interpret the printout, and deal with power surges and computer downtimes.

Poor telephone lines, poor wiring, and "call-waiting" features on the telephones caused other technical problems. Occasionally, an offender's home was located too close to an FM radio station or other strong radio wave broadcaster. Some difficulties were overcome by repairing lines or wires or by using radio frequency filters. A few programs have encountered unanticipated costs for extra telephone lines, special interconnections, underestimated long-distance charges, and supplies.

In order to be effective, electronic monitoring must have a purpose. Its use needs to be linked with an identifiable program, with stated goals and objectives. Failure to do so imposes major limitations for success. Adequate training of personnel is crucial. Although politically popular, the ultimate level of utilization remains unclear.

Assessing Community-Based Corrections

As we move toward the twenty-first century, community-based corrections is undergoing transformation. As discussed at the beginning of this chapter, community corrections is based on a philosophy of developing ties between offenders and the community. The idea is to involve the community and offenders in a mutually beneficial partnership directed toward reintegrating offenders with normal, productive roles.

However, recent ''get tough'' and punishment approaches have resulted in substantial prison overcrowding. As our nation searches for ever more places and ways to punish offenders, the community has been increasingly seen as an alternative. Through such newer approaches as house arrest, electronic monitoring, and intensive supervision, offenders are for most intents and purposes being imprisoned within their home community. So-called community programs are often ''sold'' and approved mainly on the basis of the restriction and punishment they provide. This is in direct contrast to the philosophy of community-based corrections. The philosophy of community corrections based on reintegration is in a process of being taken over by a philosophy of retribution, custody, and punishment.[20] The extent to which this occurs erodes the original purpose and meaning of what community-based corrections was conceived to be.

Summary

Meaningful community-based corrections must involve local communities. Significant ties need to be developed between offenders and the community if persons are to be reintegrated into normal community living. This implies change on the part of offenders as well as on the part of the community in order to accept and assist these persons.

Several factors account for the development of community-based corrections. One major reason has been a dissatisfaction by many with prisons. Too often they are overcrowded, unsafe, violent, and have little deterrent value. Community programs offer more humane surroundings. In many instances community-based programs can be provided for less money than institutionalization. They also offer possibilities for greater cooperation between police, courts, and correctional components of the justice system. Offenders may become involved in programs at any stage of the criminal justice process.

Community support is vital for successful community-based correctional programs. Ideally, all segments of a local community can become involved. Participation by courts and police, as well as employers, is especially important. Policymakers and planners need to develop strategies and public relations programs to gain community support and participation. Forming citizen advisory boards is one such strategy.

Examples of community-based programs and intermediate sanctions abound. Georgia has developed a range of alternatives to incarceration that provide punishment, public protection, and rehabilitation opportunities. In some areas electronic monitoring of offenders in the community is utilized, although the eventual degree of usage is unclear.

Some programs have been developed and sold on punishment and strict supervision, rather than based on reintegration, but are also often considered by some to be community-based corrections. Such approaches run counter to the original purpose of community-based corrections.

Discussion Topics

1. Which philosophy offers the better hope for long-term effectiveness, reintegration, or retribution? Why?
2. What are some ideas to help gain local citizens' support for intermediate sanctions?
3. Do you think boot camps have a place in corrections? Explain.
4. How extensively should electronic monitoring be used? Why?

Additional Readings

McShane, Marilyn D. *Community Corrections.* New York, NY: Macmillan Publishing Company, Inc., 1981.

Smykla, John Oritz and William L. Selke. *Intermediate Sanctions: Sentencing in the 1990s.* Cincinnati, OH: Anderson Publishing Company, 1995.

Zvekic, Ugljesa, ed. *Alternatives to Imprisonment in Comparative Perspective.* Chicago, IL: Nelson-Hall Publishers, 1994.

Notes

1. The President's Commission on Law Enforcement and the Administration of Justice, *Task Force Report: Corrections* (Washington D.C.: Government Printing Office, 1967), 30.
2. Ibid., 7.
3. Robert M. Carter, Richard A. McGee, and E. Kim Nelson, *Corrections in America* (Philadelphia, PA: J. B. Lippincott Company, 1975), 103–104.
4. National Advisory Commission on Criminal Justice Standards and Goals, *Corrections* (Washington, D.C.: Government Printing Office, 1974), 222.
5. Chris Eskridge, Richard Seiter, and Eric Carlson, "Community Based Corrections," in *Critical Issues in Corrections: Problems, Trends, and Prospects,* eds. Roy R. Roberg and Vincent J. Webb (St. Paul, MN: West Publishing Company, 1981), 186.
6. James McSparron, "Community Corrections and Diversion," *Crime and Delinquency* (April 1980): 230–231.
7. Peggy McGarry and Madeline M. Carter, eds. *The Intermediate Sanctions Handbook: Experiences and Tools for Policymakers* (Washington, D.C.: United States Department of Justice, National Institute of Corrections, October, 1993).
8. Doris Layton MacKenzie, "Boot Camp Prisons in 1993," *National Institute of Justice Journal* (November 1993): 21.

9. Claire Souryal and Doris Layton MacKenzie, ''Can Boot Camps Provide Effective Drug Treatment?,'' *Corrections Today* (February 1994): 54.
10. Merry Morash and Lila Rucker, ''A Critical Look at the Idea of Boot Camp as a Correctional Reform,'' *Crime and Delinquency* (April 1990): 204–222.
11. Ibid., 218.
12. National Institute of Corrections, *Residential Corrections Facilities: A Current Practice and Policy Issues* (Washington, D.C.: August, 1992), 2–3.
13. Sherry Haller and F. G. Mullaney, *Marketing Community Corrections* (Washington, D.C.: National Institute of Corrections, January, 1988).
14. National Institute of Justice, Michael N. Castle, *Alternative Sentencing: Selling it to the Public* (Washington, D.C.: Government Printing Office, September, 1991), 2.
15. Ibid.
16. Michael C. Musheno et al., ''Community Corrections as an Organizational Innovation: What Works and Why,'' *Journal of Research in Crime and Delinquency* (May 1989): 163.
17. Adopted from ''Probation's Role in a Balanced Approach to Corrections,'' Georgia Department of Corrections, Undated (Circa 1988).
18. Ibid., 9.
19. Adopted from Annesley K. Schmidt, ''Electronic Monitoring of Offenders Increases'' (Washington, D.C.: National Institute of Justice, January/February, 1989), 2–5.
20. Peter J. Benekos, ''Beyond Reintegration: Community Corrections in a Retributive Era,'' *Federal Probation* (March 1990): 52–56.

Chapter 11

Probation

Learning Objectives

After studying this chapter students should be able to:

1. Define probation.
2. Illustrate characteristics of the probation population.
3. Describe antecedents of present-day probation.
4. Understand the legal basis for probation.
5. Recognize the importance and purpose of the presentence investigation report.
6. Display increased awareness of the time demands necessary to write an accurate presentence investigation report.
7. Distinguish between the granting of probation and the administration of probation.
8. Explain typical conditions of probation.
9. Define and describe revocation and identify procedural due process rights involved.
10. Distinguish between a casework and brokerage approach to service delivery.
11. Distinguish between role categories performed by probation officers.
12. Recall contrasting results from research on outcomes of probation.
13. Define and describe the practice of split sentencing.

Key Terms

Bail
Benefit of clergy
Brokerage approach
Casework approach
Community resource management teams (CRMT)
Conditions of probation
Filing of cases
Helping role
Investigative role
Judicial reprieve
National Probation Act
Outcomes of probation
Presentence investigation (PSI)
Probation
Probation/parole officers
Recognizance
Revocation of probation
Role conflict
Shock probation
Split sentences
Surveillance/enforcement role
Technical violation

Overview

Probation developed as an effort to mitigate the harsh punishment of early law. A primary aim was to shield first-time offenders who had committed less serious law violations from the negative effects of confinement in prison. It is considered a judicial function and is regarded as being a substitute for incarceration. Probation is utilized within federal, state, and local jurisdictions.

Care should be taken not to confuse probation with parole, the latter of which occurs *after* a period of incarceration and is not a function of the judiciary. Numerous aspects of probation are discussed in this chapter, including its development, legal considerations of granting and revoking, preparation of presentence investigation (PSI) reports, and roles of probation officers.

Time Line

An overview of the development of probation is depicted by the time line in table 11.1. Narrative discussion of the events are referred to throughout this chapter.

Definition

Probation is a sentencing alternative available to judges except in those cases in which mandatory incarceration is specified by law. Depending upon jurisdiction, some or all of the following characteristics are considered as constituting the operation of probation:

- There is a sentencing alternative that postpones final adjudication, a judicial function wherein the court retains authority
- Incarceration is avoided
- Probationer remains at liberty in the community subject to court imposed conditions
- Probationer is placed under administrative supervision of a probation officer
- Probation may be revoked if conditions are violated; final sentence is then issued

Probation Population

The proportion of adults on probation from both state and federal court constitutes nearly 60 percent of persons under correctional supervision in the United States, as illustrated in figure 11.1.

Table 11.2 breaks down the large probation population by region and jurisdiction. This table also provides information regarding the probation rate per 100,000 adult residents.

Table 11.1

Time Line of Selected Events in the Development of Probation

Date	Developmental Events
From English common law—13th–19th century	Benefit of clergy Judicial reprieve Recognizance
1836	Massachusetts law providing for release on recognizance
1841	John Augustus begins informal "probation" service
1878	Massachusetts law authorized Boston to appoint a paid probation officer
1884	*People ex rel. Forsythe v. Court of Sessions*
1898	Massachusetts law extends probation system to include state superior courts
1899	Probation development
1916	*Killits* case
1925	Federal probation act passed Administrative responsibility placed with the Bureau of Prisons
1939	Administration of federal probation transferred to the Administrative Office of the Courts
1965	Shock probation utilized
1967	By this year statutory provisions for adult probation were provided in all states
1967	*Mempa v. Rhay*
1970	Development of probation standards by the American Bar Association
1973	*Gagnon v. Scarpelli*
1973	Development of probation standards by the National Advisory Committee
1977	Development of probation standards by the American Correctional Association
1982	Intensive Probation Supervision (Georgia)

Antecedents of Present-Day Probation

Although probation developed mainly in the United States, some antecedents may be traced to British common law. These practices were utilized to avoid the severe punishments (often a death sentence) given to offenders.

Benefit of Clergy

Under the original practice, dating back to the thirteenth century, members of the clergy accused of a crime have their cases heard in a church court rather than a secular court, a practice known as **benefit of clergy.** A common sentence of death found in the secular court could thus be avoided and a less severe punishment instituted by a church court.

Figure 11.1 Relative proportions of jail, prison, conditional release, and probation populations

Source: Bureau of Justice Statistics, Press release 9.11.94, Washington, D.C.

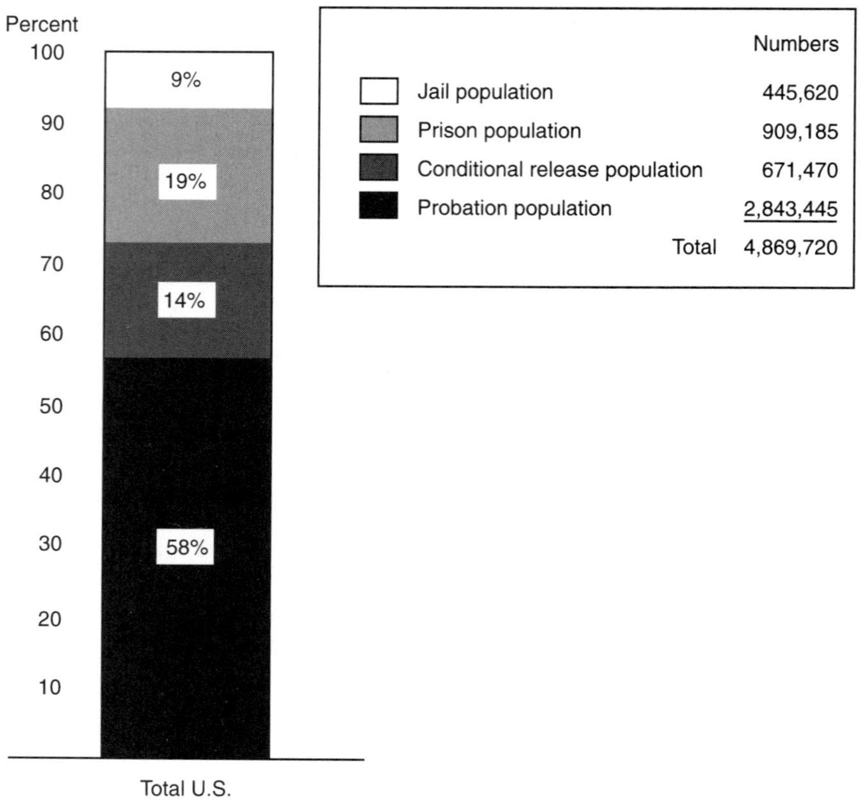

By the fourteenth century, the practice was to hear cases in secular courts but to provide benefit of clergy. In such cases only testimony favorable to the defendant was presented, which usually resulted in acquittals or light sentences.

Benefit of clergy was gradually extended to any defendant who could read as well as to additional classes of accused. As a result, its special meaning became lost and was not in use past the mid-1800s. Scholars differ regarding the impact benefit of clergy had on present probation. It made for less severe sentencing but did not incorporate essential features of current probation practices.

Judicial Reprieve

An early practice of English courts was to grant **judicial reprieve.** This practice temporarily suspended the imposition or execution of a sentence. This allowed an opportunity for convicted individuals to seek a pardon, during which

Table 11.2

Adults on Probation

Region and Jurisdiction	***Probation Population, 12/31/93***	***Number on Probation Per 100,000 Adult Residents***
U.S. total	2,843,445	1,490
Federal	43,095	23
State	2,800,350	1,458
Northeast	487,861	1,252
Connecticut	50,904	2,036
Maine	8,650	928
Massachusetts	47,150	1,021
New Hampshire	4,122	490
New Jersey	109,576	1,831
New York	156,617	1,141
Pennsylvania	88,180	961
Rhode Island	16,604	2,170
Vermont	6,058	1,402
Midwest	603,391	1,341
Illinois	73,550	852
Indiana	82,804	1,951
Iowa	14,505	697
Kansas	24,083	1,304
Michigan	139,682	2,003
Minnesota	74,186	2,256
Missouri	32,916	850
Nebraska	14,594	1,249
North Dakota	1,954	422
Ohio	98,211	1,193
South Dakota	3,781	746
Wisconsin	43,125	1,167
South	1,155,168	1,745
Alabama	33,721	1,084
Arkansas	17,527	980
Delaware	15,571	2,966
District of Columbia	10,434	2,254
Florida	199,275	1,896
Georgia	145,230	2,861
Kentucky	11,458	407
Louisiana	32,434	1,063
Maryland	80,208	2,154
Mississippi	9,943	527
North Carolina	86,212	1,645
Oklahoma	25,689	1,088
South Carolina	38,855	1,444
Tennessee	40,463	1,056
Texas	378,523	2,946
Virginia	23,619	482
West Virginia	6,006	433

(continued on next page)

Table 11.2 (continued)

Region and Jurisdiction	Probation Population, 12/31/93	Number on Probation Per 100,000 Adult Residents
West	553,930	1,365
Alaska	3,214	784
Arizona	36,815	1,285
California	280,749	1,241
Colorado	35,494	1,351
Hawaii	10,100	1,157
Idaho	4,749	620
Montana	4,107	677
Nevada	8,826	851
New Mexico	7,673	676
Oregon	37,902	1,585
Utah	7,325	613
Washington	114,018	2,952
Wyoming	2,958	891

Source: Bureau of Justice Statistics, Press release 9.11.94, Washington, D.C.

time they usually remained at liberty. However, during this time, no special conditions of behavior were imposed by the court nor was any supervision provided. In some instances the procedure led to a permanent or indefinite suspension of sentence.

Recognizance

Recognizance is a personal pledge or promise entered into by a defendant before a court. This binds a person to adhere to certain conditions of behavior and to avoid future misbehavior while remaining at liberty. A recognizance is entered into for a specified period of time. If a defendant violates the promise, the state may then take enforcement action by such means as incarceration or loss of money paid for bond.

By the nineteenth century, recognizance was used mainly as a measure to deter additional law violations. It was often employed with young offenders who had committed less serious offenses. This practice was first written into law in Massachusetts in 1836. Other states have followed, and in more recent times recognizance has also been applied at the adult level for a variety of criminal offenses.

Recognizance incorporated a number of features associated with present probation, such as the offender's remaining at liberty, conditions being imposed on behavior, and possible incarceration if conditions are violated. Absent, however, was any supervision by an officer of the court.

Bail

Bail is a system of providing a security bond in the form of cash or property for the purpose of guaranteeing appearance in court at a future date. This allows an accused person to remain at liberty during the intervening time. Bail may be used with or without recognizance.

Typically, bail was actually paid by someone other than the defendant. Those providing bail payment (usually for a fee), called sureties or bondspersons, obviously had an interest in seeing that those defendants who were "out on bail" adhered to the court's requests. Thus, we see a fragment of supervision arising; although not court sponsored, it was condoned by the court.

Filing of Cases

The **filing of cases** was carried out in Massachusetts during the nineteenth century and was unique to this state. After guilt was established, the imposition of sentence was suspended and the case "filed". Such action required the consent of the defendant and prosecutor and allowed the court to impose conditions for the defendant. A case could be reopened at any time. In some instances, though, a case might be kept on file indefinitely by the judge, which had the effect of freeing the defendant.

John Augustus

At Boston, Massachusetts the voluntary activities of John Augustus earned him the title "Father of Probation." He was a bootmaker by trade and at the age of fifty-seven (in 1841) became interested in the operation of the courts. Particularly, he was sensitive to the plight of "common drunkards" who came before the court. He reasoned that ". . . many offenders required no more than the sincere interest of another human being to be able to straighten out their lives."[1]

Augustus was the first to use the term probation with reference to corrections, a term of Latin derivation meaning a period of testing or trial of a person's conduct or character. He considered probation as assistance and the support he provided as treatment. His probation work was informal in nature and not based on legal statutes. He worked as a volunteer, and releases for offenders were based on informal arrangements between Augustus and the judges. By the time of his death in 1859, he had assisted nearly 2,000 offenders.

In addition to paying their bail, Augustus housed some of his charges in the large house he occupied with his family; other probationers had homes to which they could return. Augustus fed and clothed the needy and required those who should work to seek and keep at it. Eventually he offered bail for women and youths also.[2]

It should also be noted that Augustus carefully screened those whom he volunteered to help. In his words:

> Great care was observed of course, to ascertain whether the prisoners were promising subjects for probation, and to this end it was necessary to take into consideration the previous character of the person, his age and the influences by which he would in future be likely to be surrounded.[3]

Since the successful efforts of John Augustus took place in Massachusetts, it is no surprise that some of the first laws to place probation on a formal, legal foundation were passed in Massachusetts. For example, in 1878 the Massachusetts legislature enacted authorization for the mayor of Boston to appoint a paid probation officer. This marked the first case of a paid probation officer. The officer was an authorized agent of the court, although a member of the police force. But a law passed in 1890 stated that henceforth probation officers were not to be active members of a police force. Legislation passed in 1898 extended probation to the superior courts of Massachusetts, thereby providing probation services on a statewide basis.

The development of probation nationwide received a boost with the establishment of the first juvenile court at Chicago, Illinois in 1899. The juvenile court movement and philosophy were consistent with the idea of probation. The delivery of probation services was considered essential for effective juvenile court operation. As other states established courts at the juvenile level, authorized probation services also increased.

Legal Basis

The legal basis for probation at the state level is provided by authorization passed by individual state legislatures. By 1925 probation for juveniles was authorized in the forty-eight states. At the adult level, it was not until 1967 that probation was authorized by all the states and Puerto Rico.

Prior to 1925, attempts to establish probation at the federal level were not successful. During 1925 the passage of the **National Probation Act** allowed federal district courts (outside Washington, D.C.) to hire a probation officer.

There was opposition to the concept of probation, and its growth was not without struggle. A most formidable challenge involved a legal issue. Could the court legally indefinitely suspend sentence? By practice, sentence was suspended when a defendant was placed on probation. Successful completion of probation implied a permanent suspension of sentence. Opinion was divided on whether this was proper legal practice. In New York a law has been enacted authorizing suspension of sentence. However, opponents argued this infringed on executive power of pardon. In 1894 the legality of indefinite suspension was challenged in the case of *People ex rel. Forsyth v. Court of Sessions*.[4] The decision in this case upheld the authority for indefinite suspension since New York had a law authorizing such action. Yet this very same issue later arose in another case, this time to the United States Supreme Court in 1916. This involved what has become known as the Killits case.[5] In this instance the Supreme Court held that federal courts (and state courts by implication) did *not* have the authority on their own to indefinitely suspend sentence. However, this case also ruled that indefinite suspension of sentence would be a legal practice if legislatures passed laws authorizing such a practice (as had been done in New York). Following the Killits case, many legislative bodies then enacted laws authorizing the courts to indefinitely suspend sentence. Solving this legal problem led to expansion and increased acceptance of probation.

Rationale

You will recall from chapter 10 that probation is traditionally considered to form the backbone of many community-based efforts. Rationales for probation are essentially the same as those on which the development of any community-based corrections program are based. Included are dissatisfaction with prisons, reduced costs to the criminal justice system, and more humane treatment alternatives.

In addition, the American Bar Association provided its reasons for probation by approving the following standard:

> Standard 1.2. Desirability of probation
>
> Probation is a desirable disposition in appropriate cases because
>
> (1) it maximizes the liberty of the individual while at the same time vindicating the authority of the law and effectively protecting the public from further violations of law;
>
> (2) it affirmatively promotes the rehabilitation of the offender by continuing normal community contacts;
>
> (3) it voids the negative and frequently stultifying effects of confinement that often severely and unnecessarily complicate the reintegration of the offender into the community;
>
> (4) it greatly reduces the financial costs to the public treasury of an effective correctional system; and
>
> (5) it minimizes the impact of the conviction upon innocent dependents of the offender.[6]

The Presentence Investigation

A major purpose of **presentence investigation** (PSI) reports is to provide information to aid the court in deciding whether to place an offender on probation. Information contained in the report may also be utilized

- by probation or parole officers to aid in supervision;
- by correctional institutional personnel for classification and treatment;
- by the parole board as a decision-making aid;
- for research purposes.

Probation officers spend a high proportion of their time preparing presentence reports.

Variation exists throughout the United States in terms of (1) the extent to which such reports are prepared and (2) the depth and scope of information contained in those that are prepared. In some instances preparation of a presentence report is mandated by law. Some judges rely heavily on the information and sentencing recommendations contained in the report. At the other end of the scale, some jurisdictions do not prepare such information.

Standards call for the authorization of PSI reports in every case, coupled with the recommendation that preparation of the report should not begin until *after* adjudication of guilt, except as may be specifically authorized with adequate precautions.[7]

Preparation of PSI Reports

Preparing PSI reports is a major activity of persons employed as probation officers. This task is demanding and time-consuming since it involves searching records, verifying school attendance, checking employment history, and interviewing family and others who know the defendant. Since reports normally include an all-important recommendation concerning whether to institutionalize or place a defendant on probation, it is imperative that this document be as accurate and meaningful as possible.

Some items to keep in mind as the report is developed include the need for accuracy. Only true and accurate facts should be included. It is also important to include a diagnostic and supervision plan. The report should be as objective as possible, free from biases and stereotypes. Finally, it should be remembered that this written document will probably never be destroyed. It may be used numerous times by a variety of individuals. A well-written report reflects positively on the officer who developed the document.

As shown by figure 11.2, contents of a PSI interview form and report typically include information about

- the offense committed and prior record;
- chemical dependency;
- educational background;
- employment history;
- family history and financial information;
- military service;
- mental health;
- victim impact statement;
- verification information; and
- evaluation summary and a recommendation whether or not to place on probation.

Disclosing Information Contained in PSI Reports

A long-standing issue is whether information contained in PSI reports should be disclosed. There are arguments for and against disclosure. Those favoring no disclosure do so on the grounds that sources of information will be unwilling to cooperate in the future, sentencing may be delayed, and supervision of probationers may be compromised. On the other hand, advocates feel that disclosure of information may aid the sentencing and reintegration process, provides for a more adequate defense counsel representation and that is in and of itself consistent with the idea of fundamental fairness.

Disclosure of information in actual practice is mixed and lacks uniformity. In many jurisdictions trial judges are permitted wide discretion. In other jurisdictions, statutes provide for full disclosure (to defendant and counsel) routinely or upon request. In other instances only partial disclosure is required by law. Sometimes substantive information may be disclosed without identifying the source. Where disclosure has been practiced, there is no evidence of significant negative effects. Such experience prompted the National Advisory Council to recommend disclosure of information to both prosecution and defense counsel.[8]

Granting Probation

Authority to grant probation resides ultimately with the court. This decision is a sentencing alternative, as was discussed in chapter 4. It has been established that a person does not have a right to be placed on probation. Rather, probation

Figure 11.2 Presentence/postsentence investigation interview form

Source: Reprinted by permission of the State Probation Office, Frankfort, Kentucky.

COMMONWEALTH OF KENTUCKY
CORRECTIONS CABINET
DIVISION OF PROBATION AND PAROLE

PRESENTENCE/POSTSENTENCE INVESTIGATION INTERVIEW FORM

RE: ____ DATE INTERVIEWED __________
LAST FIRST MIDDLE

I. DATA ON FACE SHEET CC:1076:

II. PRIOR ARRESTS AS JUVENILE OR ADULT:

III. OCCUPATION, JOB SKILLS, AND ECONOMIC STATUS:

(a) PRESENT EMPLOYMENT STATUS:

(b) PRESENT OR MOST RECENT EMPLOYMENT:

DATES	NAME & ADDRESS OF EMPLOYER	WAGES	NATURE OF WORK	REASON LEFT	
STARTED					
ENDED					

(c) SECOND MOST RECENT EMPLOYMENT:

STARTED					
ENDED					

(d) SUMMARY OF ANY OTHER EMPLOYMENT:

(e) ANY EMPLOYMENT TRAINING OR SKILLS:

(f) PRESENT TAKE-HOME SALARY:

(g) OTHER SOURCES OF INCOME AND AMOUNT:

(h) OWNED PROPERTY AND VALUE:

(i) RENT OR MORTGAGE PAYMENTS:

(j) OTHER DEBTS:

(k) CHILD SUPPORT:

(l) CIVIL JUDGEMENTS:

RE: ____

IV. SOCIAL HISTORY

(a) BIRTHDATE AND PLACE OF BIRTH:

(b) FATHER'S NAME AND OCCUPATION:

(c) MOTHER'S NAME AND OCCUPATION:

(d) NUMBER OF BROTHERS AND SISTERS:

(e) SUBJECT'S DESCRIPTION OF CHILDHOOD:

(f) RELIGIOUS AFFILIATION AND PRESENT ATTENDANCE:

(g) CHILDHOOD PROBLEMS:

(h) FIRST SCHOOL ATTENDED, LOCATION AND YEAR:

(i) OTHER SCHOOLS ATTENDED, LOCATIONS, AND YEARS:

Figure 11.2 (*continued*)

(j) G.E.D., LOCATION, AND YEAR:

(k) MILITARY BRANCH AND DATE ENTERED:

(l) TYPE DISCHARGE AND DATE:

(m) DUTY STATIONS AND DECORATIONS:

(n) AGE AND REASON LEFT PARENTAL HOME:

(o) DATE(S) MARRIED, LOCATION(S), MAIDEN NAME OF SPOUSE(S), AND CHILDREN:

(p) DIVORCE(S), SEPARATION(S), DATE(S), LOCATION(S), AND REASON(S):

RE: ______

(q) PRESENT RELATIONSHIPS WITH PARENTS AND SIBLINGS:

(r) PRESENT RELATIONSHIPS WITH SPOUSE AND CHILDREN:

(s) COHABITAL RELATIONSHIP(S) AND DESCRIPTION:

V. FAMILY MEMBERS

FATHER, MOTHER, SIBLINGS, SPOUSE(S), CHILDREN, COHABITAL

RELATIONSHIP	NAME	AGE	ADDRESS	OCCUPATION	CRIM. RECORD

VI. PHYSICAL AND MENTAL CONDITION:

(a) DESCRIPTION OF PRESENT PHYSICAL HEALTH:

(b) PRESENT PHYSICAL ILLNESSES, IMPAIRMENTS, OR ALLERGIES:

(c) PRESCRIBED MEDICATION, CURRENT PHYSICIAN AND LOCATION:

(d) ALCOHOL USAGE AND AMOUNT:

(e) CONTROLLED SUBSTANCE OR SOLVENT ABUSE:

RE: ______

(f) HISTORY OF PSYCHOLOGICAL OR PSYCHIATRIC TREATMENT:

(g) DESCRIPTION OF PRESENT MENTAL HEALTH:

(*continued on next page*)

Figure 11.2 *(continued)*

VII. CASE SUMMATION

(a) SUBJECT'S VERSION OF OFFENSE:

(b) SUBJECT'S PLANS IF GRANTED PROBATION:

INVESTIGATING PROBATION AND PAROLE OFFICER

is a discretionary decision made by the judge, who may or may not consider recommendations, such as a presentence report. In most jurisdictions there are laws that govern eligibility for probation. For example, some state statutes specify that any offender may be considered for probation, while in other states serious offenders (for example, murderers or rapists) are not eligible. Typically, state statutes allow probation to be considered when a capital offense is not involved, a defendant is not likely to commit another offense, or incarceration is not mandatory. In general terms, probation has been more likely to be granted to first-time offenders who have committed less serious offenses, although probation is used for both felony and misdemeanor cases. Increasing numbers of felony offenders are being placed on probation, often times as a result of plea bargaining. The fact of prison overcrowding many times becomes a factor in the bargaining process and subsequent probation.[9]

The case of *People ex rel. Ward v Moran* established that deciding to grant probation would be left to the sound discretion of the trial court.[10] This was based on the feeling that the court is in a position to weigh the interests of society and the individual in each case.

The National Advisory Commission suggests that the length of probation should not exceed the length of maximum permissible incarceration, except that a one-year maximum is recommended for misdemeanants.[11] The American Bar Association recommends that supervision while on probation not extend past legislatively fixed time but should not exceed two years for a misdemeanor or five years for a felony.[12]

Conditions of Probation

Depending upon jurisdiction, a variety of **conditions of probation** may be imposed on an offender's behavior and mobility. An example of conditions imposed by federal district courts is shown in figure 11.3. A universal condition is to obey the law. Other areas commonly subject to certain conditions include rules regarding supporting dependents, owning an automobile, using of alcoholic beverages, changing place of residence, and changing jobs. Certain behaviors might be prohibited, such as liquor consumption or possession of a firearm. Other conditions, supporting dependents, for example, would likely be mandatory. Still others could be permitted, but permission must be obtained from the probation supervisor. Permission to travel or change residence would fall into this category.

The purpose of these rules is to aid the person in living a law-abiding life in the community. However, an examination of conditions imposed reveals that some are related to a treatment philosophy while others emphasize a more punitive or controlling orientation. Thus, another correctional issue arises over the best means (conditions) to achieve a desired outcome (law-abiding behavior).

While wide discretion is permitted in posing conditions, it is generally established that these rules should be reasonably related to individual needs of the offender and should not be arbitrary, cruel, or exceed reasonable possibility for compliance. The courts have upheld a number of conditions, such as travel restrictions, and have struck down some others, such as not permitting a woman probationer to have children. The American Bar Association has provided some guidance by suggesting areas where conditions can be helpful and appropriate. They are as follows:

1. Cooperating with a program of supervision
2. Meeting family responsibilities
3. Maintaining steady employment or engaging or refraining from engaging in a specific employment or occupation

Figure 11.3 Conditions of probation and supervised release

Source: U.S. Department of Justice.

Conditions of Probation and Supervised Release

UNITED STATES DISTRICT COURT

FOR THE

____

Name ______________________________________ Docket No. ______________________

Address ______________________________________

Under the terms of your sentence, you have been placed on probation/supervised release (strike one) by the Honorable ____, United States District Judge for the District of ________________________. Your term of supervision is for a period of ____________, commencing _______________.

CONDITIONS

It is the order of the Court that you shall comply with the following conditions:

(1) You shall not commit another Federal, state, or local crime;

(2) You shall not leave the judicial district without the permission of the court or probation officer;

(3) You shall report to the probation officer as directed by the court or probation officer, and shall submit a truthful and complete written report within the first five days of each month;

(4) You shall answer truthfully all inquiries by the probation officer and follow the instructions of the probation officer;

(5) You shall support your dependents and meet other family responsibilities;

(6) You shall work regularly at a lawful occupation unless excused by the probation officer for schooling, training, or other acceptable reasons;

(7) You shall notify the probation officer within seventy-two hours of any change in residence or employment;

(8) You shall refrain from excessive use of alcohol and shall not purchase, possess, use, distribute, or administer any narcotic or other controlled substance, or any paraphernalia related to such substances, except as prescribed by a physician;

(9) You shall not frequent places where controlled substances are illegally sold, used, distributed, or administered;

(10) You shall not associate with any persons engaged in criminal activity, and shall not associate with any person convicted of a felony unless granted permission to do so by the probation officer;

(11) You shall permit a probation officer to visit you at any time at home or elsewhere, and shall permit confiscation of any contraband observed in plain view by the probation officer;

(12) You shall notify the probation officer within seventy-two hours of being arrested or questioned by a law enforcement officer;

(13) You shall not enter into any agreement to act as an informer or a special agent of a law enforcement agency without the permission of the court;

(14) as directed by the probation officer, you shall notify third parties of risks that may be occasioned by your criminal record or personal history or characteristics, and shall permit the probation officer to make such notifications and to confirm your compliance with such notification requirement.

Figure 11.3 *(continued)*

The special conditions ordered by the Court are as follows:

Upon a finding of a violation of probation or supervised release, I understand that the Court may (1) revoke supervision or (2) extend the term of supervision and/or modify the conditions of supervision.

These conditions have been read to me. I fully understand the conditions, and have been provided a copy of them.

(Signed) ______________________________ ____________
Defendant Date

______________________________ ____________
U.S. Probation Officer Date

4. Pursuing prescribed educational or vocational training
5. Undergoing available medical or psychiatric treatment
6. Maintaining residence in a prescribed area or in a special facility established for or available to persons on probation
7. Refraining from consorting with certain types of people or frequenting certain types of places
8. Making restitution of the fruits of crime or reparation for loss or damage causes thereby[13]

Revocation of Probation

Once granted, the court may take back or withdraw the disposition of probation. Such action constitutes the **revocation of probation.** Probation may be revoked for two general reasons: (1) breaking the law and (2) breaking the rules, in which case a technical violation is said to have occurred. For example, if a probationer breaks a rule that requires permission from the probation officer to travel outside the jurisdiction, then a **technical violation** has occurred. Such behavior is not a violation of law but a violation of one of the rules of probation. In most instances, revocation is more likely to be requested and granted when another crime has been committed, such as a drug violation, rather than for a technical violation alone.

Revocation represents an important decision since the probationer will then likely face a term of incarceration. The hope for successful reintegration has been dashed, and the state is faced with increased cost of imprisonment.

Since the mid-1960s considerable attention has been focused on procedural due process rights that apply during the stage of revocation. These concerns resulted in several benchmark court cases dealing with due process issues. One major issue centered on whether defendants had the right to counsel at revocation hearings. The United States Supreme Court in *Mempa v. Rhay*[14] in 1967 ruled that probationers should be afforded the right to counsel in cases where sentence had been suspended in lieu of placing the defendant on probation. Left unclear, however, by this decision was the technical point regarding whether this right also applied to probationers in those jurisdictions where sentence was considered passed (rather than suspended) at the time of trial. There was some confusion and differing interpretations. Some courts extended this right in all cases, while others extended it only in cases of suspended sentences.

This question was addressed in the important case of *Gagnon v. Scarpelli*[15] in 1973. This case, which involved a probationer, ruled that probationers (and parolees as well) have a conditional right to counsel at revocation hearings. The court held that

1. a decision regarding the right to counsel should be made on a case-by-case basis (rather than the right being automatically extended in all cases);
2. counsel should be provided when complexities would make it difficult for the defendant to present the case; and
3. if a request for counsel is refused, the reasons for refusal should be written into the court record.

The importance of this case is reflected by the fact that most jurisdictions today allow probationers the right to counsel at probation revocation hearings. Most jurisdictions did not allow counsel prior to this decision. *Gagnon v. Scarpelli* was also important in another respect since it also held that revocation hearings for probationers were to be conducted according to due process procedures outlined the year earlier for parolees in the case of *Morrissey v. Brewer.*[16] In that landmark case, the plaintiff Morrissey was convicted of check fraud and sentenced to seven years. He was paroled from the Iowa State Penitentiary in June of 1968. Seven months after his parole, he was arrested, and subsequently, his parole was revoked on the basis of information that he had violated his conditions of parole by buying a car under an assumed name and failing to report his place of residence to his parole officer.

Morrissey alleged that he had been denied due process because his parole had been revoked without a hearing. The district court for the southern district of Iowa found that the state's failure to accord a hearing prior to parole revocation did not violate due process. The court of appeals affirmed that parole revocation required no hearing. The Supreme Court reversed the appellate court's decision and laid out these due process standards for parole revocation hearings:

- Written notice of violation
- Disclosure of evidence
- Right to confront and cross-examine
- Neutral and detached hearing
- Written reason for revocation

Service Delivery Strategies

Probation agencies manifest a particular strategy or approach in providing service and supervision to probationers. The traditional strategy has been the casework approach. An alternative orientation, identified as the brokerage approach, has been recommended and increasingly utilized.

Casework Approach

The **casework approach** emphasizes a one-to-one relationship between the probation officer and the probationer in an effort to individualize treatment. It is considered important for probationers to participate in developing their own probation plan. The caseworker assumes a treatment role in attempting to change the offender's behavior. The authority inherent in the treatment role is often seen as an important component of this process. With this strategy, the officer attempts to meet most of the probationer's needs.

Brokerage Approach

Contrasted to the casework approach is the **brokerage approach,** wherein the role of the probation officer is considered to be that of a community resource manager. Such a strategy de-emphasizes a close one-to-one relationship between probation officer and probationer. This orientation does not consider the officer as a primary agent in changing client behavior as does the casework approach. Instead, successful linkage between the officer and community service agencies is considered most important. The officer functions to locate and refer clients to existing service agencies in the community and to conduct proper follow-up.

In this role the probation officer is seen as a manager of resources available in the community. Considerable time is spent in locating and developing an effective working relationship with the various community agencies.

In many instances, services that probationers need will not be available in the community. This results since a wide range of services are often needed. Under these conditions, some officers assume an advocacy role. They work with the community in an effort to develop the needed services rather than try to provide these services by themselves.

Community Resource Management Team (CRMT)

The development of **community resource management teams** (CRMT) provides an example of a service delivery arrangement that stresses brokerage, advocacy, and the role of the community. CRMT also stresses that a team effort can more effectively provide services than an individual effort because of the wide diversity of client needs. Under the team concept, an officer may become a specialist in only a few areas. For example, one officer may work with community agencies in the area of employment assistance; another may concentrate in drug and alcohol abuse; another in educational programs. Thus, a client with multiple needs could be assisted by several officers. Such an arrangement is known as a pooled caseload. Where services do not exist, team members assist in their development. The overall concept and function of CRMT has been outlined as follows:

> In servicing an offender, the CRMT Probation Officer will assess the needs, identify the services available, contact the appropriate resource and work in a coordinated team effort to assist the probationer in obtaining these services and finally will provide a follow-up on the programming. In those areas where needed services fail to exist, staff will be responsible for mobilizing the needed resources and services. The underlying rationale is that by pooling the resources of the CRMT and the community a large number of probationers and the diversity of their needs will be serviced in a more effective manner.[17]

It is clear that CRMT follows a brokerage strategy. The role tasks and relationship between officer and probationer are very different from those inherent in the casework approach. The concept of a pooled caseload is a major departure from traditional methods and is subject to debate.

Delivery of services in most probation agencies reflects a mix, or blend, of strategies. Approaches, in many instances, will seldom be ''pure'' casework or brokerage. Although one approach is followed, some features of the other will likely be included.

Probation/Parole Officers

Although probationers are released to supervision rather than being incarcerated, and supervision of parole occurs after incarceration, both groups are usually supervised by officers employed as **probation/parole officers.** It is appropriate to discuss the formal operations of probation and parole in separate chapters. However, the day-to-day supervision of probationers or parolees is similar and often carried out by the same officer. One of the more significant differences of supervision is the ultimate authority. Jurisdiction over the probationer resides with the court, while the parolee jurisdiction resides with paroling authorities in the executive branch of government.

Federal, state, and local levels of government employ such officers. In nearly all instances, officers work in departments, which vary greatly in size. Most have at least some division of labor and bureaucratic organization.

Probation/parole officers typically perform a number of tasks in each of the following categories:

1. **Investigative role**
2. **Helping role**
3. **Surveillance/enforcement role**

As was discussed earlier in this chapter a major investigative activity is the preparation of presentence reports. Another major activity is providing help to offenders. This entails identifying needs and working to meet those needs. This may be accomplished by a casework or brokerage approach, as part of a team or on an individual basis. In some instances, volunteers are utilized to assist in the delivery of services.

Protecting the public also constitutes a major responsibility, a duty commonly identified as law enforcement. In this role the officer is expected to monitor the offender's behavior. If violations occur, the officer must decide the best course of action for all concerned. In some instances counseling may be sufficient while in another, revocation may be recommended.

Most departments also have a demand for administrative personnel. Typically, a person in the position of chief officer oversees departmental operations. Duties of this position entail a number of tasks including hiring professional and clerical staff, coordination and supervision of activities, management of the budget and building facilities, and training of personnel. Some departments also have a position of assistant chief and other positions designated as supervisors.

Differing Roles

Frequent attention is drawn to the nature of **role conflict** that a person in a position of probation officer is likely to experience. Role conflict occurs when incompatibility exists between two or more roles that an individual is expected to perform—performing one role interferes with or is antagonistic to the other. For example, probation officers are expected to provide help and support to probationers but are also expected to enforce the law and rules (which could lead to revocation).

In approaching such situations it has been pointed out that:

> . . . officers are permitted considerable discretion in the way they perform their work and in the kinds of information they report to their superiors. The manner

in which they exercise this discretion and reconcile their often conflicting job demands varies substantially from officer to officer.[18]

Some officers come to see their mission primarily as enforcers, placing greater emphasis on control, coercion, and revocation. Others see their role more as treatment oriented and toward helping to reintegrate offenders into the community.

Historically, little has been done to help minimize such dilemmas and role conflicts. However, some efforts have attempted to deal with this problem by using a Risk Prediction Scale to improve classification of clients and more clearly define an appropriate role for the officer. A Risk Prediction Scale provides a numerical rating by which to determine the amount of control or supervision required in a given case. Some cases may require a high degree of control, while others need only minimum supervision. A related aspect of this approach is that officers may specialize in handling one or the other type of cases. This has the effect of separating helping and enforcement roles, which should work to the benefit of the officer, the client, and ultimately to the benefit of society. Although this dilemma is likely to remain, future evaluations will indicate to what extent benefits will come about as a result of such techniques.[19]

In another statement regarding what roles are appropriate, it has been said that:

> The answer will depend upon what are believed to be the overall goals of probation and a subjective assessment of the most effective means of achieving those goals. It is most likely that every probation agency will develop, over time, a tendency to emphasize one or more goals over other goals, and this tendency will be a product of many diverse influences, not all of which can be controlled by the agency. Until we can agree on the proper goals of probation, their relative importance, and the best means of achieving them, we will find this to be a troubling question.[20]

Outcomes of Probation

Considerable research has been done in an attempt to evaluate the **outcomes of probation.** Any thorough evaluation process is costly and time-consuming. Ideally, individuals should be studied during the time they are on probation and for some period of time beyond expiration of their probationary period. Only through such efforts can reliable conclusions be drawn.

Probation is often judged as successful or unsuccessful depending upon the level of recidivism among probationers as compared to those receiving a different sentence, usually a prison term. However, results of such studies vary due to differences in the definitions and measurements of recidivism and to differences in follow-up periods which range from several months to many years. It has been suggested that:

> recidivism has no universally accepted meaning among criminal justice researchers. Different studies have defined it variously as a new arrest, a new conviction, or a new sentence of imprisonment, depending on the kinds of data

Table 11.3

Manner and Proportion of Adults Leaving Probation	
Characteristic	***Percent of Those Persons with a Known Status***
Adults leaving probation	100%
Successful completions	81
Discharged absconders	3
Discharged to detainer/warrants	2
Returned to prison	
With new sentence	3
With the same sentence	9
Other type of exit	2
Death	>.5

Source: Adapted from Bureau of Justice Statistics, Correctional Populations in the United States, 1986 *(Washington, D.C.: Government Printing Office, February, 1989), p. 20.*

they had available and their project goals. As a result, it is exceedingly difficult and complex to make comparisons among their results.[21]

Research attempting to adequately measure outcomes needs to consider several key interrelated variables. Probation is a process involving different persons, various strategies, and numerous programs. The fact that someone is placed in a status of probationer does not mean his or her experiences are similar to other probationers. In actuality, such experiences are likely to be quite varied. Individuals on probation have differing backgrounds, criminal histories, and personal resources. Approaches and role performance by probation officers vary greatly. Agency and community goals also vary from jurisdiction to jurisdiction. Such factors affect the quality and quantity of probation experiences, and useful evaluation efforts must weigh the impact of such interrelated variables, but they seldom do. Random sampling is seldom done, and control over extraneous variables often is lacking.

It is therefore not surprising that divergent results are reported. Some studies indicate that a significant proportion, perhaps as high as 81 percent,[22] successfully complete probation as shown in table 11.3.

Contrasting results were reported by Petersilia in California. During forty months of follow-up, 65 percent of the total probation population were arrested, 51 percent were convicted, and 22 percent were incarcerated at state prisons.[23] Other research has demonstrated specific programs to be successful.[24] As Shover and Einstadter[25] suggest, there is no way to resolve these discrepant findings. Furthermore, Shover and Einstadter point out that apparent successes of probation over harsher penalties may have little to do with programs at all and more to do with making good guesses about who will be successful.

Costs associated with community supervision are usually considered to be less than those for incarceration. Just how much less is shown by examples given in table 11.4. In each instance community supervision costs are less than

Table 11.4

Costs of Incarceration and Supervision

Section 5E1.2(i) of the Sentencing Guidelines provides that in addition to the general fine imposed under Section 5E1.2(c), the court shall impose an additional fine that is at least sufficient to pay the costs to the government of any imprisonment, probation, or supervised release ordered. This memo updates the figures, based upon the cost data for fiscal year 1989.

Incarceration

The Bureau of Prisons reports that the average per capita costs in Bureau operated facilities during FY 1989 were:

Federal Prison Facility	
Daily:	$46.54
Monthly:	$1,415.56
Annually:	$16,986.72
Contract Halfway House	
Daily:	$31.47
Monthly:	$957.27
Annually:	$11,487.18

Supervision (Pertaining to Probation and Supervised Release)

The Financial Management Division of the Administrative Office of the U.S. Courts reports that per capita supervision costs including separate average cost figures for drug/alcohol aftercare during FY 1989 were:

Supervision	
Daily:	$3.20
Monthly:	$96.66
Annually:	$1,160.00

Supervision costs are calculated based on a probation officer's average salary at the JSP 12/3 level and a clerical average salary at the JSP 7/3 level during FY 1989. In addition to average salaries, the cost figures also include annual costs associated with benefits, travel, telephones, supplies, equipment, furniture and facilities.

Average Drug/Alcohol Aftercare Costs	
Daily:	$4.76
Monthly:	$142.91
Annually:	$1,715.00

These figures are based upon annual expenditures for contract aftercare services divided by the number of clients receiving services. Costs for drug/alcohol aftercare are calculated independently from supervision costs. The average cost for an offender under supervision with aftercare treatment would thereby be $7.96 ($3.20 + $4.76).

Source: Donald L. Chamlee, "Memorandum to All Chief Probation Officers," Administrative Office of the United States Courts, 30 April 1990, pp. 1–2.

prison costs. The combined per capita community supervision and drug/alcohol treatment costs were 17 percent of what prison costs would be.

Split Sentences

In some jurisdictions probation is combined with a short period of incarceration, often known as a **split sentence.** There is no standard name for this practice which is referred to by various terms including shock probation, boot camp, split sentence, and shock parole. Such procedures permit judges to incarcerate offenders for a short period of time (usually thirty to sixty days) and then place them on probation. The term **shock probation** is inaccurate since probation does not involve incarceration. It is more technically correct to call this shock incarceration. Indications are that about 6 percent are placed on probation in this manner.[26]

A major intent of this program is to shock offenders who have never been incarcerated. It is suggested that this experience could dramatize the harsh realities of prison life and impress offenders with the seriousness of their law violations.

This practice has great public and media appeal, and some benefits have been claimed, but there is no evidence of long-term advantage for split sentencing.[27] The practice itself has generated some long-standing criticism, since the aim of probation has been viewed as avoiding incarceration in an attempt to maintain persons in the community. With this purpose in mind, the National Advisory Commission argued that the use of shock incarceration should be discontinued. The reasoning presented was that:

> This type of sentence defeats the purpose of probation, which is the earliest possible reintegration of the offender into the community. Short-term commitment subjects the probationer to the destructive effects of institutionalization, disrupts his life in the community, and stigmatizes him for having been in jail. Further, it may add to his confusion as to his status.[28]

Summary

Probation is a sentencing alternative utilized for juveniles and adults. Reintegration is stressed in lieu of incarceration. Probation should not be confused with parole, which occurs *after* imprisonment.

Some aspects of probation have been traced to common law practices such as benefit of clergy and judicial reprieve. In the United States, Massachusetts led the way in authorizing the use of probation. The volunteer efforts of John Augustus and the establishment of juvenile courts did much to aid the development and acceptance of probation.

Two-thirds of adults under correctional supervision are on probation. Granting probation is at the discretion of the trial court.

Presentence investigations comprise an important part of the operation of probation. Preparing an effective report is demanding and time-consuming. Reports typically contain such information as offense committed, family history, employment and education, finances, health, and sentencing recommendation. The results can determine whether one remains in the community or is imprisoned.

Probation operations are subject to some due process requirements. This has been especially noticeable at the stage of revocation. Two important cases applicable to revocation proceedings are *Gagnon v. Scarpelli* and *Morrissey v. Brewer*.

Probation officers perform investigative, helping, and enforcement roles. Service delivery may occur according to a casework or a brokerage approach. Due to the nature of their job, officers very likely have to manage role conflicts.

Such conflict is likely to be felt between the helping and enforcement roles and between the demands of the bureaucracy and needs of probationers.

Probation outcomes are often judged on the basis of recidivism. Results often vary due to different definitions of recidivism and differences in the quality of the experience. Only when researchers agree on what constitutes success will discrepant findings move toward resolution. Supervision on probation costs less than prison.

By a practice known as split sentencing, some jurisdictions combine short-term incarceration with probation for some offenders. About 6 percent enter probation in this fashion. This procedure has generated criticism and produced questionable results even though it has popular appeal.

Discussion Topics

1. Which service delivery strategy do you think could be the most effective? Explain.
2. What could be done to reduce role conflict for probation officers?
3. If you were conducting a research project how would you define and measure recidivism?
4. Split sentences are too confusing for offenders. Do you agree or disagree? Why?
5. Should efforts be made to inform the public of the cost advantages of probation as compared to prison?
6. Write up a practice PSI report to be critiqued by a classmate.

Additional Readings

Champion, Dean J. *Probation and Parole in the United States.* Columbus, OH: Merrill Publishing Company, 1990.

Cromwell, Paul F., Jr., et al. *Community Based Corrections.* 3d ed. St. Paul, MN: West Publishing Co., 1994.

Notes

1. David Dressler, *Practice and Theory of Probation and Parole.* 2d ed. (New York: Columbia University Press, 1969), 24.
2. Ibid., 24.
3. John Augustus, *John Augustus, First Probation Officer* (1852; reprint, New York: National Probation Association, 1939), 4–5.
4. People ex rel. Forsyth v. Court of Sessions 141 New York 288, 36 N.E. 386 (1894).
5. Ex parte United States, 242 United States 27–53 (1916).
6. American Bar Association Project on Standards for Criminal Justice, *Standards Relating to Probation* (Approved Draft, 1970), 27.
7. American Bar Association, *Standards Relating to Probation,* 11, 13; National Advisory Commission, *Corrections,* 186, 576; American Correctional Association, *Manual of Standards for Adult Probation and Parole Field Services* (College Park, MD: American Correctional Association, 1977), 39.
8. National Advisory Commission, *Corrections,* 188–189.
9. Dean J. Chapman, ''Felony Plea Bargaining and Probation: A Growing Judicial and Prosecutorial Dilemma,'' *Journal of Criminal Justice* (July–August 1988): 300.
10. People ex rel. Ward v. Moran, 54 Illinois 2d 552, 301 N.E. 2d 300 (1973).
11. National Advisory Commission, *Corrections,* 158.
12. American Bar Association, *Standards Relating to Probation,* 9.
13. Ibid., 45.
14. Mempa v. Rhay, 389 United States 128, 88 S. Ct. 254, 19 L Ed 2d 336 (1966).
15. Gagnon v. Scarpelli, 411 United States 778, 93 S. Ct. 1756, 36 L Ed 2d 655 (1973).

16. Morrissey v. Brewer, 408 United States 471, 92 S. Ct. 2593, 33 L Ed 2d 484 (1972).
17. West Texas Regional Adult Probation Department, *Community Resource Management Team for Adult Offenders* (El Paso, TX: West Texas Regional Adult Probation Department, January 1977), 6.
18. Neal Shover and Werner Einstadter, *Analyzing American Corrections* (Belmont, CA: Wadsworth Publishing Company, 1988), 130.
19. Franklin H. Marshall and Gennaro F. Vito, ''Not Without the Tools: The Task of Probation in the Eighties,'' *Federal Probation* 46, no. 4 (December 1982): pp. 37–40; and Stephen Gettinger, ''Separating the Cop from the Counselor,'' *Corrections Magazine*, 7 (April 1981): pp. 34–41.
20. United States Department of Justice, *Critical Issues in Adult Probation: Issues in Probation Management* (Washington, D.C.: Government Printing Office, 1979), 57–58.
21. Joan Petersilia et al., *Granting Felons Probation* (Santa Monica, CA: Rand Corporation, 1985), 20.
22. Bureau of Justice Statistics, *Correctional Populations in the United States, 1986*. (Washington, D.C.: Government Printing Office, February, 1989), p. 20.
23. Joan Petersilia, ''Probation and Felony Offenders,'' *Federal Probation* (June 1987): 56–61.
24. Michael Eisenberg and Gregory Markley, ''Something Works in Community Supervision,'' *Federal Probation* (December 1987): 28–32.
25. Shover and Einstadter, *Analyzing American Corrections*, 125.
26. Bureau of Justice Statistics, *Correctional Populations in the United States*, 1986 (Washington, D.C.: Government Printing Office, February, 1989), 20.
27. Dale K. Sechrest, ''Prison 'Boot Camps' Do Not Measure Up,'' *Federal Probation* (September 1989): 15–20.
28. National Advisory Commission, *Corrections*, 321.

Chapter 12

Parole

Learning Objectives

After studying this chapter students should be able to:

1. Define parole.
2. Distinguish between parole authority and parole agency.
3. Identify and describe release categories.
4. Recall the historical origins of practices that lead to parole.
5. Describe release eligibility.
6. Describe the nature of parole board hearings.
7. Discuss various problems and challenges faced by offenders.
8. Explain typical conditions of parole.
9. Identify and describe tasks parole officers perform in working with released offenders.
10. Identify due process requirements for parole revocation hearings.
11. Recognize divergent parole outcomes.
12. Explain reasons for and give examples of contemporary changes in parole.

Key Terms

Amnesty
Commutation
Conditional release
Conditions of parole
Determinate sentencing
Discretionary release
Disparity in sentences
Early parole
Elmira Reformatory
Emergency release
Enforcing
Executive clemency
Good time
Helping
Indeterminate sentence
Institutional parole officer (IPO)
Interstate compact
Interviewing
Investigating
Mandatory release
Meritorious good time
Pardon
Parole
Parole agency
Parole authority
Parole board
Parole hearing
Parole officer
Parole outcomes
Parole plan
Penal transportation
Release eligibility date
Reprieve
Revocation of parole
Risk assessment guidelines
Sentence reduction
Statutory good time
Stigma
Supervision
Technical violation
Ticket of leave
Unconditional release

Overview

Parole occurs after a period of incarceration and represents another form of correctional supervision in the community. Although parolees face the problem of returning to the community after incarceration, they are often supervised by the same officers who supervise probationers. They now find themselves in the community where they are to be monitored, often participating in some form of community-based correctional program, such as halfway houses, substance abuse programs, and job training.

Conditional releases of inmates from prisons have been granted in the United States since the 1870s. Both adult and juvenile correctional systems have mechanisms for community release. The transition from incarceration to the "free" world is often difficult for the ex-offender. The **parole officer** is the correctional worker charged with the responsibility of both easing the offender's reentry and monitoring behavior in the community.

The initial months of a community release are crucial in terms of parole success and the avoidance of a revocation. The courts have frequently intervened in the parole revocation process and other aspects of conditional release to help shape the present operation of the nation's systems. Coupled with court intervention and major change movements, parole has received major challenges and undergone alterations over the years.

Parole may be defined as the conditional release from incarceration in a correctional institution by a paroling authority after having served less than a full sentence. Parolees are then placed under the supervision of a correctional agency empowered with authority to reincarcerate upon determining that conditions of parole have been violated. Care should be taken not to confuse parole with probation, the latter occurring in place of incarceration.

Administrative Organization

There are shared responsibilities for determining when parole will be granted. Legislatures assume responsibility for establishing criminal status and penalties. The role of the judiciary is to make an initial determination on the need for incarceration and to set some limits on the length of imprisonment. A **parole authority** (a component of the executive branch of government) is assigned the task of interpreting and applying a specific length of sentence to each individual. Although simplified, this indicates the traditional shared responsibility between parole and sentencing.

Parole is a function of the executive branch with quasi-judicial responsibilities. A primary task is to oversee administration of the agency, but because parole authorities make decisions on release of individuals, they are considered to have some limited judicial functions.

Each state, the federal government, and the District of Columbia have some form of paroling authority. In forty-three jurisdictions (83 percent) authorization is granted by statute and in nine jurisdictions (17 percent)—Alaska,

Florida, Georgia, Idaho, Nebraska, Ohio, Oklahoma, South Dakota, and Texas—authorization is contained in the state constitution.[1]

Parole authorities share some commonalities but at the same time operate in many diverse ways. Examples of diversity include how parole board members are chosen, factors considered when debating granting parole, how hearings are conducted, what constitutes parole violations, and revocation considerations.

A **parole agency** is a "correctional agency which may or may not include a parole authority, of which the principal functions are the supervision of adult or juveniles placed on parole."[2]

The distinction between a parole agency and a parole authority is not always clear-cut. Typically, the parole board or body responsible for deciding whether an inmate should be released from incarceration is referred to as the parole authority. The parole agency is the administrative department within a jurisdiction that maintains responsibility for parolees.

Interstate Compact

In 1934 the United States Congress enacted a Crime Control Act. Among other provisions, this act permitted states to enter into compacts for mutual aid in fighting crime and regulating the movement of probationers and parolees.

Interstate compact members have agreed to cooperate in the exchange of parolees and probationers from another state. For example, if an inmate in Oregon indicated that Georgia was his preferred place of residence on parole, the compact system would be utilized. The Oregon compact coordinator (one exists in each state) would contact his or her counterpart in Georgia to initiate the plan. If parole was granted in Oregon, the inmate would reside in Georgia under the supervision of the local parole agency. Although it becomes bogged down in bureaucratic red tape at times, the interstate compact system has been a reasonably successful cooperative venture in an area where cooperation is often limited.

Method of Release from Prison[3]

More than 85 percent of those released from prison receive supervision in the community. Prisoners enter parole supervision either by a **discretionary release** or by fulfilling the conditions for a **mandatory release.**

In many jurisdictions the parole board has discretionary authority to release prisoners to conditional supervision in the community based on statutory or administrative determination of eligibility. Usually prisoners must serve some portion of their sentence before becoming eligible for parole.

In other jurisdictions, primarily those with determinate sentencing statutes, inmates receive a **conditional release** when they have served their original sentence minus time for good behavior or program participation; this type of release is referred to as supervised mandatory release. In both discretionary

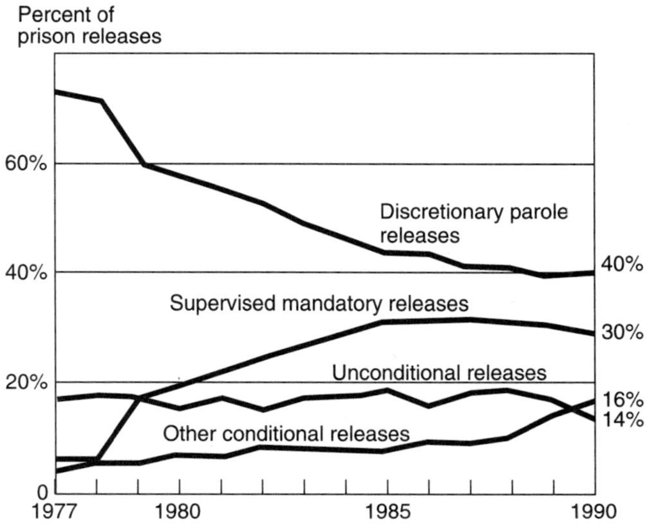

Figure 12.1 Method of prison release, 1977–1990

Source: Adapted from Bureau of Justice Statistics, Probation and Parole, 1990 *(Washington, D.C.: Government Printing Office, November, 1991), p. 6.*

release and supervised mandatory release, conditions of the release are supervised by a parole officer, and rule violations or new crimes may result in a return to prison for the balance of the unexpired sentence.

By contrast, with an **unconditional release,** an offender's obligation to serve a sentence has been fully satisfied. Expiration of term, for example, refers to a release from prison after full service of a sentence or after reductions for earned credits. No further conditional supervision in the community is required. Figure 12.1 illustrates trends in the method of releases during recent years. The percent of supervised mandatory releases from prison increased fivefold over the past 12 years, from about 6 percent of all releases in 1977 to nearly 30 percent in 1990. The number of prisoners released by a parole board decision declined from almost 72 percent of all releases in 1977 to about 41 percent in 1990. The percentage of prisoners released after their term had expired changed only slightly after 1977. Depending on the year, between 13 percent and 17 percent of all prisoners were released unconditionally after completing their sentence.

As shown in table 12.1, the total adult parole population in the United States is near 700,000. A striking feature is the great diversity in parole rate per 100,000 population.

Historical Perspective

The origins of conditional release can be traced back to early acts of leniency. These attempts to relieve the miseries of confinement were sporadic in their occurrence and not designed with any long-term plan.

Table 12.1

Adults on Parole

Region and Jurisdiction	Parole Population, 12/31/93	Number on Parole per 100,000 Adult Residents
U.S. total	671,470	352
Federal	44,156	23
State	627,314	329
Northeast	166,806	428
Connecticut	824	25
Maine	34	4
Massachusetts	4,370	95
New Hampshire	624	74
New Jersey	35,775	598
New York	52,186	380
Pennsylvania	72,100	786
Rhode Island	538	70
Vermont	555	128
Midwest	81,042	180
Illinois	24,177	280
Indiana	2,891	68
Iowa	1,887	91
Kansas	7,141	387
Michigan	14,015	201
Minnesota	2,094	64
Missouri	13,643	352
Nebraska	815	70
North Dakota	91	20
Ohio	6,997	85
South Dakota	676	133
Wisconsin	6,615	179
South	257,188	388
Alabama	7,284	234
Arkansas	3,940	220
Delaware	914	174
District of Columbia	6,591	1,424
Florida	17,326	165
Georgia	20,790	410
Kentucky	4,144	147
Louisiana	14,463	474
Maryland	13,858	372
Mississippi	2,003	106
North Carolina	17,284	330
Oklahoma	2,503	106
South Carolina	5,671	211
Tennessee	11,819	309
Texas	116,021	903
Virginia	11,504	235
West Virginia	1,073	77

Table 12.1 (continued)

Region and Jurisdiction	*Parole Population, 12/31/93*	*Number on Parole per 100,000 Adult Residents*
West	122,278	301
Alaska	685	167
Arizona	4,017	140
California	87,018	385
Colorado	2,731	104
Hawaii	1,604	184
Idaho	837	109
Montana	708	117
Nevada	3,398	328
New Mexico	1,281	113
Oregon	13,687	608
Utah	2,213	185
Washington	3,720	96
Wyoming	379	114

English Experience

The English system of **penal transportation** was the first systematic and sustained form of conditional release. As has often been the case in correctional change, the initial goals of penal transportation were directed not at benefiting the offender but at addressing current governmental or societal needs. Under penal transportation, convicted individuals were sent to English colonial areas such as the thirteen colonies, Canada, and Australia. Here they worked in penal camps or were sold into indentured servitude. Although harsh, under both arrangements offenders were provided opportunity to gain pardons with special conditions after a period of service.

Alexander Maconochie

It was at one of the penal colonies on Norfolk Island far off the coast of Australia in the 1840s that Alexander Maconochie developed a forerunner of modern parole. He was a naval officer assigned by the English government to an administrative position at the Norfolk Penal Colony. Maconochie was critical of sentencing that did not provide some reward if a prisoner changed to accept more responsibility. In an effort to provide incentive, he developed a five-step procedure he called a ''mark system.'' Prisoners could progress through these stages eventually earning full freedom:

1. Strict custody
2. Labor in government work gangs
3. Limited areas of freedom

4. Tickets of leave for conditional release
5. Full freedom[4]

The **ticket of leave** (first instituted by the English in 1790), one component of a mark system, evolved into parole. These tickets of leave were conditional pardons granted to imprisoned offenders. Maconochie's changes were not well accepted, and he was relieved of his duties. However, his ideas survived and were advanced by Sir Walter Crofton.

Sir Walter Crofton

In the country of Ireland, Sir Walter Crofton further refined Maconochie's ideas, developing the ticket of leave into three stages of penal servitude; strict imprisonment, indeterminate sentence, and ticket of leave. Upon release from prison, Crofton's Irish system called for community supervision. In urban areas in and around Dublin a position of Inspector of Released Prisoners was created to carry out supervision. The first inspector was James P. Organ, a civilian who became a prototype for modern-day parole officers.[5]

In rural areas the task of supervision was assigned to local police forces, thus establishing a dichotomy that still exists in parole between surveillance and a helping aspect of supervision.

In 1853 the English Parliament enacted a Penal Servitude Act that gave legal status to the ticket of leave system. This act also established minimum time to be served prior to eligibility for release. Legislatures and correctional officials have continued to grapple with similar issues ever since.

Experience in the United States

Howard Abadinsky, discussing the origins of parole release in the United States, cites three primary factors that lead to the implementation of the form of conditional release:

1. The decline of prison industries
2. The expense of constructing fortress prisons
3. Prison escapes and overcrowding[6]

The use of conditional release in the United States began in the 1870s, although groups such as the Philadelphia Society for the Alleviation of Miseries of Prisons had provided for community supervision in the 1700s. Correctional officials and reformers read with interest about the developments in England. Then in 1870 in Cincinnati, Ohio, Sir Walter Crofton was a featured speaker before a group meeting which would become the forerunner of the American Correctional Association.[7] His appearance helped promote the use of conditional release in the United States.

Zebulon R. Brockway

The individual credited with popularizing conditional release in the United States was Zebulon R. Brockway. While working at the Detroit House of Detention, Brockway experimented with a number of innovative programs, but it was not until 1876 that his efforts gained widespread attention. In that year, New York State opened a new correctional facility at Elmira under the direction of Brockway. The **Elmira Reformatory** became the model for United States prisons, and by 1901 twenty-five states had established similar facilities and parole programs.[8]

Brockway's program at Elmira incorporated many aspects of the English system and established some unique features. A reformatory system was designed for young (thirty years or less), first-time offenders. Programs at Elmira were structured on a graded system with heavy emphasis placed upon education, military training, and physical fitness. Individuals sent to Elmira were given indeterminate sentences with zero fixed minimum and five years maximum term. It was left to prison officials to determine when to release an individual.

Upon release from the institution, community supervision was provided by volunteers known as guardians. Releasees were required to report on a regular basis to their assigned guardian.

Conditional release underwent many modifications both at Elmira and in other states that adopted the model. Since establishment at Elmira, conditional release of imprisoned individuals has remained a component of corrections in the United States.

Discretionary Granting of Parole

Each year thousands are released from long-term correctional facilities in the United States. These individuals go through a variety of release processes. This section provides a framework that can be useful in understanding various components. A major purpose of the release process is to facilitate a bridge or smooth transition for the inmate in the move from incarceration to the community. Primary participants in this process include the inmate, institutional parole officer, hearing officer, parole boards, and parole officer.

An **institutional parole officer** (often referred to as the IPO) has primary responsibility for working at the correctional institution with the inmate and the parole authority. This officer is often viewed as a liaison between the two groups, who assists in the exchange of information, parole planning, and completion of the required paperwork.

Release Eligibility

Early in an individual's incarceration (typically within the first three months), a **release eligibility date** is established. This date of eligibility can be determined by a variety of means and can have vastly different meanings. Release eligibility in a parole system refers to the date, usually established by statute, when an inmate with a particular sentence first becomes eligible for consideration for release on parole. If, for example, the state's statutes require that the individual serve one-third of the original maximum sentence prior to any parole consideration, then the following formula would apply:

10 Years × 1/3 = 3 Years and 4 Months
(120 Months)

So an inmate who entered prison on July 7, 1995, would not be eligible for parole consideration until November 7, 1988.* Table 12.2 presents a typical chart for determining parole eligibility.

In some jurisdictions, a determination of a parole probability range is also made. A new prison arrival is not only informed of their fixed eligibility date but also of the date(s) when release on parole is most likely to occur. A probability range is a span of time during which the inmate may reasonably expect serious parole consideration. This range is calculated based upon discretionary paroles granted in the previous six months to inmates with similar offenses, sentences, and personal characteristics.[9]

It is important to recognize that in none of these cases does the inmate have a guaranteed parole date. The release eligibility date is only the first time parole consideration is given. Usually the offenders, particularly repeat offenders, are not released at the first parole hearing. The probability range is just that—a time period when parole release is probable but not assured. Parole is discretionary and can be granted throughout the statutorily permissible period of a sentence.

In a determinate sentencing jurisdiction, an exact release date subject only to good time provisions could be established at the time of the individual's sentencing. This date would constitute a guaranteed release time.

Good Time

Good time, which is provided in all but four states, is a major factor in establishing an individual's release date. The first good time law was passed in New York in 1817. Good time statutes were originally established as a tool for prison administration to use in controlling inmate behavior. Depending on the jurisdiction, there is a formula that awards days to be removed from an individual sentence in exchange for trouble-free days served by that inmate. For example, if a state's **statutory good time** provided for one good day for each five trouble-free days served, then the following yearly formula would apply:

365 Trouble-Free Days ÷ 5 = 73 Good Days per year

*This figure does not account for any jail served prior to prison incarceration or good time earned, which in both cases would reduce the time incarcerated in prison prior to a first hearing.

Table 12.2

Computing Parole Eligibility	
Length of Sentence	***Time to Serve before Meeting the Parole Board the First Time***
1 year to 1.9 years	4 months
2 years	5 months
2.5 years	6 months
3 years	7 months
3.5 years	8 months
4 years	10 months
4.5 years	11 months
5 years	1 year
5.5 years	1 year + 1 mo. (13 months)
6 years	1 year + 2 mo. (14 months)
7 years	1 year + 5 mo. (17 months)
8 years	1 year + 7 mo. (19 months)
9 years	1 year +10 mo. (22 months)
10 years	2 years
11 years	2 years + 2 mo. (26 months)
12 years	2 years + 5 mo. (29 months)
13 years	2 years + 7 mo. (31 months)
14 years	2 years + 10 mo. (34 months)
15 years	3 years
16 years	3 years + 2 mo. (38 months)
17 years	3 years + 5 mo. (41 months)
18 years	3 years + 7 mo. (43 months)
19 years	3 years + 10 mo. (46 months)
20 years	4 years
21 years	4 years + 2 mo. (50 months)
22 years	4 years + 5 mo. (53 months)
23 years	4 years + 7 mo. (55 months)
24 years	4 years + 10 mo. (58 months)
25 years	5 years
26 years	5 years + 2 mo. (62 months)
27 years	5 years + 5 mo. (65 months)
28 years	5 years + 7 mo. (67 months)
29 years	5 years + 10 mo. (70 months)
30 years	6 years
31 years	6 years + 2 mo. (74 months)
32 years	6 years + 5 mo. (77 months)
33 years	6 years + 7 mo. (79 months)
34 years	6 years + 10 mo. (82 months)
35 years	7 years
36 years	7 years + 2 mo. (86 months)
37 years	7 years + 5 mo. (89 months)
38 years	7 years + 7 mo. (91 months)
39 years	7 years + 10 mo. (94 months)
39.1 years to Life	8 years
Persistent Felony Offender	10 years

Source: Reprinted by permission of the Commonwealth of Kentucky Department of Corrections, Frankfort, KY.

In some jurisdictions the ratio is one trouble-free day equals one day of good time. "Characteristically, good time credits are large—between one-third and one-half of the sentence."[10]

A trouble-free day is typically defined as one where no major institutional violation occurs or when the inmate is not being held in some form of special punitive detention. Most jurisdictions provide for discretionary reduction or elimination of earned good time for certain major violations of institutional rules. Good time can typically be divided into two categories: statutory and meritorious. Statutory good time, as described in the preceding paragraph, is available to all inmates in exchange for trouble-free time served. **Meritorious good time,** available in some jurisdictions, allows for additional reductions in an inmate's sentence above and beyond statutory good time for program participation or exemplary behavior. This form of good time might be granted for heroic behavior in an emergency or outstanding work in a prison work assignment. In the past, participation in medical experiments has also been the basis for awarding meritorious good time.

In both forms of good time, the days earned are subtracted from the inmate's maximum sentence, thus reducing the time to be served and also shortening the period required before parole eligibility consideration begins. Good time is one of the few positive reward mechanisms available to the prison administration and staff to control inmate behavior. It provides a tangible reward for good behavior in prison.

It has been reported by sixty percent of paroling jurisdictions that accrual of good time credit can affect the date of the first parole hearing. Earning credits may make possible an earlier hearing date while loss of credit may move back a hearing date. In thirty-seven percent of jurisdictions, the date of the first parole hearing is not affected and Pennsylvania and Utah do not provide earning for good time.[11]

Parole Plan

In jurisdictions where there are requirements for administrative approval of the release, some steps are typically taken prior to the parole hearing. The first of these is to develop a **parole plan.**

A few months prior to an inmate's fixed eligibility date, a preparole interview is conducted. This interview conducted by the institutional parole officer begins the parole release consideration process. Typically, an inmate has already become fully informed as to the various parole procedures through discussions with other inmates or correctional staff members. Nevertheless, the officer will briefly review the steps and answer any questions an inmate may have. Inmates are given a parole planning form to be completed and returned to the officer prior to any further consideration.

In some cases the officer or other institutional staff members have the opportunity to provide valuable guidance to inmates who are attempting to formulate a parole plan. This is a time when an individual is faced with the reality of returning to the free community and making sometimes difficult decisions about his or her future. In some cases even the hardest of old cons can be reached by a correctional staff member during this phase of incarceration.

The two most important aspects of the parole plan are a place of residence and employment. Both of these factors require contacts on the streets, which may be limited or unavailable to some inmates who have been incarcerated for many years.

When jurisdictions were asked to rank the amount of influence of (1) parole plan, (2) overcrowding, and (3) community resources had on a release decision, a parole plan was ranked as most important.[12]

Preliminary Hearing

In many jurisdictions, hearing officers working for the parole authority complete a preliminary review or hearing of a pending case. A preliminary review determines if all required documents are available for consideration by the parole board. A hearing officer may utilize a rating scale or other type of evaluation instrument to determine the probability of parole. These instruments are generally developed through a statistical analysis of past parole board decisions in similar cases.

A hearing officer will report findings to the parole board. Typically this report includes a recommendation on whether or not to grant parole. In some states, a hearing officer's favorable recommendation for parole can be acted upon by the parole board without holding a formal hearing. In a few states there are no requirements that a parole hearing be held in the presence of the inmate. Generally any denial of parole must be preceded by a formal hearing. The use of hearing officers and preliminary parole reviews has reduced parole board work loads and proven cost-effective in some states.

Parole Boards and Hearing

Parole boards may be described by a variety of characteristics. Some of these are as follows:

- Most members are appointed by the Governor
- Nearly seventy-five percent are confirmed by the House of Representatives and/or Senate
- Most boards have five or seven members
- Most states require some criminal justice or related experience
- The most common term lengths are four, five, and six years
- Twenty-five percent of members are females
- Sixty-six percent of members are white
- Twenty-four percent of members are black[13]

A **parole hearing** is a formal, quasi-legal proceeding where an inmate appears before the parole agency representatives empowered with authority to make a determination regarding release from incarceration. Different jurisdictions have varying procedures, but commonly a hearing does not include all parole board members. Generally a panel of two or three members hears a case.

Members have an opportunity to review an inmate's institutional file, parole plan and institutional parole officer's report, the hearing officer's report, and any supporting information submitted by the inmate. In reality, many parole board members have very little time to review the abundance of documentation that accompanies each case. Most parole hearings are completed in less than fifteen minutes. The parole hearings and the subsequent decision-making processes have angered many inmates and correctional officials. Major controversy centered around alleged arbitrariness of decisions. Such controversy led to modifications and even elimination in some jurisdictions.

There is an old phrase among inmates regarding parole board decisions: crime and time. When they say crime and time, they are referring to the parole board's tendency to base their decision on the particular crime committed and the time or length of sentence. This phrase is certainly reflective of the basis for many parole board decisions, particularly in first-appearance cases. Decisions made near the end of an individual's sentence are greatly influenced by the parole plan.

Discretionary parole decisions are based on a variety of variables involving the offense that resulted with incarceration, as well as previous events and other factors. Figure 12.2 presents recent research results showing the relative importance paroling jurisdictions attached to a variety of factors in the parole decision process.[14]

The scoring process ranged from a low of zero, meaning the factor was not considered at all, to two hundred and eight, meaning it was considered very important. As can be seen, the results were very clustered with the nature (presumably seriousness) or the present offense, ranking as most important, but by a tiny margin ahead of a prior violent crime and a prior felony record. Following closely in descending order were use of a firearm, previous parole adjustment problem, prison disciplinary record, psychological report, and victim input. Future researchers should build on these data with a multivariate analysis to determine the interrelatedness between these factors as well as the possible impact of other factors such as adequacy of parole plan, community resources, and likelihood of benefitting from treatment in a community setting.

There is no constitutional right to have counsel at parole hearings. Each state must establish its own policy on this issue. The inmate is generally offered an opportunity to make a statement in their own behalf.

There are three basic outcomes that could result from the parole board hearing: parole, deferral, or continuance. Granting parole is a tentative approval for release pending a parole plan investigation. A deferral is a temporary delay in making a decision to acquire additional information (such as a psychological report, community investigation, or other supporting documentation). A continuance, commonly referred to as a flop, is a denial of parole and the establishment of a future hearing date. The continuance can be of any length

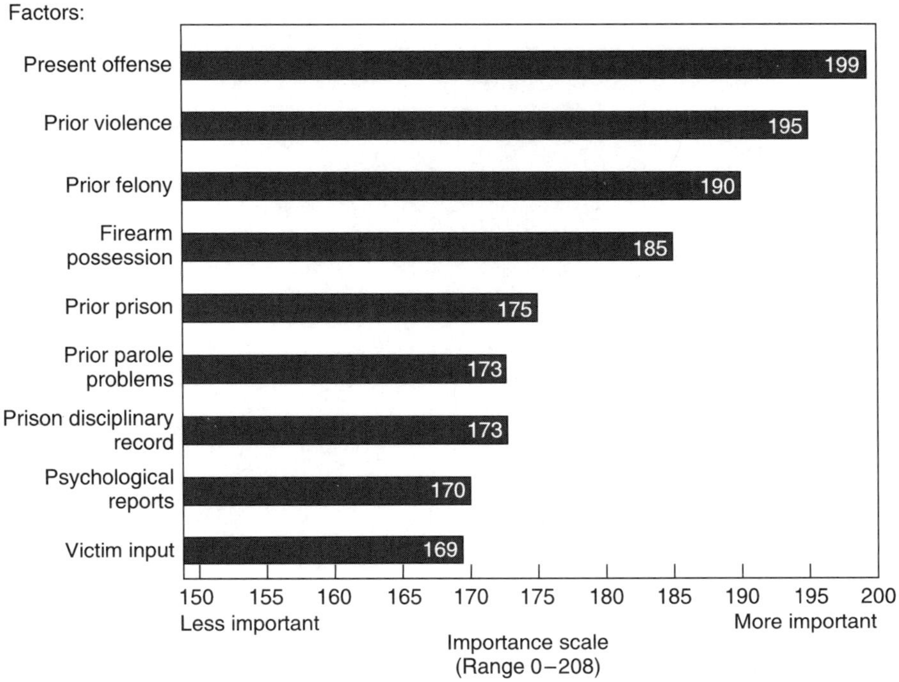

Figure 12.2 Relative importance of selected factors in the parole decision process

Source: From The Practice of Parole Boards, *Copyright © 1994 Association of Paroling Authorities International.*

depending upon the local policy and state laws. Generally, a continuance is not for less than six months or more than five years. A favorable recommendation for parole is followed by a community investigation to verify the parole plan. A local parole officer in the jurisdiction where an inmate plans to live typically conducts the investigation. If the investigation uncovers no discrepancies or other problems, then release on parole follows.

Court Impact on Granting Parole

In *Greenholtz v. Inmates of Nebraska Penal and Correctional Complex,*[15] Chief Justice Berger concluded "that prison inmates have no constitutional or inherent right to be conditionally released prior to the expiration of their sentence."[16] The release hearing is a discretional function of the local jurisdiction and has thus not been subject to the same due process requirements for a formal legal hearing.

The inmate being considered for release is entitled to a written notice of the reason for denial of parole (*King v. United States*).[17] Often these notices are designed along a checklist format with a standard reason for denial and space provided for comments relating to the particular case in question.

The courts have not granted automatic reviews of parole board decisions. The responsibility for review of decisions relating to conditional release has

Parolees often face many adjustments and uncertain futures when returning home.

been left with the administrative agency that oversees the process. Court review generally has been provided when there has been evidence of arbitrary and capricious actions or abuses of authority.

An inmate should receive a written reason of the board's decision and in many cases is told the outcome of a hearing by the panel the same day he or she appears. Parole decisions can be appealed through administrative channels. Most jurisdictions have a standard procedure that leads eventually to a full parole board hearing to review the case. Because parole decisions are discretionary in nature, challenges in court have generally not proven successful, unless an inmate can prove that existing laws or policies have been violated in the hearing process.

Return to the Community

From the offender's perspective, parole and other conditional releases represent more than just legal restriction since this helps to define his or her social status. There is clearly a **stigma** attached to being a parolee. The label identifies an individual as a ''loser,'' one who must be watched carefully and not be trusted. Not every individual released from prison encounters such a negative reaction, but a stigma may remain in both the minds of the public and the parolee.

The support of family, friends, and other significant individuals in the community is also a crucial factor in parole success. The individual who returns to the community without established ties is faced with yet another problem. Everyone needs the support and caring of others, but often the parolee must face a new life without these key resources. Maintaining regular community ties with significant others while incarcerated may be one of the most important activities in which an inmate can become involved.

The stresses and challenges facing the parolee in returning to the community are numerous. The individual's self-image and attitude in many cases help him or her to make it past those first months of parole. Numerous studies have indicated that the highest rates of parole revocation occur during the first six months after release.[18] Parolees may find it especially difficult to adjust to a wife and family upon release from prison. After an absence of several years, a family has probably become more independent and learned to function without the incarcerated individual.

During these early months, many adjustments must be made by the parolee. Typically during this period, the individual would be under maximum supervision and have the greatest contact with the parole officer. At this stage in the parolee process, a PO can provide the support and encouragement along with supervision that may make the difference for some parolees.

Employment and Parole

Many parole systems require that an individual have a promise of or an actual job prior to release. These initial jobs are often manual labor or only token forms of employment. It has been reported that the jobs secured while incarcerated are not crucial to parole success, but that employment in general is very important.

The issue of employment and the first job upon release from prison were the focus of a ten-year follow-up study.[19] The study follows the lives of sixty-five parolees who entered jobs with the APEX Trust upon release from prison. Among the findings of the study were that one-third of the parolees were reincarcerated within one year of release. An additional one-third of the releasees were reconvicted by the criminal courts some time during the ten-year follow-up period.

Nine individuals maintained their employment with APEX in excess of one year. Interestingly, none of these nine were reconvicted during the ten years of follow-up. The conclusion drawn by the researchers was that finding "just any job" for a parolee immediately upon release did not insulate that individual from the likelihood of reconvictions. The researchers determined that the pressing task was to find work commensurate with the offender's skills.

There appears to be a strong positive correlation between acquiring and maintaining a job and successful completion of parole. If an individual is to "make it" in the community, he or she must have a means of support. The most common legal means of support is a job. Without employment, the parolee's chances of living a crime-free life are greatly diminished.

Conditions of Parole

Conditions of parole established for parolees' behavior upon release vary greatly among jurisdictions. Figure 12.3 is an example of one set of standard conditions. Many states have considerably more conditions. The special

Figure 12.3 Statement of parole agreement

Source: Courtesy State of Ohio Department of Rehabilitation and Correction.

STATE OF OHIO
Department of Rehabilitation and Correction
ADULT PAROLE AUTHORITY
STATEMENT OF PAROLE AGREEMENT

The Members of the Parole Board have agreed that you have earned the opportunity of parole and eventually a final release from your present conviction. The Parole Board is therefore ordering a Parole Release in your case.

Parole Status has a two-fold meaning: One is a trust status in which the Parole Board accepts your word you will do your best to abide by the Conditions of Parole that are set down in your case; the other, by Ohio law, means the Adult Parole Authority has the legal duty to enforce the Conditions of Parole even to the extent of arrest and return to the institution should that become necessary.

The following Conditions of Parole are in effect in your Parole Release:

1. Upon release from the institution, report as instructed to your Parole Officer (or any other person designated) and thereafter report as often as directed.
2. Secure written permission of the Adult Parole Authority before leaving the State of Ohio.
3. Obey all municipal ordinances, state and federal laws, and at all times conduct yourself as a responsible law-abiding citizen.
4. Never purchase, own, possess, use or have under your control, a deadly weapon or firearm.
5. Follow all instructions given you by your Parole Officer or other officials of the Adult Parole Authority and abide by any special conditions imposed by the Adult Parole Authority.
6. If you feel any of the Conditions, or instructions are causing problems, you may request a meeting with your parole officer's supervisor. The request stating your reasons for the conference should be in writing when possible.
7. Special Conditions:

I have read, or have had read to me, the foregoing Conditions of my Parole. I fully understand them and I agree to observe and abide by my Parole Conditions.

Witness: ______________________ Parole Candidate: ____________

Date: ____________

conditions, item 7, may be any individualized stipulation established by the parole authority. A typical special condition might be a requirement that the parolee seek assistance in dealing with a substance abuse problem. Each parolee must agree to the conditions detailed in the parole agreement prior to being released.

There was a time when parole authorities were given almost unlimited discretion to establish conditions of parole. With the advent of increased civil rights for prisoners and parolees, limits have been placed on the range of conditions of parole. A general rule is that conditions of parole must be reasonable and have some justifiable application to an individual situation. In most, conditions of parole that limit free speech, require church attendance, or mandate surgery are considered unreasonable by present standards.

One typically unwritten condition of parole relates to the parolee's protection against search and seizure. The courts have ruled that a search that would be unlawful if directed against an ordinary citizen may be proper if conducted against a parolee (*Santos v. New York State Board of Parole*).[20] The courts have generally required a lower standard of proof to justify warrantless searches and arrests in the case of parolees.

Final Release

Whether one examines the repeat offender or the first-time parolee, crucial factors help to determine parole outcomes. The most positive possible outcome is final release. Final release received generally in two years or less ends an individual's legal obligation for a conviction. The term *parole success* generally refers to receipt of final release and thus the termination of parole.

Parole Officer Supervision

Parole officers are the representative of the criminal justice system responsible for the **supervision** of conditionally released offenders. Supervision is the broad term used to describe the various tasks that a parole officer is required to perform. The tools available to parole officers (POs) include the legal authority of the office, communication and counseling skills, community contacts, and other resources.

A relationship that develops between a parolee and the PO is largely dependent upon two factors: the attitudes of the parolee and the parole officer. There are different styles and approaches taken by POs. The same range of differences can be found among parolees. A relationship can range from close and trusting to hostile and distant. From the parolee's perspective, the type of relationship that develops can be important in determining parole success. Tasks of officers may be divided into four basic categories: interviewing, investigating, enforcing, and helping. Each category of task is present in most parole officers' assignments, although the relative importance varies greatly by jurisdiction and among personnel. The same officers often supervise both probationers and parolees as discussed in chapter 11.

Interviewing

A primary communication tool between releasees and POs is **interviewing.** Interviews may be described as a formal discussion with a purpose. Officers conduct interviews not only with the releasee but also with many other individuals, including the family of the client, neighbors, employers, and representatives of various community agencies.

The first interview may occur before the inmate is released from incarceration, but more frequently it will occur after release. In many jurisdictions it is common practice to require the new releasee to see his or her parole officer within twenty-four hours of leaving the institution.

Typically, parolees and officers meet for regularly scheduled interviews throughout the supervision period. Although interview styles differ greatly among officers, the initial interview and the impressions established at the time are crucial to the development of the relationship. Both the officer and releasee are checking each other out and making initial decisions about how to approach the relationship.

Interviews have a purpose, and in the parole setting, that purpose generally centers around the supervision plan. The supervision plan is the formal or informal guideline for the individual's behavior during the supervised period. The parole officer may be seeking information to establish the supervision plan, to determine if the releasee is abiding by the plan, or to assist the individual in achieving one of the objectives of the plan. The level of supervision varies greatly between and among releasees. Differential levels of supervision are generally associated with such factors as the length of parole, individual's characteristics, and the PO's caseload. Interviews and the parole officer's skills in conducting effective interviews are crucial elements of supervision.

Investigating

Investigations by parole officers are conducted prior to and throughout a supervision period. **Investigating** is nothing more than a careful examination of a particular situation or incident.

The first investigation in most parole cases is conducted prior to release and in some jurisdictions even prior to a parole hearing. Parole officers are called upon to verify information contained in the inmate's parole plan, such as place of residence and potential employer. In some states, officers check with the sentencing judge and prosecuting attorney to elicit their reactions to the inmate's possible release. The information gained from such prerelease investigations is returned to the releasing authority and is retained by the local parole officer as background data in the event that release is granted.

Investigations are conducted throughout the actual supervision period to determine if the releasee is abiding by the supervision plan. In the event that a violation of the release condition occurs, a formal investigation is conducted prior to any decision regarding a change in status. The term follow-up is often synonymous with investigation. The parole officer may place a follow-up call to a community agency where the releasee had agreed to seek assistance. These follow-ups can serve two purposes: to assist the officers in monitoring the releasee and to inform the community agency that the officer is concerned about the individual. Often the follow-up, a seemingly insignificant act, can be very important in helping the parolee receive needed services.

Enforcing

Enforcing is a check of the releasee's behavior by the parole officer. In the vast majority of cases, the actual physical surveillance of a parolee by an officer does not occur, although some use this tool to a greater extent than others. Actual surveillance involves talking with others in the community and is usually prompted by some suspicion of misbehavior and is generally short-term.

Three basic outcomes that can result from the surveillance of a releasee: maintain the status quo, alter supervision, or initiate the revocation process. If the surveillance uncovers no wrongdoing, the supervision plan remains intact. If the surveillance indicates that additional supervision is needed, a change in the individual's plan might result. If the PO determines that major technical or criminal violations have occurred, the initiation of the parole revocation process could result. Limitations of time, personnel, and resources reduce the enforcement role of most parole officers, but it remains an important component of the supervision of conditional releasees.

Helping

A parole officer also provides **helping** services to the releasee and the community. Many officers view themselves as community resource agents. A community resource agent approach focuses upon referrals. Most referrals are made to various community resources, such as the employment service, health department, mental health agency, and other governmental and private agencies. Since it is often difficult to locate services within a community, a parole officer can provide a crucial link between a newly released offender and the community.

Parole officers also provide direct services in the form of individual and group counseling, assistance in securing employment, and other related areas. A variety of factors influence the parole officer's capability to provide direct services, but it is clear that many parolees have special needs that cannot be met by the individual officer. A combination of direct services and referrals is the approach taken by many parole officers. This combination of services seems most effective.

Revocation of Parole

An unsuccessful parole occurs with a **revocation of parole,** and the releasee is returned to incarceration for a parole violation. The violation of parole can be criminal or technical. A criminal violation occurs when the releasee is found guilty of breaking the law. A **technical violation** occurs when it has been determined that a parolee has violated a condition of parole that would not be considered a criminal conviction for a nonparolee. An example of a technical violation would be a parolee who is restricted from leaving the state and found to be visiting a relative in a nearby state without the permission of the parole officer. Traveling between states is not against the law for a ''free'' citizen, but it violates that condition of parole and thus is a technical violation.

A technical or criminal violation can result in a termination of parole and a return to prison. Such violations do not always result in a revocation of parole, but the option is available to the parole authority. Revoking parole is a discretionary decision, which is ultimately the responsibility of the paroling authority. Court cases in the late 1960s and early 1970s brought more uniformity to the revocation process, but diversity remains. For example, sixty-nine percent of jurisdictions consider an arrest for a felony or misdemeanor a parole violation, while twenty-nine do not; two percent consider only a felony arrest a violation. Nearly all jurisdictions consider a conviction for a felony or misdemeanor as a parole violation.[21]

The area of court intervention that has had the greatest impact on the operation of parole has been revocation. The court has established limited due process guidelines for parole revocation hearings. The court differentiated between the parole release hearing and the revocation hearing. The release hearing was viewed as a discretionary function of the correctional agency, but the revocation hearing was seen in a quasi-judicial or legal setting. The difference between the two rests on the fact that revocation of parole results in a significant deprivation of liberty, while the parole hearing was viewed as an administrative decision-making process to determine whether or not to continue a status as an inmate who had no liberty interest at stake.

Morrissey v. Brewer

The landmark case regarding revocation was *Morrissey v. Brewer*.[22] In 1967, the plaintiff Morrissey was convicted of check fraud and sentenced to seven years. He was paroled from the Iowa State Penitentiary in June of 1968. Seven months later he was arrested. His parole was revoked on the basis of information that he had violated his conditions of parole by buying a car under an assumed name and failing to report his place of residence to his parole officer.

Morrissey alleged that he had been denied due process because his parole had been revoked without a hearing. The district court for the southern district of Iowa found that the state's failure to accord a hearing prior to parole revocation did not violate due process. The court of appeals affirmed that parole revocation required no hearing. The Supreme Court reversed the appellate court's decision and laid out these due process standards for parole revocation hearings:

- A written notice of the claimed violations of parole
- A disclosure to the parolee of evidence against him
- The opportunity to be heard in person and to present witnesses and documentary evidence

Table 12.3

Manner and Proportion of Adults Leaving Parole	*Percent of Those Persons with a Known Status*
Adults leaving parole	100%
Successful completions	57
Discharged absconders	2
Discharged to detainers/warrants	2
Returned to prison	
With new sentence	11
With parole revoked	19
With revocation pending	5
With charges pending	1
Transferred to another state	2
Death	1

Source: Adapted from Bureau of Justice Statistics, Correctional Populations in the United States, 1986 *(Washington, D.C.: Government Printing Office, February, 1989), p. 80.*

- The right to confront and cross-examine adverse witnesses (unless the hearing officer specifically finds good cause for not allowing confrontation)
- A neutral and detached hearing body such as a traditional parole board whose members need not be judicial officers or lawyers
- A written statement by the fact finders as to the evidence relied on and the reasons for revoking parole

A revocation process involves two hearings. First is a preliminary hearing, which is conducted by a parole agency to determine probable cause. A second is conducted by a paroling authority to determine whether or not to revoke parole. In both, the listed due process safeguards apply.

Gagnon v. Scarpelli

The court in *Morrissey v. Brewer* did not address whether an attorney must be present, but in a latter case, *Gagnon v. Scarpelli,*[23] the court ruled on this issue. The court held that in revocation hearings a right to counsel does not exist in all cases, rather on a case-by-case basis.

Parole Outcomes

Parole outcomes shown in table 12.3 indicate the manner and proportion in which adults left parole jurisdiction during a given year. Some were successful, others were not. This data indicated a success rate of 57 percent. Other studies show varying rates of success, some higher, some lower.[24]

Table 12.4

Amount and Proportion of Time Offenders Who Successfully Completed Parole Spent Incarcerated and on Parole

	Mean Time under Correctional Supervision			
Most Serious Offense	***Total***	***Jail and Prison***	***Parole***	***Percent of Total Time Spent on Parole***
All offenses	46 months	27 months	19 months	41%
Violent offenses	60	36	24	40
Murder	115	77	38	33
Manslaughter	63	37	26	41
Kidnapping	58	35	23	40
Rape	74	48	26	35
Other sexual assault	54	34	20	37
Robbery	60	35	25	42
Assault	46	28	18	39
Other violent	41	25	16	39
Property offenses	37	21	16	43
Burglary	38	21	17	45
Larceny/theft	33	18	15	45
Motor vehicle theft	42	25	17	40
Arson	47	28	19	40
Fraud	38	22	16	42
Stolen property	34	19	15	44
Other property	33	16	17	52
Drug Offenses	42	24	18	43
Possession	34	18	16	47
Trafficking	46	27	19	41
Other drug	37	21	16	43
Public-order offenses	34	20	14	41
Weapons	43	26	17	40
Other public order	29	16	13	45
Other offenses	37	20	17	46

Note: Data on offense distribution and mean time served are based on the 41,514 successful parole releases who had entered prison with sentences of more than a year. Data include those on supervised release even if not technically termed "parole."

Source: Adapted from Stephanie Minor-Harper and Christopher A. Innes, Time Served in Prison and on Parole, 1984 *(Bureau of Justice Statistics, Washington, D.C.: Government Printing Office, December 1987), p. 7.*

Table 12.4 shows the amount of time various offenders who successfully completed parole spent incarcerated and on parole. Overall the successful parolees spent about 60 percent of their total time under correctional supervision incarcerated and 40 percent on parole in the community. Murderers served a smaller percentage of their sentences on parole, but overall, the proportion of sentences served on parole did not vary very much among offenses.

Figure 12.4 provides information concerning those who were unsuccessful on parole. The average amount of time completed on parole prior to revocation was eighteen months. Serious violent offenders, such as murderers and rapists,

Figure 12.4 Average number of months on parole prior to revocation

Source: Adapted from Stephanie Minor-Harper and Christopher A. Innes, Time Served in Prison and on Parole, 1984 *(Bureau of Justice Statistics, Washington, D.C.: Government Printing Office, December, 1987), p. 8.*

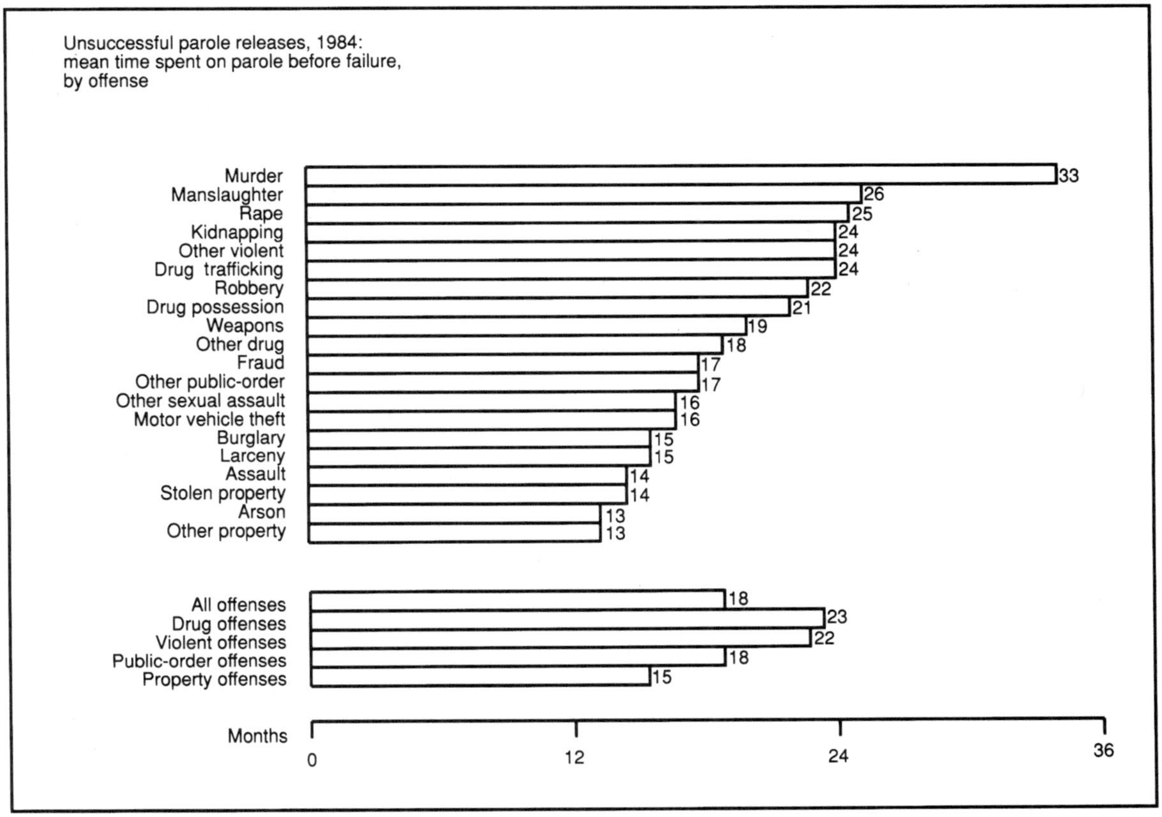

were on parole longer prior to revocation than were those convicted of such offenses as arson, stealing property, and assault.

Measuring parole outcomes raises the same issues as with probation. There is no universally agreed upon definition of recidivism. Until there is, research results will vary. Definitions of recidivism vary according to the length of follow-up time and the nature and seriousness of the offense. Random sampling and controlling extraneous variables remains a problem.

The degree of success is also linked to the selection process. Some jurisdictions may be better at determining who will be successful. Finally, in important ways, parole is a continuation of correctional supervision and likely reflects positive and negative experiences of inmates while in prison. Success or failure on parole might be linked as much or even more to a carry over of institutional experiences as to parole per se. Research to date has not sorted out this important dimension.

Changes in Parole

Over the years, efforts to change the nation's parole systems have unfolded at both the state and federal level. This was brought about by court intervention, changing public attitudes, prisoner grievances, and research findings. Perhaps the most dramatic has been the abolishing of parole boards' releasing authority at the federal level of jurisdiction. As discussed in chapter 4, a determinate sentencing guideline is in place at the federal level that provides for good time credit but not for discretionary release by a parole board. Originally the role of parole authorities was to "determine whether a person sentenced to a term of imprisonment could be released from prison before expiration of the term without undue risk that the offender would commit further criminal offenses."[25] With time and ample opportunities, it became apparent that parole boards could not or had not been accurately predicting recidivism or rehabilitation. As the limitations of the parole board's predictive powers became clear, there was a shift in emphasis from estimates of likelihood of recidivism to alleviation of **disparity in sentences** and control of prison population.

Coupled with this changing emphasis was a growing body of research that indicated parole supervision had a questionable impact upon recidivism. These and other factors led to a reappraisal of parole and many calls for change. One point of agreement among all parties in the debate over parole was that similar offenders who shared similar sets of circumstances should serve similar prison sentences. The arguments that emerged over parole focused on how to develop a more equitable system. As a result, about a dozen states and the federal government have abolished parole boards, more determinate sentencing legislation has been enacted, and various forms of parole guidelines have been established.

At the same time, it should be made clear that the federal government has not given up on the concept of community supervision for released inmates. Although prisoners receive no shortening of sentence through a parole board's action, most are subject to supervised release in the community, just like parole supervision under jurisdiction of the sentencing judge.

Determinate Sentencing

Determinate sentencing can be viewed as one attempt to change the parole system. Parole is premised upon the **indeterminate sentence,** which allows the correctional system to establish the actual date of release. Critics who advocated the abolishment of parole tend to favor establishment of a determinate sentencing system.

The determinate sentence or "flat time" is set by the sentencing judge based on legislated guidelines. The offender is given a definite sentence and, once calculations for good time provisions are made, has a fixed release date. In most jurisdictions, the inmate knows how long he or she must serve from the day the sentence is handed down.

Thirty-eight states or about 75 percent of parole boards currently maintain traditional discretionary release authority. Fourteen other jurisdictions operate under determinate sentencing or some form of restriction on discretionary release power by the parole board. These are: The federal government, Alabama, Delaware, Illinois, Indiana, Maine, Minnesota, Mississippi, New Mexico, North Dakota, Oregon, Rhode Island, Vermont, and Washington.[26]

Parole Guidelines: Risk Assessment

A principal change has been changes made in **risk assessment guidelines** by which to make parole decisions. A goal is to increase fairness and reduce disparity. Some guidelines seek to develop predictions on a statistical basis. That is, given a certain set of risk factors, what is the probability for success on parole? The severity of criminal history is one major factor taken into consideration. Second, risk factors such as early crime, age at present offense, drug use, and early incarceration are also evaluated.

These changes themselves have generated some criticism and brought forward some problems. For example, some of the statistical prediction techniques are complex and sometimes parole board members do not have sufficient training necessary for effective application. Others have questioned the value of statistical risk assessment in attempting to make a prediction for a given individual. Statistical prediction works better at the group level than at the individual level. This was described so very well some years ago by Stanley in pointing out that:

> A parole board may know that of 100 offenders with a certain set of characteristics, 80 will probably succeed and 20 will fail on parole. But the board members do not know whether the person who is before them belongs with the 80 or the 20.[27]

Nevertheless, guidelines have probably reduced arbitrary decision making. Developing and applying guidelines have media appeal and are popular with government officials and the public. Some form of risk assessment instruments are used by about 50 percent of parole boards.[28]

Maintaining Parole

It has been stated that:

> Parole in the United States is experiencing a major transition the outcome of which is not yet clear. The incessant politicization of crime control policy and the pronounced loss of faith in the capacity of the correctional system to change criminal behavior have fostered a growing reliance on mandatory sentencing provisions, sentencing, enhancements, and other punitive measures, all under the rubric of truth-in-sentencing. This trend poses a serious challenge to parole.[29]

Some states have chosen to maintain their system of parole with only minor modifications. These states have retained the indeterminate sentence and have shunned parole board guidelines that limit discretion. Parole reform legislation is pending in many states that have not changed the parole system.

Support for maintaining parole and the indeterminate sentence can be found at two levels: individual and systemwide. Many supporters of parole point to individual cases where sentencing disparity has resulted in inequitable treatment. Another view is that parole provides the opportunity to recognize and reward significant changes in individual behavior and attitude. To some, parole provides both a punishment and a reward system, rather than the exclusively negative sanctions available under determinate sentencing provisions.

The systemwide level of support for parole concerns itself with the need for a prison population control mechanism (a safety valve). Parole has long been used as leverage by prison personnel to maintain order in correctional institutions. There is concern among some institutional workers about the loss of this control tool.

Special Forms of Early Release

There are several special forms of release in addition to those already discussed. Because prison overcrowding is a major problem many states have authorized release of inmates earlier than usual to control prison populations.[30] Three types of early release are generally designated as

- emergency release;
- sentence reduction; and
- early parole.

Emergency release permits the release of inmates who are approaching the end of their sentences. Typically, those within four to five months of discharge may be released in states having this provision.

Some states use **sentence reductions** to lower prison populations. This approach usually requires a formal declaration that the prison system is above its authorized capacity. If this occurs, sentences of selected inmates (such as first offenders and nonviolent offenders) may be reduced by up to ninety days.

Other states allow **early parole** by advancing release dates to be advanced for certain categories of offenders when prisons are overcrowded. Such programs may involve a period of more intensive supervision by a parole officer or participation in special community-based correctional programs.

Executive Clemency

As has already been pointed out, parole is a function of the executive branch of government. Though seldom used, the executive branch also has clemency powers that can be utilized to release incarcerated individuals. Provisions for the use of **executive clemency** differ among jurisdictions. Some states have a special board or commission to review and recommend the use of these executive privileges. Other states utilize the existing parole authority to make recommendations to the chief executive. In some states there is no established format for the use of executive clemency, and decisions rest exclusively with the chief executive. The following paragraphs detail common forms of executive clemency.

Pardon

A **pardon** can be granted to an individual by a chief executive. It exonerates the person from blame and protects him or her from future criminal prosecution for the acts in question. A pardon can be granted at any time in the criminal justice process (pretrial or posttrial). Typically, a pardon is granted after conviction and frequently after the individual has served his sentence.

Amnesty

Amnesty is a form of pardon but applies to a group or class of individuals rather than a single person. It has been common practice in the United States to grant amnesty to soldiers who deserted, once the war has ended. Amnesty precludes prosecution.

Reprieve

A **reprieve** is a temporary stay of execution of a sentence. The use of reprieves was common prior to the 1960s when there was extensive use of capital punishment. The granting of a reprieve provided additional time to seek further court reprieve or reconsideration. With an increasing number of states reintroducing capital punishment provisions, the reprieve may soon return to common use.

Commutation

A final form of executive clemency to be discussed is **commutation.** Commutation provides for a modification of sentence to the benefit of the offender. A typical example of the use of a commutation would be in the case of an individual sentenced to a life in prison. A commutation could be granted reducing the sentence from a straight life term to a maximum term of life. This change in sentence would make the inmate eligible for parole and possible release.

In some jurisdictions commutations are regularly granted to individuals who have served long sentences, thus opening up the possibility of release. In some states, commutations are only given in recognition of meritorious service. Commutation and other forms of executive clemency can provide additional safeguards and protection against harsh and excessive sentences.

Summary

Parole is one major form of correctional supervision that takes place in the community. Individuals are placed on parole after serving some time incarcerated.

About 10 percent of all persons under correctional supervision are on parole. Offenders may be placed on parole as a result of a discretionary release granted by a parole board decision or by a mandatory release based on earned good time credit. During recent years, discretionary release of parole boards has declined due in large part to the increased use of determinate sentencing. However, during the same time, the parole population has increased because of major growth in the prison population.

Present-day parole evolved from a series of earlier practices and individual efforts. Notable among these was the ticket of leave first instituted by Alexander Maconochie at the Norfolk Penal Colony, Sir Walter Crofton's development of community supervision, and Zebulon Brockway's programs instituted at the Elmira Reformatory.

Granting parole on a discretionary basis involves a series of processes and several individuals. Often inmates are assisted by an institutional parole officer in developing a parole plan. An inmate's records are reviewed by a parole board, and a decision made on whether or not to grant parole.

Offenders paroled for community supervision face a variety of adjustments. They most certainly will encounter some degree of negative labeling. They need employment and must work to develop and maintain relationships with family and other community members.

Parole officers have broad responsibilities supervising parolees. Their jobs call for a variety of tasks in the categories of interviewing, investigating, enforcing, and helping.

Parole may be revoked by breaking the law again or by committing a technical violation. Courts have established some limited due process guidelines for revocation hearings. Significant cases that had an impact were *Morrissey v. Brewer* and *Gagnon v. Scarpelli.*

Some leave parole successfully while others do not. Measuring the success of parole raises a number of complex issues such as the way recidivism is defined and the manner in which prior life experiences and prison experiences may be linked to parole outcomes.

Discretionary release of parole boards has resulted in considerable disparity in sentences. In efforts to develop a more equitable system, some parole boards have been abolished, more determinate sentencing legislation has been passed, and various forms of parole guidelines have been instituted.

Discussion Topics

1. Do you favor discretionary or mandatory release of inmates? Explain.
2. Parole board members should be selected based upon qualifications. Do you agree or disagree?
3. How might a community offer help to a new parolee in an effort to reduce stigma?

4. How important do you feel employment is for parolees?
5. In your view, what is most important for a parole officer to try to accomplish with a parolee?
6. Do you think abolishing parole boards can result in less disparity in sentences? Why or why not?

Additional Readings

Abadinsky, Howard. *Probation and Parole.* 5th ed. Englewood Cliffs, NJ: Prentice-Hall, 1994.

Association of Paroling Authorities, International. *The Practice of Parole Boards.* Lexington, KY: Host Communications Printing, 1994.

Ellsworth, Thomas, ed. *Contemporary Community Corrections.* Prospect Heights, IL: Waveland Press, Inc., 1992.

Notes

1. Association of Paroling Authorities, International, *The Practice of Parole Boards* (Lexington, KY: Host Communications Printing, 1994), 1.
2. United States Department of Justice, *State and Local Probation and Parole Systems* (Washington, D.C.: Government Printing Office, 1979), 9.
3. This section adapted from Bureau of Justice Statistics, *Probation and Parole 1990* (Washington, D.C.: Government Printing Office, November, 1991), 5–6.
4. Rodney J. Henningsen, *Probation and Parole* (New York: Harcourt Brace Jovanovich, 1981), 55.
5. Louis P. Carney, *Probation and Parole: Legal and Social Dimensions* (New York: McGraw-Hill, 1977), 138–140.
6. Howard Abadinsky, *Probation and Parole: Theory and Practice,* 3d ed. (Englewood Cliffs, NJ: Prentice-Hall, 1987), 136–143.
7. Oscar D. Shade, ''The Demise of Wisconsin's Contract Parole Program,'' *Federal Probation 45,* no. 1 (March 1981): 34.
8. Henningsen, *Probation and Parole,* 58.
9. *Parole in Wisconsin, 1981* (Madison: Department of Health and Social Services, 1981), 7.
10. Andrew von Hirsch and Kathleen Hanrahan, ''Determinate Penalty Systems in America: An Overview,'' *Crime and Delinquency 27,* no. 3 (July 1981): 299.
11. Association of Paroling Authorities, International, *The Practice of Parole Boards,* 10–11.
12. Ibid., 12–13.
13. Ibid., 28–32.
14. Ibid., 12–13.
15. 99 S. Ct. 21000, 1979.
16. ''Fourteenth Amendment—Parole Release Determinations,'' *The Journal of Criminal Law and Criminology 70,* no. 4 (September–October, 1979): 468.
17. 492 F. 2nd 1337, 7th. Cir. 1974.
18. George F. Davis, ''Assessing Parole Violation Rates by Means of the Survivor Cohort Method,'' *Federal Probation 44,* no. 4 (December 1980): 26–32.
19. Keith Soothill and Jennifer Holmes, ''Finding Employment for Ex-Prisoners: A Ten Year Follow-up Study,'' *The Howard Journal 20* (1981): 29–36.
20. 441 F. 2nd 1216.
21. Association of Paroling Authority, International, *The Practice of Parole Boards,* 20–21.
22. 408 United States 471, 1972.
23. 411 United States 778, 1973.
24. Neal Shover and Werner Einstadter, *Analyzing American Corrections* (Belmont, CA: Wadsworth Publishing Company, 1988), 121–125.
25. Karen Skriveth, ''Abolishing Parole: Assuring Fairness and Certainty in Sentencing,'' *Hofstra Law Review* (Winter 1979): 209.

26. Association of Paroling Authority, International, *The Practice of Parole Boards,* 7.
27. David T. Stanley, *Prisoners Among Us* (Washington, D.C.: Brookings Institute, 1976), 56.
28. Association of Paroling Authority, International, *The Practice of Parole Boards,* 9.
29. Ibid. (unnumbered).
30. Bureau of Justice Statistics, *Report to the Nation on Crime and Justice,* 2d ed. (Washington, D.C.: Government Printing Office, March 1988), 109.

Unit 4

Juvenile Offenders

Society responds to juvenile offenders with a system separate from the adult system and with a different history, philosophy, and purpose. This unit provides a look at that system.

Chapter 13 Juvenile Law and Offenders
Chapter 14 Treatment for Juvenile Offenders

Chapter 13

Juvenile Law and Offenders

Learning Objectives

After studying this chapter students should be able to:

1. Identify major forces that brought change in how children were viewed and treated.
2. Recognize that a juvenile system was created for a different purpose than the adult system.
3. Understand how juvenile jurisdiction is based on age and behavior.
4. Differentiate and provide examples of public and status offenses.
5. Explain the waiver process and due process safeguards that apply.
6. Compare and contrast basic features of juvenile courts and criminal courts.
7. Summarize pathways of case flow in juvenile courts and criminal courts.
8. Identify, recall findings, and describe outcomes of several benchmark court cases applying to juvenile justice systems.
9. Describe the variety of illegal behavior among juvenile offenders.
10. Describe the measurement of delinquency by self-report and identify information gained by this method.

Key Terms

Caseflow
Diversion
Due process rights
Enforcement
Hidden delinquency
Intake function
Juvenile court
Juvenile jurisdiction
Juvenile justice system
Juvenile law
Juvenile legal terms
Parens patriae
Petition
Public offense
Self-report
Status offenses
Treatment
Waiver of jurisdiction

Background

Providing protection and treatment for children has been a long struggle. As Empey has made clear, before the modern concept of childhood emerged, children were treated with indifference.[1] They were neither loved nor thought to need love and emotional attachment. Such attitudes led to centuries of various abuses, which included such practices as abandonment and infanticide.

During the seventeenth and eighteenth centuries, social reformers began to criticize such practices and urged that children be safeguarded physically and morally and be given special attention in preparation for adulthood. Increased emphasis on law and the humanistic view of life, freedom, and the role of the family were stressed. The emergence of the nuclear family helped to promote more caring attitudes toward children.

Efforts in America were undertaken to provide social and governmental practices that were consistent with this emerging modern concept of childhood. Forces that brought change to other sectors of American society were the same ones that brought change in how children were viewed and treated. Chief among these forces for change were the following:

- Increasing industrialization, immigration, and urbanization during the 1800s and early 1900s created new problems for families and children.
- Efforts by concerned citizen groups known as the child savers resulted in the establishment of juvenile courts during the early 1900s.
- The Great Depression of the 1930s had a devastating economic effect on American families.
- Passage of the Social Security Act in 1935 marked a beginning of major federal funding for programs to aid children and families.
- The experiences of World War II during the 1940s helped increase public concerns for the well-being of children.
- Civil rights efforts during the 1960s helped broaden concerns for all children.
- The visionary "Great Society" of the 1960s further advanced causes for children and families by providing federal monies for attacking poverty and problems of crime and delinquency.

Linked with these major events has been the creation and continued development of a distinct set of laws and procedures designed to deal with juvenile offenders.

Overview of Juvenile Justice Systems

Laws in the United States have been enacted in such a manner that juvenile offenders are processed by a **juvenile justice system** rather than the adult criminal justice systems. In most instances, with some important exceptions, children through the age of seventeen are subject to the jurisdiction of the juvenile court. The juvenile system is organized and operates according to laws enacted specifically for handling juvenile offenders. Juvenile justice was created for a different purpose than the adult system, as will be made evident throughout this chapter. Creation of a juvenile system was directed toward providing treatment and services rather than punishment, which characterizes the adult system, although intense debate continues over the perceived merits of a punishment versus rehabilitation approach for juvenile offenders.

Although a distinct system, prominent features of the juvenile system share some basic similarities with the adult system. First, the system operates according to a body of law and democratic principles, although juveniles have only some of the rights that have been extended to adults. Second, public sentiment, funding resources, legislative decisions, and judicial decisions represent aspects of the larger society that impact juvenile justice. Third, the operational aspect of the system includes four major phases: enforcement, intake, courts, and corrections or treatment. Fourth, the system operates with a large amount of discretionary decision-making. Many agencies and individuals are involved in making such discretionary decisions, including police officers, intake officers, judges, and treatment personnel. Such decisions determine whether youths remain in the juvenile system and the path followed if they do continue in the system. Fifth, because juvenile systems operate at the local and state levels, operations are highly diverse and fragmented. Local communities develop particular responses to delinquent behavior. For example, some may rely heavily on early diversion programs, whereas others may emphasize court-ordered sanctions. Mindful of variations that do exist, the operation of juvenile justice has been described as: ". . . a myriad of different courts on different levels of government performing a variety of functions; . . . the structure of the courts vary significantly, both between states and within the same state."[2] Our focus in this chapter is to describe and explain the nature, purpose, and operation of this system.

Juvenile Law and Jurisdiction

Juvenile law is concerned with three major areas affecting the lives of children. One concern is illegal behavior that is in violation of the criminal law for adults and juveniles. A second addresses behaviors that are not criminal but have been deemed forbidden only for juveniles. The third area regards the problem of child neglect and dependency cases. It should be noted that traffic violations by juveniles are handled by the juvenile court in some cases, while in other cases, they are handled by the same court as adult traffic offenders.

In simplified form, **juvenile jurisdiction** with respect to juveniles is based on two elements:

Age
Behavior

Consideration of these two factors leads to construction of figure 13.1.

Figure 13.1 Corrections and the Justice System

Behavior	*Age*: Yes (Usually 17 or less)	*Age*: No (Usually 18 or more)
I. Criminal behavior	A Public offenses	D
II. Noncriminal behavior	B Status offenses	E (shaded)
III. Neglect/dependency cases	C	F

Age

If a given behavior involves a person of a particular age such as depicted by cells A, B, and C in figure 13.1, jurisdiction lies with the juvenile court. In other instances, the juvenile court would not have jurisdiction. The term juvenile is a status that is defined by age. In most states, a young person in violation of criminal law is considered within the jurisdiction of juvenile law until the age of eighteen, after which time the adult criminal court has jurisdiction. However, in a few states there exists what is known as concurrent jurisdiction. That is, the law in these states permits officials to decide whether to hear juvenile cases in juvenile court or adult criminal court.

Most state laws do not indicate a lower age limit for juvenile court jurisdiction. The common law presumption was that a child under the age of seven was incapable of criminal behavior. A few states have specified this age as a lower limit, but most have set no lower age limit.

Behavior

Some behavior engaged in by juveniles violates criminal law. Often this conduct is called a **public offense;** that is, this behavior constitutes illegal behavior for all citizens regardless of age or status. But due to age, the juvenile court has jurisdiction, rather than the adult criminal court (row I in figure 13.1). Examples of such offenses would include theft, assault, and rape.

Some behaviors by juveniles are not a violation of any criminal law. Rather, they are behaviors forbidden only to juveniles. For this reason, they are known as **status offenses** since they are based on the status of age (row II in figure 13.1). (Cell E is shaded since by definition there is no logical possibility other than at the juvenile level.) There are a substantial number of behaviors contained in this category). A few examples are

- truancy;
- disobedience (incorrigible);
- running away from home;

- sexual irregularities;
- using tobacco; and
- leading an idle life.

In a final instance (row III in figure 13.1), parental circumstances or behaviors that adversely affect children may be subject to a juvenile and/or adult court jurisdiction. Various laws have attempted to deal with the problem of abused children left homeless by death or those who are abandoned and do not have adequate adult care. Although terminology varies from state to state, these are usually known as neglected or dependent child statutes. Although this is an important dimension of the juvenile court, our presentation is limited to discussing the handling of juveniles who violate the law.

Waiver of Jurisdiction

Most states have statutory provisions for changing jurisdiction of a juvenile case from the juvenile court to the adult criminal court. This process is technically known as a **waiver of jurisdiction.** Among others, terms used to identify this process are transfer, bind-over, and certification.

A decision to try a juvenile defendant in adult court has significant consequences, most notably that the child no longer has the protective nature of the juvenile court to guide decision-making toward what is in the best interests of the child. Rather, the advisarial and punitive characteristics of criminal prosecution will not affect the child, although some increase in protection of due process rights is likely gained. There is considerable variation throughout the United States regarding the reasons for which jurisdiction may be waived. In general terms, the most frequently considered factors include age, seriousness of the present offense, and previous record. Usually, the older the juvenile, the more serious the present offense; and the more serious any previous record, the greater the chances for a waiver of jurisdiction.

In some jurisdictions, only one factor might be utilized, but in others a particular combination might be used in making a waiver decision. In still other jurisdictions, waiver is permitted without considering such items as age or offense.

Prior to 1966, most waivers were accomplished on an arbitrary basis with little or no regard for due process procedures. However, in 1966 this arbitrary nature of waivers was reviewed by the United States Supreme Court in *Kent v. United States.*[3] The court indicated the crucial importance attached to such decisions because of eventual consequences such as sentencing and available treatment. After reviewing this case, the court set forth four basic due process safeguards for waiver procedures. They are the following:

1. The juvenile is entitled to a hearing on the question of waiver.
2. The juvenile is entitled to representation by counsel.

3. The juvenile's attorney must have access to the juvenile's social records on request.
4. If jurisdiction is waived, the juvenile is entitled to a statement of reasons.

In some states, however, these due process safeguards have been circumvented. This has been accomplished by passing statutes allowing broader discretion to prosecutors for trying juveniles in adult court.[4] Trying increasing numbers of juveniles in adult courts with resulting incarceration in state prisons erodes juvenile justice as distinct and separate and marks a return to an incarceration practice deemed undesirable as long as 100 years ago.

Juvenile Legal Terms

A particular set of words, or terms, has been developed to describe legal procedures at the juvenile level. These terms were chosen to be used because it was thought that terms used at the adult level carried undue harshness and stigma for children. While this aim has really not been achieved, special terminology continues to be applied at the juvenile level. Some of the more commonly used **juvenile legal terms** and the corresponding adult terms are the following:[5]

Juvenile Terms	***Adult Terms***
Adjudication	Conviction of guilt
Aftercare	Parole
Commitment	Sentenced to prison
Detention	Holding in jail
Dispositional hearing	Sentence
Hearing	Trial
Petition	Accusation or indictment
Probation	Probation (same term used)
Taking into custody	Arresting

Juvenile Court

The American legal institution established to administer juvenile justice is the **juvenile court.** The year 1899 represents a very significant date in the history of American juvenile justice. On July 1 of that year, legislation was passed that created the first juvenile court located at Cook County (Chicago), Illinois.[6] By 1928 juvenile courts had been established in all but two states. In 1945, with the development of a juvenile court in Wyoming, all states had separate court systems for juveniles.

The juvenile court was founded on the concept of ***parens patriae,*** a Latin phrase that means "the father of his country." This idea was based on English common law wherein the king, being "father of the country" was granted power to protect the interests of children whose parents for one reason or another could not or would not provide this protection. Application of this

Table 13.1

Comparison of Juvenile Court and Adult Criminal Court

Characteristic Feature	*Juvenile Court*	*Adult Court*
Purpose	Protect/treat	Punish
Jurisdiction	Based mainly on age	Based on offense
Responsible noncriminal acts	Yes (status offense)	No
Court proceedings	Less formal/private	Formal/public
Proceedings considered to be criminal	No	Yes
Release of identifying information to press	No	Yes
Parental involvement	Usually possible	No
Release to parental custody	Frequently	Occasionally
Plea bargaining	Less frequently, open admission of guilt more common	Frequently
Right to jury trial	No (McKeiver case)	Yes
Right to treatment Fourteenth Amendment	Yes	No
Sealing/expungement of record	Usually possible	No

philosophy in the United States provided a rationale for the states, through the juvenile court, to extend this doctrine to also include delinquent children.

Consistent with this philosophy, the juvenile court was not designed as a forum for deciding guilt or innocence. Due process was not addressed. The proceedings, informal in nature, were to be in the best interests of the child. Any measures taken were not to be punitive but rather were to be helpful and rehabilitative. The court was considered to be an advocate for the child. Two primary considerations were that children were not totally responsible for their behavior and that the state had some duty to help ''socialize'' its children. Table 13.1 provides a quick reference to differentiate between selected characteristic features of juvenile courts and adult criminal courts.

Caseflow

The processing of juvenile offenders is not entirely dissimilar to adult criminal processing, but there are crucial differences in procedures. Figure 13.2 illustrates, in summary form, pathways of **caseflow** typically found in juvenile justice systems. Operationally, cases may be thought of as processing through four phases: enforcement, intake, court, and corrections. Many cases enter the system as a result of law **enforcement** efforts, but some others are referred by school officials, social service agencies, neighbors, and even parents for behavior or conditions that are determined to require intervention by the formal systems for social control.

An **intake function** is a key decision-making point in juvenile courts. The task of intake officers is to screen all cases and decide whether each case should

Figure 13.2 Caseflow in Juvenile Justice Systems

Source: Adapted from The Challenge of Crime in a Free Society *(A report by the President's Commission on Law Enforcement and the Administration of Justice, Washington, D.C.: Government Printing Office, 1967), pp. 8–9.*

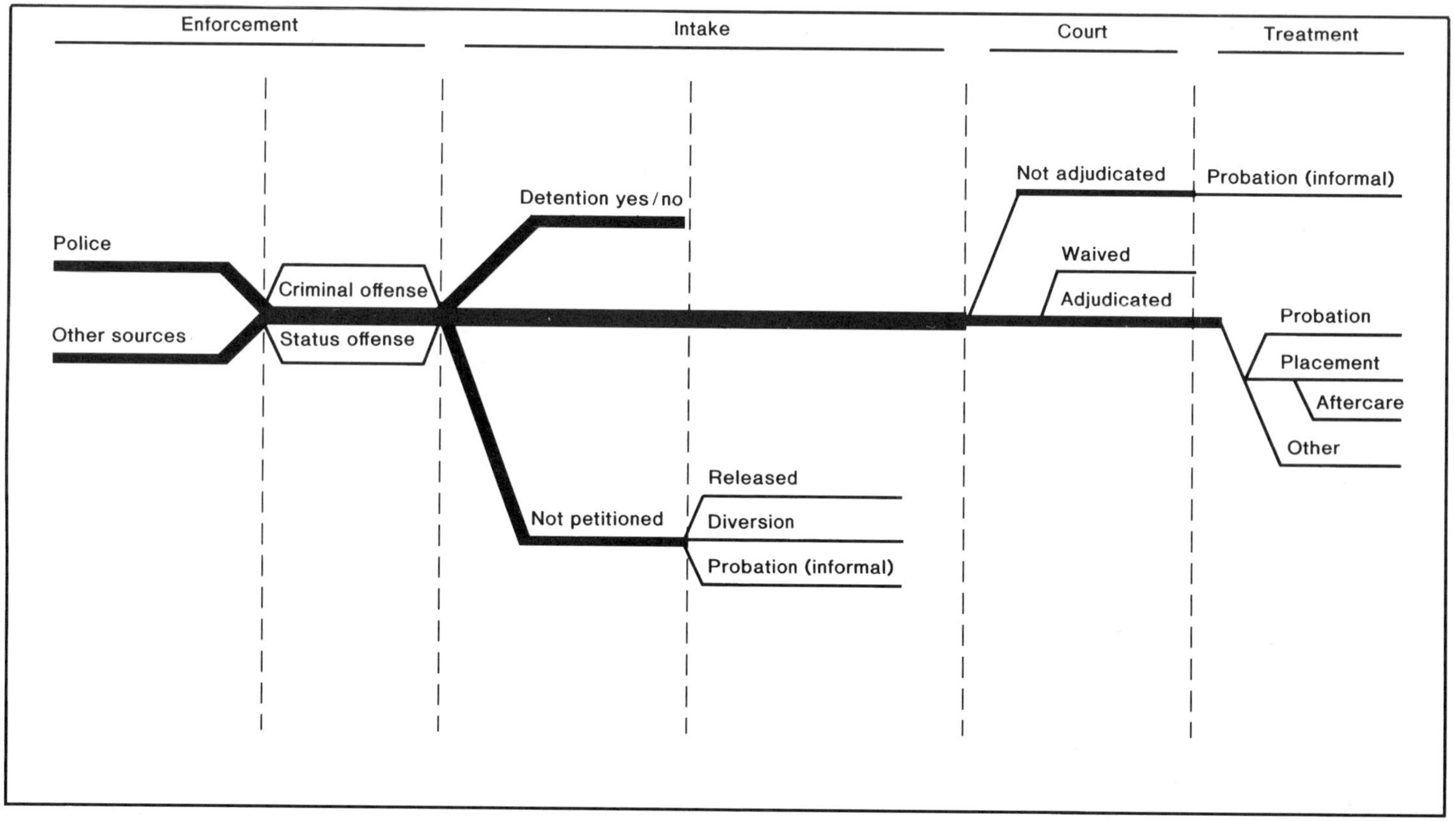

be petitioned to juvenile court (official handling) or not petitioned and handled in some manner other than involving the court (unofficial handling). This decision is based on information gained through conducting investigations and interviews. Officers have wide discretion in deciding whether to refer cases to the court. In most jurisdictions, it is also decided at this time whether a juvenile should be placed in a secure detention facility. This marks one of the more controversial points in the system. Many juveniles are secured in adult jails, which are often in poor condition and lack a trained staff to deal with children. Juveniles do not have a right to release on bail, and it is felt by many that significant numbers are needlessly jailed.

Most youth not petitioned to the juvenile court are released. Others are placed on probation or in some diversion program. **Diversion** is a process whereby an alleged delinquent offender's case is "turned away" from further movement through the formal juvenile justice system. In so called "true" diversion an alleged offender could just walk away. However in actual practice, the person is most likely diverted to an alternative community program. Diversion is consistent with *parens patriae* and with avoiding the stigmatization suggested by labeling theory.[7]

If a **petition** for an adjudicatory hearing is accepted, the juvenile may be brought before a juvenile court quite unlike the court with jurisdiction over adult offenders. At the **treatment** phase, juvenile courts usually have far more discretion than adult courts. In addition to such options as probation, commitment to correctional institutions, restitution, or fines, state laws grant juvenile courts the power to order removal of children from their homes for placement in foster homes or treatment facilities. Juvenile courts may also order participation in special programs aimed at shoplifting prevention, drug counseling, or driver education. They may also be waived to criminal court for trial as an adult.

Due Process Rights

Because the juvenile court philosophy emphasized informality and protection, little attention was given to providing constitutional guarantees of **due process rights.** Over the years it became more and more evident that this was a serious omission. Too often juveniles were deprived of liberty and freedoms on an arbitrary basis without the benefit of procedural due process as provided by the Constitution. It is not surprising, therefore, that test cases were brought before the United States Supreme Court in an effort to provide increased due process in juvenile courts.

Several cases concerning different aspects of this issue were heard between 1966 and 1971. The previously mentioned case of *Kent v. United States* marked an important beginning. The court demonstrated a willingness to consider this issue and with the decision interjected some measure of due process into juvenile court proceedings. It was only one year later when a most significant case affecting juvenile justice was heard, the case of *in re Gault*.[8]

Gerald Gault, age fifteen, was adjudicated a juvenile delinquent, charged with making an obscene phone call. He was committed to the State Industrial School until he reached twenty-one, a sentence of six years. Interestingly, the maximum penalty if Gerald had been eighteen or older would have been a fine of $5 to $50 or a jail term of no more than two months.

The adjudicatory hearing had been completely informal. No notice of charges was given, Gerald did not have legal counsel, and none of the testimony was sworn. Yet, Gerald was adjudicated a delinquent and sentenced to an institution for six years.

Upon reviewing this case, the Supreme Court ruled in favor of Gault and outlined due process procedural requirements considered essential for adjudicatory hearings. The procedures enumerated were the following:

1. The child and parents must be given a written notice of charges.
2. The child and parents have the right to be represented by counsel.
3. The child has the right to remain silent under provisions of the freedom from self-incrimination clause of the Fifth Amendment.
4. The child has a right to confront and cross-examine any witnesses against him.

During 1970 the Supreme Court considered a case concerning the strength of proof necessary before a juvenile could be judged delinquent when in violation of criminal law. This case was *in re Winship*.[9] The test of evidence to convict an adult is that the proof must be "beyond a reasonable doubt." Prior to *Winship* the test of evidence in juvenile court was less strict, requiring proof only by a "preponderance of evidence." The court ruled that for criminal violations juveniles should have the same safeguards as adults and thus required proof beyond a reasonable doubt in juvenile court.

Impact of *in re Gault*

The *Gault* case carried great potential for change in juvenile courts. A juvenile court that operated with increased procedural fairness was now a possibility. In many juvenile courts, this has been achieved to a considerable degree by adhering to the due process requirements specified by *Gault,* including the right to counsel. However, it has been well documented by Feld and others that in many jurisdictions juveniles do not have a lawyer.[10] In some instances, there are even great county-by-county differences in the rate of representation by counsel within the same state. Thus, many juveniles are subject to court proceedings without benefit of an attorney, some of whom receive a secure confinement disposition.

Related research on the impact of *Gault* reveals an outcome in many courts that most certainly could be considered an unintended consequence. That is, there is evidence that at least in some cases, juveniles represented by counsel are actually receiving more severe dispositions, even when controlling for the

Table 13.2

Arrests by Offense Charged among Persons under 18 Years of Age		
Offense Charged	*Number*	*Percent*
Murder and nonnegligent manslaughter	2,626	.4
Forcible rape	4,766	.8
Robbery	35,632	5.5
Aggravated assault	52,653	8.1
Burglary	109,965	16.9
Larceny/theft	369,227	59.0
Motor vehicle theft	70,659	10.8
Arson	6,940	1.1
Total	652,468	100.0[a]
Violent crime[b]	95,677	15.0
Property crime[c]	556,791	85.0
Total	652,468	100.0

[a] *Rounding error*
[b] *Murder, forcible rape, robbery, and aggravated assault*
[c] *Burglary, larceny/theft, motor vehicle theft, and arson*
Source: Adapted from U.S. Department of Justice, Federal Bureau of Investigation, Crime in the United States, 1991 *(Washington, D.C.: Government Printing Office, 1992), pp. 223–224.*

type of offense. For example, Bortner reported that nearly 36 percent of juveniles represented by counsel received out-of-home placements compared to only 10 percent of those without counsel.[11] The dynamics causing this outcome await additional research. It has been suggested that such variables as inexperienced lawyers, co-optation by the judge of court-appointed counsel, and traditional *parens patriae* philosophy, might alter lawyers effectiveness.[12]

What started with great promise for juvenile justice has brought mixed results. Youths in some jurisdictions are receiving counsel, while many others are not. Much legislative work remains to be done to ensure that juveniles are in fact being accorded this specific constitutional right.

Juvenile Offenders

What is it that juveniles do that involves them in the juvenile justice system? As we will see, they engage in a variety of illegal behaviors. As previously discussed, some offenses are considered status offenses, while others are public offenses. Much of this illegal behavior by juveniles remains unknown or hidden, never coming to the attention of officials.

The Uniform Crime Report provides a measure of those juvenile offenses that are known to the police. Table 13.2 provides an example of the distribution for index (serious) offenses for which persons under eighteen years of age were charged. Remember that this table presents offenses not individuals. We know that in many instances the same individual accounts for multiple offenses. We note that juvenile arrests occurred within each category. However, a very high proportion of the arrests, nearly nine of every ten, are for property crime.

Self-Reports

Beginning during the 1950s, the measurement of delinquency by a method known as **self-report** has greatly enhanced our knowledge regarding the behavior of young people. When this approach is used according to appropriate research techniques, which include confidentiality of responses, reliable and valid data are produced.[13] Typically, a sample of students completes a questionnaire (sometimes in an interview setting) in which they record their illegal behaviors. Table 13.3 provides an example of data collected for selected offenses from a self-report survey. By using this kind of measure we are able to learn about illegal events even if unknown to police or juvenile officials. In this way we learn more about **hidden delinquency;** that is, illegal behavior that is otherwise unknown. It has been estimated that only about 24 percent of those committing offenses for which they could be arrested are in fact arrested.[14]

Self-report studies have provided more complete information about juvenile offenders. Previously, offenders were considered to be located within only certain segments of the population. Specifically, delinquency was thought to occur primarily among minority group males, living in lower socioeconomic areas. However, included among the findings of self-report research are such revelations as the following:

- More juveniles violate the law than previously recognized.[15]
- Offenders are located within all socioeconomic levels.[16]
- Female juveniles report a variety of offenses, although generally with less frequency and severity than males.[17]
- Self-reports provide little evidence of racial differences in offender rates.[18]
- Self-reported delinquency rates remain stable; sharp increases and decreases have not been reported.[19]
- Hidden delinquency is high. Less than one-fourth of chronic offenders are apprehended by police, and only eight percent of offenders reporting delinquent behavior are arrested.[20]

Self-reports represents a very important data source in learning about juvenile offenders. Self-report findings prompted rethinking of our theories of delinquency causation and have provided a more comprehensive picture of juvenile offenders. Views regarding how offenders are identified and handled by society have been altered.

Self-reports are an important data source for learning about juvenile offenders.

© D. Schaefer/Uniphoto Picture Agency

Table 13.3

High School Seniors Reporting Involvement in Selected Delinquent Activities in Last 12 Months

Question: "During the last 12 months, how often have you:		
Delinquent Activity	*Class of 1982 (N = 3,678)*	*Class of 1992 (N = 2,690)*
Hit an instructor or supervisor?		
Not at all	97.3	96.7
Once	1.6	1.9
Twice	0.5	0.5
3 or 4 times	0.1	0.3
5 or more times	0.4	0.6
Gotten into a serious fight in school or at work?		
Not at all	82.7	81.1
Once	11.2	11.5
Twice	3.0	4.0
3 or 4 times	1.8	1.8
5 or more times	1.3	1.7
Taken part in a fight where a group of your friends were against another group?		
Not at all	81.9	78.7
Once	10.0	11.5
Twice	4.2	4.4
3 or 4 times	2.4	3.2
5 or more times	1.5	2.2
Hurt someone badly enough to need bandages or a doctor?		
Not at all	88.6	87.2
Once	7.5	7.3
Twice	2.1	2.9
3 or 4 times	0.8	1.6
5 or more times	0.9	1.1
Used a knife or gun or some other thing (like a club) to get something from a person?		
Not at all	97.7	95.7
Once	1.6	2.2
Twice	0.3	1.0
3 or 4 times	0.2	0.5
5 or more times	0.2	0.5
Taken something not belonging to you worth under $50?		
Not at all	68.7	67.4
Once	13.9	14.2
Twice	6.9	7.5
3 or 4 times	5.4	5.6
5 or more times	5.2	5.2
Taken something not belonging to you worth over $50?		
Not at all	93.1	89.5
Once	3.7	5.5
Twice	1.1	1.7
3 or 4 times	0.9	1.5
5 or more times	1.2	1.7

Table 13.3 (continued)

Question: "During the last 12 months, how often have you:		
Delinquent Activity	***Class of 1982 (N = 3,678)***	***Class of 1992 (N = 2,690)***
Taken something from a store without paying for it?		
Not at all	71.2	69.6
Once	12.5	12.6
Twice	5.8	6.7
3 or 4 times	4.6	5.2
5 or more times	5.9	5.9
Taken a car that didn't belong to someone in your family without permission of the owner?		
Not at all	95.9	94.0
Once	2.4	3.1
Twice	0.6	1.4
3 or 4 times	0.5	0.7
5 or more times	0.7	0.9
Taken a part of a car without permission of the owner?		
Not at all	94.3	93.9
Once	3.4	3.2
Twice	1.2	1.2
3 or 4 times	0.4	1.0
5 or more times	0.7	0.8
Set fire to someone's property on purpose?		
Not at all	98.8	97.2
Once	0.7	1.6
Twice	0.2	0.4
3 or 4 times	0.1	0.4
5 or more times	0.2	0.4
Damaged school property on purpose?		
Not at all	87.6	85.3
Once	6.9	7.9
Twice	3.2	3.5
3 or 4 times	1.2	1.2
5 or more times	1.2	2.1
Damaged property at work on purpose?		
Not at all	94.2	94.0
Once	3.1	2.7
Twice	1.5	1.3
3 or 4 times	0.6	1.0
5 or more times	0.7	1.0
Gotten into trouble with police because of something you did?		
Not at all	78.3	77.8
Once	13.7	11.9
Twice	5.4	5.2
3 or 4 times	1.9	3.0
5 or more times	0.7	2.2

Sources: Jerald G. Bachman, Lloyd D. Johnston, and Patrick M. O'Malley, Monitoring the Future 1981, *pp. 100–102;* 1983, *pp. 100–102;* 1985, *pp. 99–101;* 1987, *pp. 103–105 (Ann Arbor, MI: Institute for Social Research, University of Michigan); Jerald G. Bachman, Lloyd D. Johnston, and Patrick M. O'Malley,* Monitoring the Future 1980, *pp. 100–102;* 1982, *pp. 100–102;* 1984, *pp. 99–101;* 1986, *pp. 102–104;* 1988, *pp. 103–105 (Ann Arbor, MI: Institute for Social Research, University of Michigan); and data provided by the Monitoring the Future Project, Survey Research Center, Lloyd D. Johnston and Jerald G. Bachman, Principle Investigators. Table adapted by* Sourcebook *staff. Reprinted by permission.*

Summary

Before the modern concept of childhood emerged, children were treated with indifference. Through reform efforts, attitudes and practices changed toward promoting greater safeguards and nurturing for children. Government became increasingly involved by establishing distinct laws and procedures to deal with juveniles, including those who violate the law.

The juvenile justice system has both similar and contrasting features when compared to the adult justice system. The juvenile system was designed consistent with the doctrine of *parens patriae,* supportive of treatment rather than punishment as characterizes the adult system. Some due process rights have been extended to juveniles by *Kent v. United States* and *In re Gault* as examples.

Juveniles commit a variety of acts that are illegal. Illegal behaviors are categorized as being public or status offenses. Many offenses remain unknown or hidden but have been measured by researchers utilizing self-report methods. Information gained by self-report studies has provided a more comprehensive picture of juvenile delinquency and its treatment.

Discussion Topics

1. Should status offenders be dealt with by the juvenile court system? Why or why not?
2. What cases, if any, would you waive to adult court? Why?
3. What are some diversion program activities you would like to see developed?
4. What should state and local governments do to better implement provisions of the *Gault?*
5. What are some ways in which self-reports have caused us to ''rethink'' juvenile delinquency?

Additional Readings

Cox, Steve M. and John J. Conrad. *Juvenile Justice: A Guide to Practice and Theory.* 4th ed. Madison, WI: Brown and Benchmark Publishers, 1996.

Jones, Michael A. and Barry Krisberg. *Images and Reality: Juvenile Crime, Youth Violence and Public Policy.* San Francisco, CA: National Council on Crime and Delinquency, June 1994.

Platt, Anthony. *The Child Savers: The Invention of Delinquency.* Chicago: University of Chicago Press, 1970.

Notes

1. LaMar T. Empey, *American Delinquency: Its Meaning and Construction* (Homewood, IL: Dorsey Press, 1982), 3–74.
2. Mark M. Levin and Rosemary Sarri, *Juvenile Delinquency: A Comparative Analysis of Legal Codes in the United States* (Ann Arbor, MI: National Association of Juvenile Corrections, 1974), 37.
3. Kent v. United States 383 U.S. 541 (1966).
4. Ira M. Schwartz, (In) *Justice for Juveniles* (Lexington, MA: D.E. Heath and Co., 1989), 10–11.
5. Adapted from Ruth S. Cavan, *Juvenile Delinquency,* 2d ed. (New York: J.B. Lippincott Co., 1969), 367; and National Advisory Commission on Criminal Justice Standards and Goals, *Corrections* (Washington, D.C.: United States Government Printing Office, 1973), 248.
6. Sophia M. Robinson, *Juvenile Delinquency* (New York: Holt, Rinehart & Winston, 1964), 250–51.

7. James O. Finckenauer, *Juvenile Delinquency: The Gap Between Theory and Practice* (Orlando: Academic Press, Inc., 1984), 134.
8. In re Gault 376 U.S. 1 (1967).
9. In re Winship 397 U.S. 358 (1970).
10. Barry C. Feld, ''*In re Gault* Revisited: A Cross-State Comparison of the Right to Counsel in Juvenile Court,'' *Crime and Delinquency* 34 (October 1988): 393–395.
11. M. A. Bortner, *Inside a Juvenile Court: The Tarnished Ideal of Individualized Justice* (New York: New York University Press, 1982), 139–140.
12. Feld, ''*In re Gault* Revisited,'' 419–420.
13. Michael Hindelang, Travis Hirschi, and Joseph Weis, *Measuring Delinquency* (Beverly Hills, CA: Sage, 1981), 114.
14. David Huizinga and Delbert Elliott, ''Juvenile Offenders: Prevalence, Offender Incidence, and Arrests by Race,'' *Crime and Delinquency* 33 (April 1987): 208.
15. Martin Gold, ''Undetected Delinquent Behavior,'' *Journal of Research in Crime and Delinquency* 3 (January 1966): 27–46.
16. Huizinga and Elliott, ''Juvenile Offenders,'' 206–223.
17. *Ibid.*
18. Jay Williams and Martin Gold, ''From Delinquent Behavior to Official Delinquency,'' *Social Problems* 20 (Fall 1972): 209–229.
19. Rosemary Sarri, ''Gender Issues in Juvenile Justice,'' *Crime and Delinquency* 29 (July 1983): 381–397.
20. Franklyn Dunford and Delbert Elliott, ''Identifying Career Offenders Using Self-Reported Data,'' *Journal of Research in Crime and Delinquency* 21 (February 1984): 57–86.

Chapter 14

Treatment for Juvenile Offenders

Learning Objectives

After studying this chapter students should be able to:

1. Identify and describe four competing treatment philosophies.
2. Explain the concept of blended philosophies.
3. Describe and provide examples of the three phases of deinstitutionalization during the past one hundred and fifty years involving juveniles.
4. Understand the nature of dispositions for juveniles.
5. Understand the range in the type of facilities utilized for juvenile offenders.
6. Understand the variety of residential and nonresidential programs for juvenile offenders.
7. Identify several treatment methods commonly in use for juvenile offenders.
8. Summarize the current state of treatment outcomes.

Key Terms

Aftercare
Blended philosophies
Crime control
Deinstitutionalized model
Due process
Group counseling
Indeterminate dispositions
Individual counseling
Institutional model
Least restrictive alternative
Nonintervention
Nonresidential programs
Private facilities
Probation
Public facilities
Rehabilitation
Residential programs
Right to treatment
Training schools
Transitional adjustments
Treatment outcomes

Competing Treatment Philosophies

Whenever a society seeks to intervene to treat inappropriate behavior by juveniles, it is likely that different viewpoints will emerge, just as we noted for adult corrections. Although the concept of *parens patriae* with accompanying emphasis on treatment marked the founding philosophy, several other philosophies have evolved and are competing for dominance. Seigel and Senna suggest that these philosophies can be assembled into four main categories:

- rehabilitation
- due process
- nonintervention
- crime control[1]

Rehabilitation Philosophy

Juvenile treatment was founded upon the philosophy of **rehabilitation** which was consistent with *parens patriae* and with efforts designed to be in the best interest of the child. This approach assumes people have a capacity to learn and to change their behavior. Juveniles who misbehaved were seen more as victims of circumstances and environment, rather than offenders per se. The aim of rehabilitation was to support and offer treatments on an individual basis in a nonpunitive atmosphere. The court structure resulting from this approach was more informal and confidential. This was the prevailing philosophy until the 1960s.

Due Process Philosophy

During the 1960s concerns for constitutional procedural rights for juveniles emerged. The founding rehabilitative philosophy focused almost entirely on establishing a benevolent and informal system. Overlooked in these efforts were **due process** rights. As a result, many youths were losing freedom and liberties when they were institutionalized, but without due process safeguards, including the right to counsel, freedom from self-incrimination, and the right to confront and cross-examine witnesses. The courts extended some important due process rights to juveniles beginning in the mid-1960s in the benchmark cases of *Kent* (1966) and *in re Gault* (1967). Granting rights was consistent with the idea of fairness which was implicit in the rehabilitation philosophy, but as attorneys became more involved, proceedings became less informal and more adversarial.

Nonintervention Philosophy

A **nonintervention** approach, which emerged during the 1970s, sought to avoid or minimize the stigma of being labeled a delinquent. This concern has been raised by labeling theory, which reasoned that extended penetration into the system might make matters worse. In order to minimize such risk, this philosophy suggested intervention should be limited in order to avoid negative labels and stereotyping. Any intervention deemed necessary should be according to the principle of the least restrictive alternative; that is, treatment in community-based facilities was the desired environment, and placement in a secure facility should be utilized only as a necessary last resort. Such a philosophy promotes deinstitutionalization and diversion programs.

Crime Control Philosophy

Classical philosophy, under the heading of **crime control,** was reexpressed during the more conservative 1980s. This approach is located at the opposite end of the continuum from rehabilitation. It basically rejects the premise that people have the capacity to change behaviors. Punishment and incarceration represent the desired policies. As a result, increasing numbers of juveniles have been transferred to adult criminal court and sentenced to state prisons in some jurisdictions.

Blended Philosophies

Elements from all four of these philosophies may be found at the juvenile level. There are no "pure" types, as these categories do overlap and are not always mutually exclusive. They form what would be called **blended philosophies.** As the overall political climate changes, a particular philosophy may gain in popularity for a period of time then replaced later by a different approach. Rehabilitation, for example, does not have the degree of dominance it once had but is still very much in evidence. Due process considerations have brought fundamental change in many courts but still have not been implemented in many others. Extension of rights to juveniles has been focused and limited. The courts have denied some rights accorded adults and not extended others such as a trial by jury.

Some notions advocated by the crime control philosophy, carried to the extreme, could essentially destroy juvenile justice as a system distinct and different from the adult system. With the blending from other viewpoints, such a change is not likely to be complete.

Categorizing philosophies in this manner serves as an analytical tool. Any program, agency policy, legislative proposal, or court decision may be evaluated in terms of its philosophical thrust and the direction in which it leads. It is likely that elements of all four will continue to provide a mix of policies and approaches for treatment of juvenile offenders.

Background

The setting for treating juveniles in America may be analyzed on a continuum from a traditional or **institutional model** to a nontraditional or **deinstitutionalized model.**[2] Although there are multiple meanings, features of traditional institutional facilities include usage of an isolated structure of significant size, housing a sizable number of persons. Activities, including work and eating, are conducted according to rules and schedules, often in large groups. Possibilities for treatment are at least compromised and may even be non-existent because of institutional constraints. The dehumanizing aspects of institutions have been documented for some time.[3]

By contrast, activities within a deinstitutionalized environment will be conducted with a much smaller number of persons within a less isolated facility in a community-like setting. Because of smaller numbers, fewer institutional restraints, and a more humanizing atmosphere, care and treatment possibilities may be enhanced.

During the past 150-year period, there have been three phases of deinstitutionalization involving juveniles.[4] Prior to about 1900, most children who were considered to be in need of attention resulting from neglect and/or law violations were institutionalized in facilities with adults. Children served time in adult prisons as well as being placed in institutional settings with adults, such as county poor houses and almhouses. Such practices marked the first phase of institutional use.

During the latter portion of the 1800s and early 1900s, the practice of institutionalizing children in the same facility as adults was coming under closer scrutiny and increasing public disapproval. Also during this time, in 1899, the first juvenile court in America was established. For the first time, there was a separate court and set of laws that dealt with the needs of children. These concerns also resulted in the establishment of separate, state-supported public institutions for children, often identified as reform schools. In addition, many children became residents in private institutions that began to be developed. Thus a second phase, institutions for children that were separate from adult institutions, was beginning.

A third phase was replacing traditional juvenile institutions with nontraditional or deinstitutionalized facilities. Beginning in the mid-1930s, many children in the neglect/dependent category were removed from institutional settings. Far less change has been realized for delinquent youths. Only a few states, including Kentucky, Massachusetts, Utah, West Virginia, Michigan, Alabama and Pennsylvania have achieved significant deinstitutionalization for adjudicated delinquents. This has been accomplished without endangering public safety.[5] In most states traditional institutions, often known as **training schools,** coupled with a large number of private institutions continue to house large numbers of juveniles. In addition, a growing number of juveniles nationwide are in adult prisons as a result of being waived and sentenced by adult courts.

Example of Deinstitutionalization

The Commonwealth of Kentucky provides a good example of transition through the phases of deinstitutionalization just described.[6] As shown in figure 14.1, removing children from adult institutions began in 1897 with the establishment of a separate institution, housing both males and females, known as the houses

Figure 14.1 Transition from Institutional to Deinstitutional Model Commonwealth of Kentucky

Source: Richard W. Snarr, A History of Services for the Commonwealth's Children, *Department for Social Services, Frankfort, KY, 1989, p. 3.*

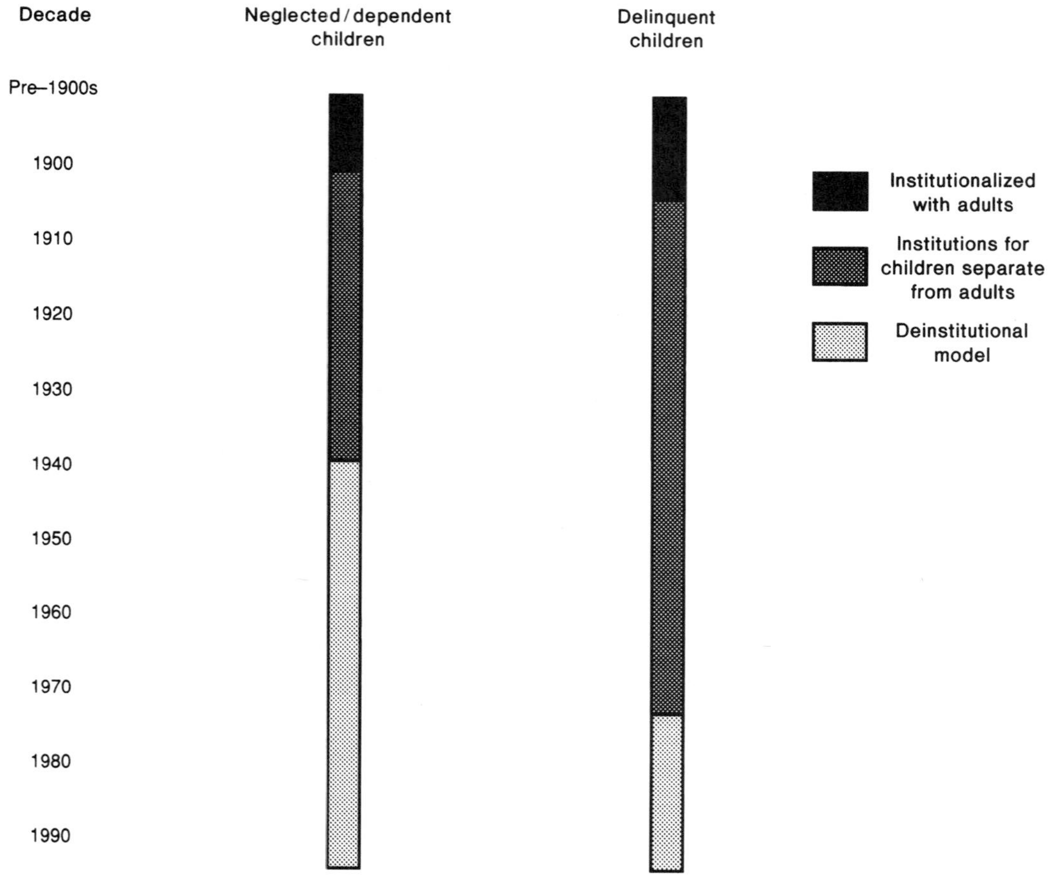

of reform. Placing dependent children (some of whom were also in trouble with the law) in a community setting began during the mid-1930s. This was accomplished in large measure by developing programs that assisted families to be able to provide for children in their own homes. Monies became available as a result of the passage of the Social Security Act.

Leadership efforts during the latter 1960s resulted in the closing of the large training school institution for delinquents in 1971, among the first in the nation to do so. Since that time delinquents have been placed in small, deinstitutionalized facilities throughout the state and in local communities in camps, group homes, and day treatment schools. There is no evidence of increased juvenile delinquency resulting from this deinstitutionalized system. In fact, research evidence indicates just the opposite is true. Only about 9 percent of those arrested in Kentucky for violent crimes are juveniles, compared to the 19 percent national violent crime ratio for juveniles. Similarly, only 9 percent

In Kentucky (since 1971) and a few other states, small residential treatment facilities for juveniles have successfully replaced large training schools.

of all persons arrested in Kentucky for all offenses, including status offenses, were juveniles. This is about one-third the 26 percent national ratio of total juvenile arrests.[7] This long-term experience suggests that placing large numbers of juveniles in training schools might not be necessary and may even be counterproductive in correcting offenders and deterring would-be offenders.

Dispositions[8]

Juvenile offenders generally receive **indeterminate dispositions;** that is, a juvenile remains under supervision until authorities deem that youth to be rehabilitated or legally an adult. In most states the juvenile treatment agency is granted this release authorization. In a few instances, this resides with the committing judge or a juvenile parole authority.

Although unable to issue death penalties or life sentences, juvenile courts nationwide have a variety of dispositional alternatives available. In many jurisdictions the range of alternatives is limited. Dispositions are commonly viewed as moving from a **least restrictive alternative** to a most restrictive alternative. Under the former, youthful offenders continue living at home while participating in a particular program—customarily classified as a nonresidential program. This arrangement is often accomplished as a part of probation, which is a frequently used form of supervision for juveniles.

Other youth may require closer supervision away from home—in residential programs. Group homes and facilities, both public and private, for youthful offenders provide varying degrees of security for thousands of young persons institutionalized in the United States.

Probation

Probation is the most widely utilized disposition for adjudicated delinquents. One estimate would indicate that around 300,000 juveniles are on probation.[9] Probation developed as described in chapter 11 and was incorporated as juvenile justice systems were created. The development of probation for juveniles was tied to the juvenile court movement during the late 1800s and early 1900s. The juvenile court was established according to the doctrine of *parens patriae* and was to act in whatever manner was in the best interests of the child. The notion of utilizing probation, rather than incarceration, so that juveniles could remain in their home communities was very consistent with the aims of the juvenile court. For these reasons probation became a very important and much-used option by juvenile courts. Probation for juveniles operates in a manner very similar to that for adults, as previously discussed. It is a dispositional alternative that allows a juvenile to remain in the community, subject to certain rules imposed by the court. Typically, such rules include staying out of trouble, attending school, and participating in a treatment program. These activities are supervised by a juvenile probation officer.

Probation provides the framework for a variety of community programs. For example, activities known as restitution, community service, or some other title are usually incorporated by some manner into ongoing probation operations and include functions performed by regular staff and probation officers.

Facilities for Juveniles

There are thousands of facilities nationwide which hold children in custody. The number of juveniles in the facilities, both public and private, is approaching 90,000.[10] There is considerable variation in the type and level of security of the facilities as exhibited by tables 14.1 and 14.2.

Public facilities operate with public funds under the administrative control of state or local governments. **Private facilities** are subject to governmental licensing but are administratively controlled by private enterprise, although many receive public funding. Some are for profit, others are nonprofit organizations.

As shown by table 14.1, facilities may be classified by type and function according to certain features and the purpose they are designed to carry out.

- Detention centers are short-term facilities that provide custody in a physically restricting environment pending adjudication or, following adjudication, pending disposition, placement, or transfer.
- Shelters are short-term facilities that provide temporary care similar to that of a detention center but in a physically unrestricted environment.
- Reception or diagnostic centers are short-term facilities that screen persons committed by courts and assign them to appropriate custody facilities.

Table 14.1

Number of Facilities and Juveniles, by Facility Type for Public and Private Juvenile Facilities

Type of Facility	
All public and private facilities	
Number of facilities	3,036
Detention centers	428
Shelters	297
Reception or diagnostic centers	39
Training schools	263
Ranches, forestry camps, or farms	260
Halfway houses or group homes	1,749
Number of juveniles	83,402
Detention centers	14,069
Shelters	2,700
Reception or diagnostic centers	1,764
Training schools	30,532
Ranches, forestry camps, or farms	13,170
Halfway houses or group homes	21,167
Public facilities	
Number of facilities	1,040
Detention centers	403
Shelters	71
Reception or diagnostic centers	21
Training schools	190
Ranches, forestry camps, or farms	104
Halfway houses or group homes	251
Number of juveniles	49,322
Detention centers	13,772
Shelters	805
Reception or diagnostic centers	1,425
Training schools	25,074
Ranches, forestry camps, or farms	5,129
Halfway houses or group homes	3,117
Private facilities	
Number of facilities	1,996
Detention centers	25
Shelters	226
Reception or diagnostic centers	18
Training schools	73
Ranches, forestry camps, or farms	156
Halfway houses or group homes	1,498
Number of juveniles	34,080
Detention centers	297
Shelters	1,895
Reception or diagnostic centers	339
Training schools	5,458
Ranches, forestry camps, or farms	8,041
Halfway houses or group homes	18,050

Source: Adapted from Joseph M. Bessette, Children in Custody, 1975–1985, *U.S. Department of Justice, Bureau of Justice Statistics (Washington, D.C.: Government Printing Office, 1989), p. 28.*

Table 14.2

Juveniles in Public and Private Juvenile Facilities, by Facility Type and Level of Security

Self-Classification and Level of Security	Number of Juveniles		
	Total	Institutional Facilities	Open Facilities
Public facilities			
All facilities	49,322	40,552	8,770
Secure	39,499	38,437	1,062
Nonsecure	9,823	2,115	7,708
Detention centers	13,772	13,741	31
Secure	13,710	13,701	9
Nonsecure	62	40	22
Shelters	805	317	488
Secure	332	272	60
Nonsecure	473	45	428
Reception or diagnostic centers	1,425	1,365	60
Secure	1,376	1,358	18
Nonsecure	49	7	42
Training schools	25,074	22,051	3,023
Secure	20,834	20,612	222
Nonsecure	4,240	1,439	2,801
Ranches, forestry camps, or farms	5,129	2,531	2,598
Secure	2,166	2,047	119
Nonsecure	2,963	484	2,479
Halfway houses or group homes	3,117	547	2,570
Secure	1,081	447	634
Nonsecure	2,036	100	1,936
Private facilities			
All facilities	34,080	7,933	26,147
Secure	5,311	3,989	1,322
Nonsecure	28,769	3,944	24,825
Detention centers	297	213	84
Secure	209	174	35
Nonsecure	88	39	49
Shelters	1,895	331	1,564
Secure	460	226	234
Nonsecure	1,435	105	1,330
Reception or diagnostic centers	339	105	234
Secure	63	63	0
Nonsecure	276	42	234
Training schools	5,458	2,311	3,147
Secure	1,543	1,326	217
Nonsecure	3,915	985	2,930
Ranches, forestry camps, or farms	8,041	2,626	5,415
Secure	1,101	1,002	99
Nonsecure	6,940	1,624	5,316
Halfway houses or group homes	18,050	2,347	15,703
Secure	1,935	1;198	737
Nonsecure	16,115	1,149	14,966

Source: Adapted from Joseph M. Bessette, Children in Custody, 1975–1985, *U.S. Department of Justice, Bureau of Justice Statistics (Washington, D.C.: Government Printing Office, 1989), p. 32.*

- Training schools are long-term facilities for adjudicated juvenile offenders typically under strict physical and staff controls.
- Ranches, forestry camps, or farms are long-term residential facilities for persons whose behavior does not require the strict confinement of a training school, often allowing them greater contact with the community.
- Halfway houses or group homes are long-term, nonconfining facilities in which residents are allowed extensive access to community resources, such as schooling, employment, health care, and cultural events.

Table 14.2 indicates facilities have important differences according to the type of onsite environment and security level they provided.

- Institutional environments impose greater restraints on residents' movements and limit access to the community. Most public or private detention centers and most public reception or diagnostic centers and training schools were classified as having institutional environments.
- Open environments allow greater movement of residents within the facilities and more access to the community. Facilities with open environments included many private facilities and many public shelters; ranches, forestry camps, or farms; and halfway houses or group homes.
- Secure facilities are institutions in which the movement of residents is controlled through staff monitoring of entrances or exits and/or through hardware such as locks, bars, and fences. Most public facilities and private detention centers were classified as secure facilities.
- Nonsecure facilities are institutions in which residents' movements are not restricted by hardware restraints such as locks, bars, and fences or by the use of staff monitoring of entrances and exits. Many private facilities and most public shelters; ranches, forestry camps, or farms; and halfway houses or group homes were classified as nonsecure facilities.

In summary we can see that thousands of juveniles are institutionalized in a variety of facilities. In only a few jurisdictions spotted across the country has any significant deinstitutionalization for handling delinquents occurred.

Programs

A feature that most marks the nature of juvenile programs is the heterogeneity of approaches.[11] Numerous therapies, varying target groups, different philosophies, types of agency involvement, and timing of the intervention serve as some examples of this diversity. Various interventions may occur at different points as juvenile cases move throughout the system. Early interventions are possible prior to actual formal court adjudication (preadjudicatory) while later interventions may not occur until after adjudication (postadjudicatory). Programs in the preadjudicatory phase represent so-called nonjudicial alternatives and are often described as diversionary and/or community-based.

Programs are often assessed by the amount of restriction placed upon participating juveniles and the corresponding nature of treatment. In many instances, youth live at home with little, if any, additional restriction—for example, nonresidential programs. At the other end of the continuum would be the secure, locked residential facilities. The National Council on Crime and Delinquency has cataloged and categorized various programs from throughout the United States, starting with the least restrictive options and moving to the more secure alternatives,[12] as is shown in the following paragraphs.

Nonresidential Programs

Young offenders living in their own homes may benefit from **nonresidential programs** established to serve all youngsters in the community, as well as from those specifically developed as alternatives to incarceration. Programs serving both offenders and nonoffenders are generally less restrictive than those serving only youths in trouble.

Job/Career Programs

Youths receive help in defining their career interests, vocational training, instruction in how to look for a job, and miscellaneous employment services.

After-School or Evening Programs

Programs include supervised recreation, special outings, informal counseling, arts and crafts, preparation for the GED exam, or classes in budgeting and other "life skills." They are usually conducted at community centers, churches, YM/YWCAs, or public schools.

Alternative Schools

Young people attend small, nontraditional schools in their communities. Some schools are designed especially for youths in trouble with the law, while some are open to any child not benefiting from the public school.

Advocacy

Staff members act on behalf of individual children and their families, negotiating with private and public agencies to secure services, to monitor the progress of youths in community programs, and to pressure for *needed change.*

Counseling

Included are individual, group, and family therapy, usually conducted in weekly sessions. Short-term therapy that focuses on specific problems or tasks is generally preferred to lengthy psychotherapy.

Mediation/Arbitration

As an alternative to the juvenile court, a panel of community residents helps resolve cases involving juveniles. Often, the youth is required to provide some form of restitution to the victim; other common stipulations are regular school attendance and a period of counseling.

Restitution

Youths repay their victims or the community through financial compensation or assigned community service work.

Intensive Service to Families

Trained youth workers provide services to families in their own homes, including crisis intervention, counseling, training in problem-solving skills, homemaking assistance, and financial planning. They often work to help families make use of the appropriate community resource and may adopt an advocacy role.

Residential Programs

Youths in **residential programs** are relocated away from their homes for close supervision and to receive intensive services. Goals include returning young people to their families or preparing them to live independently.

Wilderness Programs

Programs consist of four to eight weeks of group wilderness survival in remote, rugged areas. The best programs are followed up by services to the youngster after he or she returns to the family or other living situation.

Preparation for Independent Living

Youths are assisted in acquiring the skills and financial resources needed to be self-supporting. They may live alone, maintaining contact with the supporting agency; share an apartment with another youth under close adult supervision; or live for a period with a family or responsible adult. Funds and additional services may be provided by the sponsoring agency.

Foster Family Care

As many as three youths live with foster parents and use community schools and resources. The sponsoring agency may provide other services as needed. Foster family care is appropriate for youngsters who must be removed from their homes temporarily because of a family crisis or who need special adult support as they prepare to live on their own.

Intensive Foster Care

A two-parent family houses no more than two young people, in need of both supervision and individual attention. At least one foster parent is present at all times, and additional staff and clinical support are provided by the sponsoring agency. The youths attend public or alternative schools. Restrictiveness is gradually lessened over the course of the program.

Group Home

Six to eight young people live with houseparents or rotating staff. The home is not locked, and the youths attend community schools, make extensive use of community resources, may have jobs, and may return home on weekends.

Highly Structured Group Care

Appropriate only for youths who are dangerous to themselves or others and who are unable to control their behavior, these residences are not locked, have a high staff/client ratio, and use a structured form of treatment. School, recreational activities, and some services are provided within the residence; there is some use, with supervision, of community resources.

Secure

It may be necessary to turn to secure, locked facilities for youths who have been convicted of violent offenses and who clearly are a danger to themselves or others. Alternatives available within this category have a high staff/client ratio, house not more than fifteen youths per unit, and offer intensive services. It must be remembered that each community is different and no one jurisdiction has all of these program options. Rather, this information illustrates the nature and wide diversity of programs now operating.

Treatment Methods

The courts have generally held that juveniles have a **right to treatment** in such cases as *Inmates of the Boys' Training School v. Affleck* and *Nelson v. Heyne*.[13] This is a fundamental difference from adult corrections. However, what constitutes a minimum treatment program for juveniles has not been determined nor has the United States Supreme Court ruled whether juveniles have a right to treatment.[14]

Various treatment methods are found in both residential and nonresidential juvenile corrections settings. Some techniques involve an individual approach, while others employ a group approach. In a number of instances there may be a mix of several specific strategies, for example, utilizing both behavior modification and group counseling.

Treatments are delivered according to a variety of strategies and vehicles. For example, the training school, the family, educational institutions, recreational programs, and probation services are utilized to deliver various treatments. In some cases a treatment will be targeted toward a specific problem, such as alcohol or other substance abuse problems. Listed briefly in this section are several more commonly employed individual and group methods of treatment.

Individual Treatment Methods

Individual counseling will be found in most all institutional settings and in many nonresidential programs. This approach allows for a one-to-one relationship in dealing with personal adjustment problems. This method may be more expensive due to the number of counselors needed, and thus, the quality and quantity of treatment may be compromised by budget constraints.

A method developed by William Glasser, known as reality therapy,[15] is a frequently utilized individual approach. This technique, unlike most others, largely ignores the past and concentrates on changing future behavior. Individuals in treatment are directed toward taking responsibility for their behavior. A highly trained and trustworthy therapist is necessary for this method.

Another very frequently used method of treatment is behavior modification. A basic premise of this approach is that behavior is learned. Treatments are designed to reward acceptable behavior and to deny rewards for unacceptable behavior. The desired behaviors are often spelled out in a contract format. Typically, offenders' rewards are in the form of increasing privileges and freedoms as they progress through a series of phases or stages in the treatment process. Such programs have been utilized in both institutional and community settings.

Group Treatment Methods

Other treatment strategies involve some form of group activity, often identified by the generic term, **group counseling.** Such a setting can provide offenders an opportunity to gain help and support from others as they try to change their behavior. Group methods also are appealing since few staff and less expense may be involved.

Effective guidance for a group approach relies on well-trained staff leaders to focus the group in a positive direction and to avoid potential abuses. Some specific group therapies include positive peer culture (PPC) and group psychotherapy.

Assessment of Outcomes

Although a comprehensive discussion of this topic is beyond the scope of this section, a few salient points will be summarized. Assessing **treatment outcomes** in correctional settings has involved some spirited differences. While Martinson[16] contended that "with few exceptions, nothing works," Ross and Gendreau's research indicates that there are treatment programs that do work. Their analysis of dozens of widely different programs carried out in a variety of institutional and community settings indicated that many reported success.[17] Recognizing that many treatments have failed, their analysis suggests that reasons for this may include reliance on a single method, reliance on a single outcome, or simply not enough treatment in terms of time spent or resources committed.

It would appear that we have available treatment methods that do work, at least for some people. What does not seem to work many times is our will and commitment to deliver these treatments in an effective manner. Too often agencies and treatment personnel are unaware of techniques that have demonstrated effectiveness. Knowledge about successful treatments needs to be successfully disseminated and better delivered to those juveniles in need.[18]

Aftercare

In many jurisdictions **aftercare** (parole) is an afterthought. It is the weakest link in the system and in great need of improvement. This could be done by thinking more in terms of providing continuing treatment. Many times little or no thought and planning are given to the transition when a youth is released from an institution or residential setting and faced with living back in the community. Failures attributed to residential treatment efforts may, in fact, be more the result of what happens to juveniles after they are released. Faced with various **transitional adjustments,** including home and living arrangements, negative labeling, and schooling, it is apparent that a comprehensive and complete treatment plan should include this transitional period.

Although providing aftercare may be structured in different ways, one model for developing continuing care suggests the inclusion of a variety of specific components.

1. Home visits prior to release
2. Living arrangement for the youth upon release
3. A continuation of the treatment program and services within the community
4. Identification of community support systems to include the family and social worker
5. Availability of twenty-four-hour supervision
6. A contract to achieve specific goals
7. A gradual phasing out of services and supervision that is based on the youth's response, not on a predetermined schedule.

In addition it must be recognized that some juveniles came from a dysfunctional home environment to which they cannot return. Therefore, aftercare services will need to be involved in the development of alternative living arrangements such as foster homes and independent living programs.[19]

Aftercare represents a form of community-based treatment, incorporating ideas discussed at the beginning of chapter 10. The basic intent is to reintegrate a person's life into positive activities within the community. Such a goal must include support and involvement by the community. Therefore, efforts to improve this weak link in juvenile corrections must include a strategy for broad-based community support.

Summary

Treatment for juvenile offenders is guided from time to time by various philosophies. The founding philosophy of rehabilitation has been joined by competing philosophies of due process, nonintervention, and crime control. Elements of each may be found in juvenile corrections, sometimes in overlapping fashion.

Children were formerly incarcerated with adults. Developments around the beginning of the twentieth century resulted in separate institutions to hold only children. Only a few states have adopted a deinstitutional model for delinquents, although this has been accomplished without endangering public safety.

Dispositions for juveniles are generally indeterminate. Probation is the most widely used disposition. Dispositions are usually viewed as moving from a less restrictive alternative to a more restrictive alternative.

There are thousands of facilities, both public and private, that hold juveniles in custody. Some operate with a more open environment, while others have a secure, close environment.

Programs are characterized by a variety of approaches. Interventions may occur at different points throughout the system. Some programs are delivered in residential settings, while others are nonresidential.

The courts have held that juveniles have a right to treatment but have not indicated what constitutes a minimum treatment program. A variety of individual and group treatment methods have been used. Available evidence indicates that some treatment attempts have failed, while others have been successful. Practitioners need to become more knowledgeable about treatments that do work, and greater policy commitment to delivering these treatments appears to be necessary to meet delinquents' needs.

Aftercare is the most neglected aspect of treatment for juveniles. Continuation of treatment and services during the transition back into the community are generally lacking. Effective efforts to provide continuing aftercare rely heavily on community support and involvement.

Discussion Topics

1. Which of the four competing treatment philosophies more closely expresses your view? Why?
2. How could deinstitutionalization be accomplished in more states?

3. How could treatment personnel become more aware of successful treatment methods?
4. Develop some ideas to improve aftercare in your community.

Additional Readings

Kratcoski, Peter C. *Correctional Counseling and Treatment.* 3d ed. Prospect Heights, IL: Waveland Press, Inc., 1994.

Minor, Kevin I. "Juvenile Delinquency and the Transition to Monopoly Capitalism." *Journal of Sociology and Social Welfare,* 20, December, 1993, 59–80.

Ross, Robert R. and Paul Gendreau, eds. *Effective Correctional Treatment.* Toronto: Butterworths, 1980.

Notes

1. Larry J. Siegel and Joseph J. Senna, *Juvenile Delinquency: Theory, Practice, and Law,* 3d ed. (St. Paul, MN: West Publishing Company, 1988), 346–351.
2. Paul Lerman, *Deinstitutionalization and the Welfare State* (New Brunswick, NJ: Rutgers University Press, 1984), 3–4.
3. For example, see Donald Clemmer, *The Prison Community* (Boston, MA: Christopher Publishing Company, 1940) and Erving Goffman, *Asylums* (New York: Anchor Books, 1961).
4. Lerman, *Deinstitutionalization and the Welfare State,* 147.
5. Barry Krisberg, Ira M. Schwartz, Paul Litsky and James Austin, "The Watershed of Juvenile Justice Reform," *Crime and Delinquency* 32 (January 1986): 29.
6. Richard W. Snarr, *A History of Services for the Commonwealth's Children* (Frankfort, KY: Department of Social Services, 1989).
7. Charles Bonta, "Kentucky Statistics on Juvenile Arrests, Adjudication, and Disposition," Kentucky State Police Uniform Crime Reports, Frankfort, KY, March, 1987.
8. Materials in this section are adopted from Bureau of Justice Statistics, *Report to the Nation on Crime and Justice: The Data* (Washington D.C.: Government Printing Office, 1983), 77.
9. Ibid., 75.
10. Bureau of Justice Statistics, *Children in Custody, 1975–85* (Washington, D.C.: Government Printing Office, 1989), 28.
11. Richard W. Snarr and Jesse J. Bowe, "The Nature of Programs Utilized in Response to Delinquent Behavior Among Juveniles" (Paper presented at The Academy of Criminal Justice Sciences' Annual Meeting, Orlando, FL, 17–21 March 1986), 3–4.
12. Adopted and reprinted by permission from Margaret L. Woods, *Alternative to Imprisoning Young Offenders: Noteworthy Programs* (Fort Lee, NJ: National Council on Crime and Delinquency, 1982), 2–7.
13. *Inmates of the Boys' Training School v. Affleck* 346 F. Supp. 1354 (D.R.I. 1972) and *Nelson v. Heyne* 491 F. 2d 353 (7th Circuit 1974).
14. Siegel and Senna, *Juvenile Delinquency,* 546.
15. William Glasser, *Reality Therapy* (New York: Harper and Row, 1965).
16. Robert Martinson, "What Works? Questions and Answers About Prison Reform," *The Public Interest* 35 (Spring 1974), 22–54.
17. Robert R. Ross and Paul Gendreau, eds. *Effective Correctional Treatment* (Toronto: Butterworths, 1980), 23.
18. T. E. Backer, R. Liberman, and T. Kueknel, "Dissemination and Adoption of Innovative Psychosocial Interventions," *Journal of Consulting and Clinical Psychology,* 54 (February 1986): 111–118.
19. David W. Richart and Sandra Tiley, *More Than Prison Bars* (Kentucky: Kentucky Juvenile Justice Commission, November, 1982), 15.

Unit 5

Selected Issues

This text concludes by presenting for your thoughtful consideration several very controversial issues that involve corrections and the nation as a whole. An opportunity is represented to begin examination of these controversies and to explore ways to impact these problems as a corrections student, corrections professional, or voting citizen.

Chapter 15 Selected National Issues Involving Corrections

Chapter 15

Selected National Issues Involving Corrections

Learning Objectives

After studying this chapter students should be able to:

1. Recognize some limitations of corrections.
2. Better understand corrections as being only a portion of a whole society.
3. Identify and discuss several controversial issues involving corrections.
4. Formulate better reasoned ideas to deal with issues facing corrections and society.

Key Terms

Aggravating circumstances
Capital punishment
Guided discretion
Mitigating circumstances
National issues
Privatization
Racial discrimination
Rate of imprisonment
Retribution
Unequal justice
Vengeance

Introduction

Corrections is a complex and difficult enterprise that functions as part of a larger social, economic, and political system. A flow of persons into corrections for supervision is related to a variety of factors that in many ways are beyond the control of correctional personnel. Political pressures, new sentencing laws, and changing attitudes represent some sources of external forces and decisions that impact corrections. Once sentenced for correctional supervision, there is a very identifiable group of people. By examining this population, a nation can note such things as who is being sentenced, type and length of sentence and, in general, can form judgments about how political decisions, laws, and policies of our society are working and how financial resources are allocated.

As this occurs, different opinions, controversies and issues emerge, identified as correctional issues but having far-reaching implications for the whole society. Although corrections may be the focus of debate, such issues involve the manner in which a nation manages its crime problems. Such **national issues** abound. Several that carry deep significance and controversy were selected for review and presentation of reasoned opinions: (1) rate of imprisonment and prison overcrowding; (2) race and imprisonment; (3) privatization; and (4) the death sentence.

Rate of Imprisonment

An array of statistics and data from across the nation clearly demonstrates our heavy reliance on imprisonment. Recalling information from chapter 6, it was noted that both the **rate of imprisonment** and the number in prisons are at all-time highs and going even higher. Construction and operating costs for only a 500-bed prison over thirty years are conservatively estimated to be $225 million. Stacks of data tell the same story.

As a matter of perspective, other societies do not place such a heavy reliance on incarceration. The United States has the second highest known proportion of its people incarcerated than any other nation, surpassed only by the Russian Republic.[1] Recognizing that legal definitions and traditions vary, comparisons of several incarceration rates per 100,000 population indicate 40 in the Netherlands, 84 in France, 97 in Mexico, 558 in the Russian Republic, and 519 in the United States.[2]

Governments in countries with lower incarceration rates have adopted rationales and policies that place less reliance on imprisonment as a means of social control. Instead, emphasis is placed on the use of fines, probation, employment, and treatment in the community. Incarceration is viewed as a last resort, usually reserved for only the most serious offenders. Efforts are made to avoid dehumanizing experiences while in prison. Overcrowding and lengthy stays are generally not in evidence, and a high majority of prisoners serve less than a one-year sentence.

Increasingly in the United States, legislators have passed stricter sentencing laws. This has been especially true for drug and drug-related offenses. Generally these laws call for more mandatory prison sentences, longer periods of incarceration, and sometimes elimination of early release through parole. As a result, we have an overburdened corrections system marked by prison overcrowding and related pressures. This is often construed as a "corrections problem" when it is really a national problem.

We have tried to "build our way out of this overcrowding problem" by constructing new prisons. However, little progress has been made due to the ever-increasing number of sentenced offenders. Obviously, correctional budgets continue to soar upward for construction, operating, and personnel costs. More importantly, despite claims to the contrary, there is little valid evidence that our problems with crime are being solved or that we have a safer society. There is little evidence that reliance on incarceration contributes much to a lasting reduction of drug crimes, which have been especially targeted. Furthermore, it is simply impossible to have every drug user do prison time. Many times it appears that we incarcerate only out of frustration and as a quick response to what are more underlying social problems and causes, such as health care, housing, child abuse, education, and employment.

Examining this scenario of ever-increasing expenditures and overcrowded prisons generates some very basic questions. Reasonable and rational taxpayers might well ask: When compared to other countries, why must so many of our own citizens be imprisoned? Could money be spent more effectively in other ways to reduce crime problems? Why not use more of the alternatives that have been developed? Does prison improve people? How many prisons are enough? Although there are not quick, easy answers to such questions, being among the world's democratic and free nations it would seem that we would not forever want to continue increasing the rate of incarceration of our own people.

Race and Imprisonment

Corrections and the criminal justice system have an overwhelming impact on young black males. This issue was very clearly described in a report by the Sentencing Project and provides primary information for this discussion.[3] Findings based on this report are presented in table 15.1 and figure 15.1. Researchers studied young black, white, and Hispanic adults, ages twenty to twenty-nine, and noted that more persons than ever were under some form of correctional control—either on probation, in jail or prison, or on parole. Additionally a measure of participation in higher education was calculated. Among the sobering facts revealed were that on any given day

- one in four young black males was on probation, on parole, in jail or prison;
- one in ten Hispanic males was under correctional control; and
- one in sixteen white males was under correctional control.

Table 15.1

Criminal Justice Control Rates

Population Group 20–29	State Prisons	Jails	Federal Prisons	Probation	Parole	Total	Criminal Justice Control Rate
Males							
White	138,111	94,616	15,203	697,567	109,011	1,054,508	6.2%
Black	138,706	66,188	7,358	305,306	92,132	609,690	23.0%
Hispanic	36,302	24,357	6,155	134,772	36,669	238,255	10.4%
Total						1,902,453	8.4%
Females							
White	6,320	7,099	944	141,174	8,712	164,249	1.0%
Black	6,072	6,095	665	58,597	6,988	78,417	2.7%
Hispanic	1,509	2,036	488	29,850	3,210	37,093	1.8%
Total						279,759	1.3%

Source: Marc Mauer, "Young Black Men and the Criminal Justice System: A Growing National Problem." Copyright © 1990 The Sentencing Project, Washington, D.C. Reprinted by permission.

Related findings were that:

- direct control costs for the 609,690 black males were $2.5 billion per year;
- from 1979–1988 the number of prison inmates doubled and crime rates still increased by 2 percent;
- more *younger* black males within the twenty to twenty-nine age group were under correctional control (n = 609,690) than the total number of black males *of all ages* enrolled in college (n = 436,000). Comparable figures for white males were 1,054,508 under control and 4,600,000 enrolled in college.[4]

Regardless of the reasons these younger black males are under correctional control, it is reasonable to suggest that by "getting tough" and simply controlling and locking increasingly more of them up has been a bankrupt strategy. The situation is alarmingly out of balance and destructive to a democratic and free society when one racial segment has more people under criminal justice control than are participating in higher education. The results of such high rates of incarceration will often remain a lifetime. Most will find it very difficult to carry a label of ex-offender, and still become a productive citizen. Succinctly stated, ". . . given these escalating rates of control, we risk the possibility of writing off an entire generation of black men from having the opportunity to lead productive lives in our society."[5]

Privatization of Prisons and Jails

Should government contract with private-sector vendors to operate adult prisons and jails? This question has given rise to several issues and controversies. As governments face continued shortages of prison space, escalating costs,

Figure 15.1 Male Participation in Criminal Justice and Higher Education

Source: Marc Mauer, "Young Black Men and the Criminal Justice System: A Growing National Problem." Copyright © 1990. The Sentencing Project, Washington, D.C. Reprinted by permission.

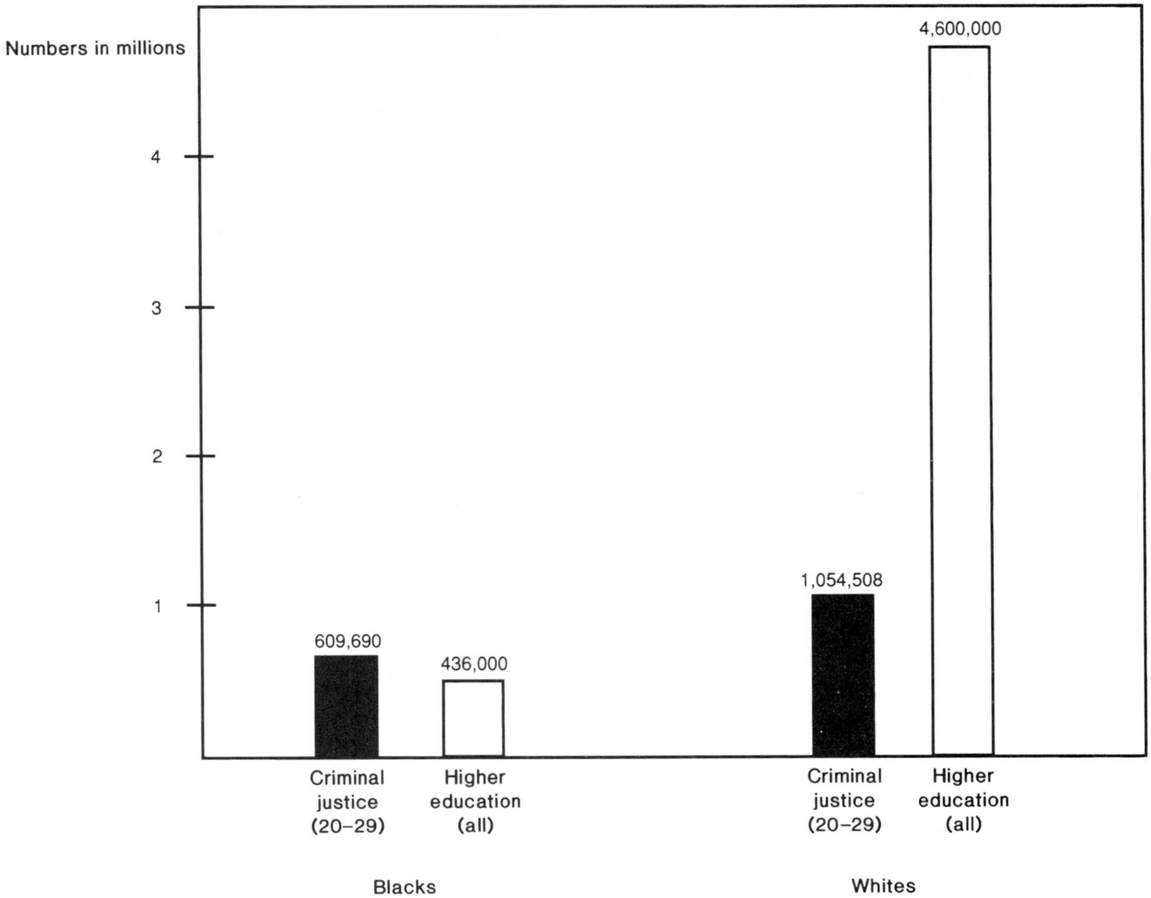

and court orders to reduce overcrowding and related problems, the idea of contracting with private vendors to help ease these pressures is increasingly considered. Under **privatization,** facilities are financed and operated by private agencies aimed at making a profit. Such agencies include Corrections Corporation of America, Behavioral Systems Southwest, United States Corrections Corporation, and Wackenhut Security Services.

Private enterprise has long provided various services and goods to minimum security institutions, especially at the juvenile level. Most contracts for adult facility operations have been for local jails and detention centers and various community-based facilities, such as halfway houses and prerelease centers. Contracting for management and operation of secure adult prisons has been limited to housing a small proportion of inmates, but is increasing. The first

Prison overcrowding has prompted construction of some private prison facilities, but not without controversy.

such facility contracted out by state government was Kentucky's minimum security Marion Adjustment Center, which began operation in January 1986.[6]

Critics charge on ideological grounds that it is improper for a private agency to be involved in the business of depriving people of certain liberties, arguing that this function should remain with the public sector, otherwise, civil and constitutional rights would be violated. Initially, such concerns posed a major question regarding whether government had a legal authority to contract for private operation of correctional facilities. Subsequent clarification indicates that such authority does exist, unless specifically prohibited by state law. For example, *Milonar v. Williams*[7] indicated that operation of a private facility did not violate Eighth Amendment rights. The facility was considered to be an extension of the state. Similarly, *Medina v. O'Neill*[8] ruled that the state could delegate the power to imprison offenders to private prisons. Despite such rulings, considerable objection on philosophical grounds is likely to remain.

Other issues relate to government cost and profit for the private contractor. Although much additional research is needed, preliminary indications are that private-sector construction and operating costs are similar to or slightly less than comparable public costs. However, providers may be earning little profit at the current level of charges.[9] Additionally, it has been suggested that total costs to private correctional firms could rise as they become more involved in legal expenses with liability issues.[10]

Some concerns have been expressed that private firms might exploit inmates in ways to earn more profit. While there is no evidence of this having happened, this potential should be adequately controlled by effective monitoring, good management, and inmate access to the courts. Private companies also are often committed to having their contracts renewed and have an interest in operating in a legal and satisfactory manner for the government.[11]

The private operation of prisons and jails continues in large measure in uncharted waters. Until actual experience accumulates, there is little to guide policy and approaches. Without a doubt, in the decades ahead additional court decisions, agency experience and model guidelines will help clarify issues and provide direction. Movement toward privatization has been cautious and careful. It remains to be seen how many inmates may eventually be housed in private facilities. Although such a population may or may not grow substantially, it seems that the vast array of public facilities will continue to operate, whatever growth may occur in the private sector.

Experts agree that if a governmental unit is considering privatization, the following measures are essential:

- There should be statutory authority in the jurisdiction.
- Contracts should be specific and clear, with appropriate insurance and indemnification clauses to cover liability issues.
- Terms of the contract and standards for the operation of the facility need to be monitored.
- Decisions affecting the liberty interests of inmates (e.g., good time, discipline, release, and transfer) should have final review with the governmental agency.
- Evaluations need to include analyses of benefits as well as costs.[12]

The Death Sentence

In some ways it is ironic that society has designated corrections to carry out executions. On the one hand we speak in such terms as correcting, rehabilitating, and supporting, and, on the other hand, authorize that same governmental unit to administer a death penalty. Most other Western, industrialized nations have abolished **capital punishment;** Great Britain, Sweden, and France provide examples, and their homicide rate is about a third as great as the United States.[13]

Probably no sentence elicits as much controversy as capital punishment—a sentence of death. A death sentence represents a very emotional issue on which citizens tend to become polarized. Frequently, people consider the same information or data and draw opposite conclusions. Arguments presented for and against capital punishment have gone on for centuries and will continue. These arguments have ranged from complex philosophical viewpoints to other considerations such as costs involved.

Since our earlier colonial days, it has been estimated that American jurisdictions have lawfully executed approximately 16,000 offenders.[14] Thirty-six

Figure 15.2 Death Penalty Laws by State

Source: Adapted and updated from Bureau of Justice Statistics, Capital Punishment in 1992 *(Washington, D.C.: Government Printing Office, December, 1993), cover page.*

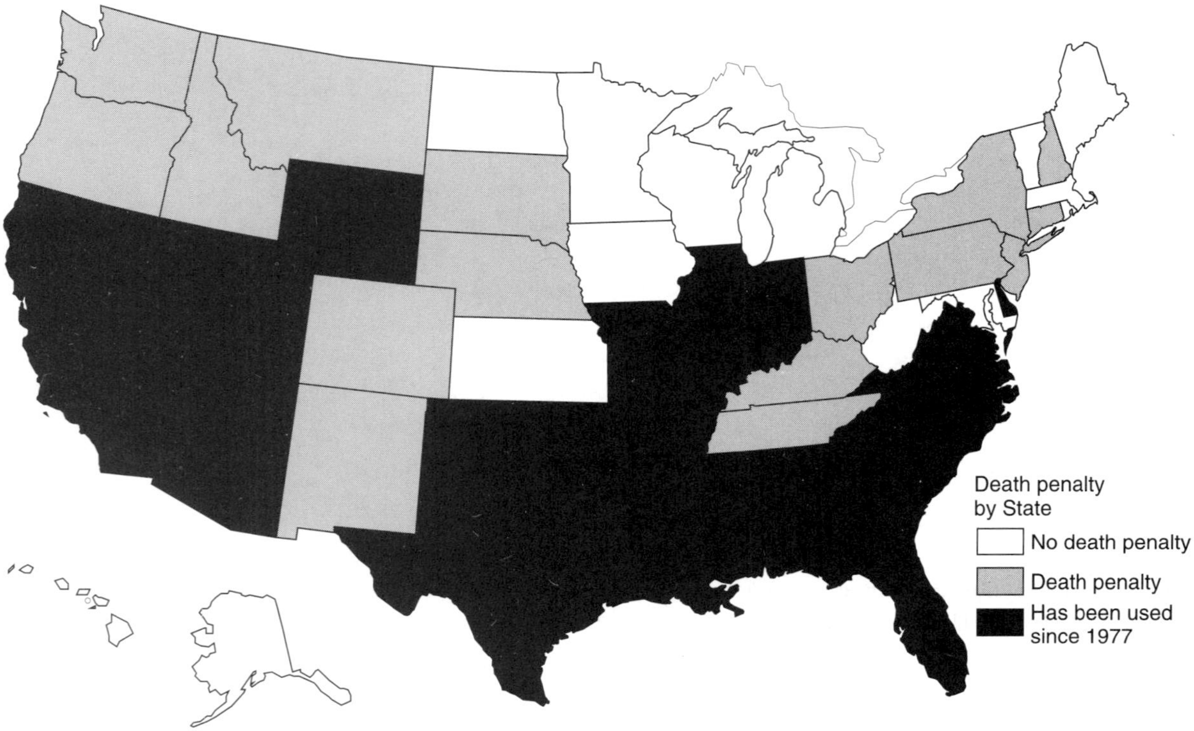

states and the federal government currently have death penalty laws in effect as shown in figure 15.2.

Each year in the United States a total of approximately 275 people receive a death sentence, but an average of only about twenty individuals are actually executed each year.[15] Allowing for the high reversal rate of death sentences, the death row population increases by about 200 each year, is approaching a total of about 3,000 persons, and is now larger than at any time in our history.[16]

Systematic records on the use of the death sentence have been available beginning in 1930. Trends since that time are depicted in figure 15.3. For three decades since the mid-1930s, the trend was downward. For ten years between 1967 and 1977, no executions occurred. Following this ten-year moratorium, the recent era of executions began with the much-publicized firing-squad death of Gary Gilmore in Utah on January 17, 1977. Since that time, close to 200 individuals have been executed in twenty states.[17]

Present-day capital punishment has been guided by several landmark Supreme Court cases beginning in 1972 with *Furman v. Georgia*.[18] In this case,

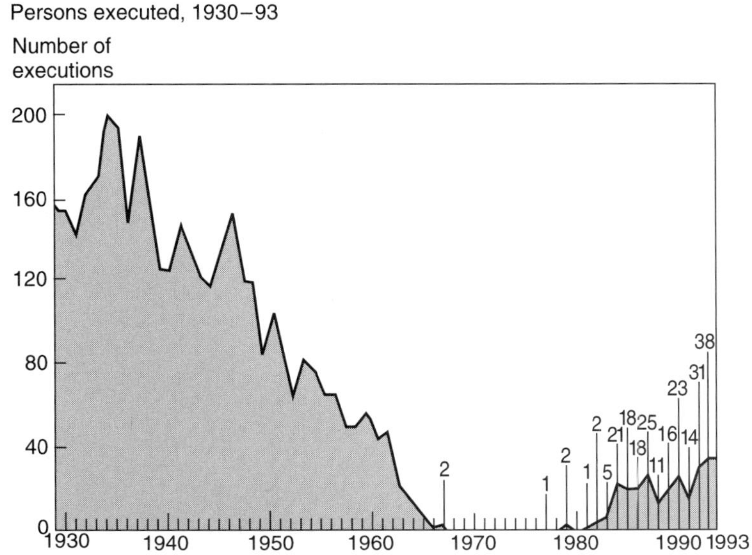

Figure 15.3 Persons Executed, 1930–1992

Source: Adapted from Bureau of Justice Statistics, Capital Punishment in 1992 *(Washington, D.C.: Government Printing Office, December, 1993), p. 9.*

the United States Supreme Court ruled that the death penalty *as applied* was arbitrary and capricious, constituting ''cruel and unusual punishment'' in violation of the Eighth Amendment. The net result of the *Furman* decision was the invalidation of all United States death penalty laws and the eventual lifting of death sentences for the more than 600 inmates who had accumulated on death row between 1967 and 1972. During subsequent years, many states enacted revised legislation designed to conform to the standards set forth in the nine separate opinions of the Supreme Court justices in the *Furman* case.

Revised laws passed in Georgia provided for **guided discretion,** whereby consideration was to be given to **aggravating** and **mitigating circumstances** by the jury at a sentencing hearing. In 1976 the United States Supreme Court in *Gregg v. Georgia*[19] upheld these death penalty statutes, finding that a death penalty does not violate *per se* the Eighth Amendment; the era of guided discretion in the use of the death penalty was launched. Subsequent cases, such as *Roberts v. Louisiana*[20] and *Lockett v. Ohio,*[21] continued to favor guided discretion, ruling against mandatory capital punishment and requiring consideration of mitigating circumstances.

It seems, however, that court rulings and new laws have done little to reduce **racial discrimination** in the application of the death penalty. It has been demonstrated that:

> In the time between the 1976 Supreme Court decisions and September 1, 1985, 46 men and one woman were executed in the U.S. . . . Only four of the 47 executed defendants had black victims (one other had both white and black victims). . . . As of August 1985 there were 1,520 men and 20 women sentenced to death in the United States. Of these 1,540 inmates, 783 (50.8%) were white, 640 (41.56%) were black, (and) 113 (7.34%) were other

> minorities. . . . Between 1930 and 1967, 50.09% of those executed for homicide were nonwhite, and in August 1985, 49.50% of those on death row were nonwhite—a difference of less than 2%. Despite all the emphasis on formal standards after *Furman*, it appears that the racial composition of the condemned population has changed very little.[22]

McClesky v. Kemp,[23] decided in 1987, held that statistics showing racial discrimination against African Americans in the imposition of a death sentence was not unconstitutional.

Whether or not a death sentence could be imposed on juvenile offenders was an issue for decades. Two recent cases, *Thompson v. Oklahoma*[24] and *Stanford v. Kentucky*[25] have clarified this issue. In 1988, *Thompson* said it was unconstitutional to impose a death sentence on someone who committed the offense at age fifteen or younger. The next year, 1989, the *Stanford* decision held that imposing a death sentence was not unconstitutional for someone who was sixteen or older when the offense was committed.

What has been striking in America during the past several decades is the change in public opinion regarding capital punishment. During the 1960s support for the death penalty declined; slightly more Americans opposed than supported the death penalty. Since that time, attitudes shifted with up to three-fourths or more now supporting capital punishment.

What seems to be uppermost among supporters is the desire for **retribution.** The appeal of other arguments, such as deterrence, has been largely replaced by a desire for what many see as **vengeance.** Arguments based on retribution are simpler and easier to make and much more widely understood. This creates an irresistible issue with high media appeal for politicians to show they are tough on crime. Critics contend, however, that retribution can be served without resorting to a death penalty. Meeting a need for retribution could be met with less excessive penalties, such as long and life imprisonment.

There are approximately 20,000 murders per year in the United States, but only about twenty executions. **Unequal justice** is being rendered. It is unequal at the outset since capital punishment is authorized for citizens under laws in some states but not in others. Furthermore, table 15.2 shows that a small number of offenders in only a few states are sacrificed to the death penalty each year, apparently satisfying vengeance needs, while all other such offenders do not pay with their life. Such an outcome in a civilized and democratic nation is beyond moral and legal sensibility. Problems are not being solved with such an approach and searches for satisfactory solutions are hindered.

Concluding Remarks

A case could be made that our nation is overloading correctional systems in ways additional to overwhelming prison populations. In the opening chapter, we learned that corrections is one major component of criminal justice, designed to supervise criminal law offenders. It seems, however, that many times we are expecting corrections to accomplish tasks and solve problems that it was not really designed nor originally intended to do.

Table 15.2

Number of Persons Executed, by Jurisdiction in Rank Order, 1930–92

State	Number Executed	
	Since 1930	Since 1977
U.S. total	4,047	188
Georgia	381	15
Texas	351	54
New York	329	
California	293	1
North Carolina	268	5
Florida	199	29
Ohio	172	
South Carolina	166	4
Mississippi	158	4
Louisiana	153	20
Pennsylvania	152	
Alabama	145	10
Arkansas	122	4
Virginia	109	17
Kentucky	103	
Tennessee	93	
Illinois	91	1
New Jersey	74	
Missouri	69	7
Maryland	68	
Oklahoma	63	3
Washington	47	
Colorado	47	
Indiana	43	2
West Virginia	40	
District of Columbia	40	
Arizona	39	1
Nevada	34	5
Federal system	33	
Massachusetts	27	
Connecticut	21	
Oregon	19	
Iowa	18	
Utah	17	4
Kansas	15	
Delaware	13	1
New Mexico	8	
Wyoming	8	1
Montana	6	
Vermont	4	
Nebraska	4	
Idaho	3	

(continued on next page)

Table 15.2 (continued)

State	Number Executed	
	Since 1930	Since 1977
South Dakota	1	
New Hampshire	1	
Wisconsin	0	
Rhode Island	0	
North Dakota	0	
Minnesota	0	
Michigan	0	
Maine	0	
Hawaii	0	
Alaska	0	

Source: Bureau of Justice Statistics, Capital Punishment in 1992 *(Washington, D.C.: Government Printing Office, Dec. 1993), p. 9.*

In addition to supervising law violators, corrections is increasingly expected to manage and essentially solve a long list of diverse social problems related to health care, illiteracy, child abuse, mental illness, drug/alcohol abuse, children of incarcerated mothers, training for jobs, and living skills. While corrections has some role in dealing with such problems, other social institutions, such as educational and health enterprises for example, are the primary institutions designed to manage such problems. The nation's institutions appear out of balance when so many young citizens are sentenced for correctional supervision who cannot read and write, have not had access to necessary health care, were abused as children, have no job, and abuse alcohol/drugs. Our country needs to recognize that many persons presenting themselves at the nation's prison doors have serious and long-standing problems that go far beyond their criminality, and it is unrealistic to expect social control institutions such as corrections to provide lasting solutions for these types of problems. That is expecting corrections to solve problems for which it was not designed.

Continuing the validity of our democratic nation demands that we move beyond a simple "get tough on crime" approach. The problems are too complex not to do otherwise.[26] Such problems should elicit our careful attention as corrections students, professionals, and also voting citizens. Perhaps a bit more tolerance, humanity, and understanding would be beneficial in approaching these problems among our own people. More reliance for decisions and management should be based on solid research and valid scientific information. Honest debate on issues should replace polemic rhetoric. There is a need for politicians and citizens with courage and leadership to move away from "locking more people up" and to work toward prevention and better solutions. The challenges and opportunities that you have at this time to impact these problems as a corrections student, corrections professional, or as a voting citizen have never been greater. You have a chance to make a difference, and each of these roles will likely be more satisfying when based on knowledge and education. Toward that end, it is hoped that this textbook and opinions expressed (whether you agree or disagree) will be useful to you.

Summary

Corrections operates as part of a larger political and social system. Legislators pass new laws, courts define changing legal boundaries, and citizens wonder how corrections spends their tax dollars. Such items often become national concerns pushing correctional topics to the forefront for consideration. Correctional students and professionals, as well as voting citizens have compelling reasons to be as knowledgeable as possible when such issues emerge.

Examples of several difficult issues were presented in this chapter for your consideration. Our nation has a high rate of incarceration, incarcerates a highly disproportionate number of young black males, is debating the merits of privatization of prisons, and inflicts a death penalty, which most other Western, industrially advanced nations have abolished.

Many people coming to prison have problems rooted in early childhood, such as lack of health care, being abused, and inadequate education. Perhaps our nation is expecting too much from corrections to manage and solve these problems when these people reach adulthood. How can our nation prevent some of these problems? Such issues pose challenges and opportunities for you as a student, professional, and citizen.

Discussion Topics

1. Select one issue identified in this chapter. Do you agree or disagree with the opinions expressed on that issue? Explain why or why not.
2. Could we have a reasonably safe and free society if we had a lower incarceration rate?
3. How can our nation prevent such a high proportion of young black males from ending up in prison?
4. If it ever became an election issue, would you vote for or against privatization of prisons? Why?
5. Is it unfair that only a tiny fraction of those found guilty of murder are executed?
6. Do you think our nation is expecting corrections to manage and solve problems it was not designed to solve?

Additional Readings

Earlie, Michael K. and Kevin I. Minor. *Prisons Around the World.* Dubuque, IA: William C. Brown Publishing Company, 1992.

Irwin, John and James Austin. *It's About Time: America's Imprisonment Binge.* Belmont, CA: Wadsworth Publishing Company, 1994.

Logan, Charles H. *Private Prisons: Cons and Pros.* New York: Oxford University Press, 1990.

Mauer, Marc. "Americans Behind Bars: The International Use of Incarceration, 1992–93." The Sentencing Project. September, 1994.

Notes

1. Marc Mauer, ''Americans Behind Bars: The International Use of Incarceration,'' *The Sentencing Project* (Washington, D.C., September, 1994).
2. Ibid., 4.
3. Marc Mauer, ''Young Black Men and the Criminal Justice System: A Growing National Problem,'' *The Sentencing Project* (Washington D.C., February, 1990).
4. Ibid., 3.
5. Ibid., 4.
6. Judith C. Hackett et al., ''Contracting for the Operation of Prisons and Jails'' (Washington, D.C.: National Institute of Justice, June, 1987), 2.
7. 691 F. 2d 931 (1982).
8. 589 F. Supp. 1028 (1984).
9. Hackett et al., ''Contracting for the Operation,'' 7.
10. Charles W. Thomas and Linda S. Calvert, ''Access to Qualified Immunity by Private Defendants in 42 Section 1983 Damage Suits: The Implications for Correctional Privatization'' (Paper presented at the annual convention of the Academy of Criminal Justice Sciences, Denver, CO, 17 March, 1990), 45.
11. Bureau of Prisons, ''Private Sector Management of Prisons'' (Washington, D.C., February, 1989), 2.
12. Ibid., 3.
13. Dean J. Champion, *Corrections in the United States* (Englewood Cliffs, NJ: Prentice Hall, 1990), 517.
14. Victor L. Streib, ''Testimony: Legislative Criteria Concerning the Death Penalty for Juveniles'' (Submitted to the Constitution Subcommittee of the United States Senate, 27 September, 1989), 4.
15. Bureau of Justice Statistics, *Capital Punishment 1992* (Washington, D.C.: Government Printing Office, December, 1993), 9–10.
16. Ibid., 2.
17. Ibid., 9.
18. 408 U.S. 238 (1972).
19. 428 U.S. 155 (1976).
20. 431 U.S. 633 (1977).
21. 438 U.S. 536 (1978).
22. Michael L. Radelet and Margaret Vandiner, ''Race and Capital Punishment: An Overview of the Issues,'' *Crime and Social Justice,* 25 (1986): 100.
23. 481 U.S. 279 (1987).
24. 487 U.S. 815 (1988).
25. 109 S. Ct. 2969 (1989).
26. Charles M. Friel, ''Intergovernmental Relations: Correctional Policy and the Great American Shell Game,'' *Criminal Justice in the 1990s: The Future of Information Management* (Proceedings of a BJS/SEARCH conference, April, 1990), 36–46.

Glossary

A

Administrative segregation A separate, very controlled section of the prison for troublemakers.

Aftercare The conditional release of a juvenile from an institution under community treatment and supervision.

Aggravating circumstances Factors that could produce a more severe sentence.

Alternative release options One of several means used instead of bail for pretrial release of defendants.

Amnesty A form of pardon precluding prosecution applying to a group rather than one individual.

Ashurst-Summers Act The 1935 federal legislation making it a criminal offense to transport in interstate commerce prison-made products for private use.

Augustus, John Considered to be the "father of probation." A bootmaker by trade in Boston, Massachusetts, who voluntarily and on an informal basis aided offenders released to him by the court.

Autocratic management style Concentrates power with emphasis on formal structure with little consideration given to human needs and employee participation.

B

Bail Any one of a number of mechanisms for pretrial release from detention of accused individuals.

Banishment The exclusion of an offender from the social group.

Benefit of clergy Original practice extended to members of the clergy in which their cases were heard in church courts where punishment was less severe than in secular courts. This was considered as one of the forerunners of probation.

Bill of Rights The first ten amendments to the United States Constitution which enumerate basic rights, such as no cruel or unusual punishment and freedom from self-incrimination.

Bondsperson A professional who pays bail for another for a fee.

Boot camp Shock incarceration units with programs patterned according to a military model.

Branches of government The separation of governmental power into legislative, executive, and judicial functions.

Brokerage approach A service delivery strategy that emphasizes the role of probation officers as community-resource managers, thereby providing a link to community agencies. This strategy deemphasizes the close one-to-one relationship characteristic of the casework approach.

C

Capital punishment The judicially ordered execution of a convicted criminal.

Casework approach A service delivery strategy in which the caseworker assumes major responsibility to change offender's behavior through a close one-to-one relationship.

Chemically dependent An individual who is physiologically and/or psychologically dependent on alcohol, opium derivatives and synthetic drugs with morphinelike properties, stimulants, and depressants.

Citizens' advisory board Utilizing local citizens in planning and operating community-based corrections.

Civil law A division of law under which private disputes are regulated.

Civil rights The legally guaranteed rights due citizens by just claim.

Classical view Considers free will and choice as major determinates in criminal behavior.

Classification An ongoing formal process concerned with identification, categorization, and assignment of inmates to various levels of security, programs, and work.

Coed prison Prison facilities holding both males and females. To be effective, both sexes should be managed and treated similarly, and the ratio of females to males should be in the range of forty females to sixty males or sixty males to forty females.

Community acceptance A necessary component for successful community-based corrections efforts.

Community-based corrections Any and all activities involving the community in efforts to reintegrate offenders.

Community service A sentencing alternative that uses a form of nonmonetary or symbolic restitution to victims or the general public.

Commutation The modification of a sentence to the benefit of an offender.

Conditional release The broad umbrella term assigned to the most widely utilized mechanisms for releasing convicted adults and adjudicated juveniles from long-term incarceration.

Conditions of probation/parole The special rules applying to those granted probation. Examples include reporting to probation officer, refraining from alcohol use, and seeking permission to travel outside jurisdiction.

Congregate system Provided for prisoner confinement in separate cells but brought the inmates together into congregate workshops (New York State Prison at Auburn).

Conjugal visits Visits that are private and unsupervised between an inmate and spouse.

Contraband Any restricted or prohibited item so designated by the correctional institution and found in the possession of an inmate or within the facility.

Convict lease Leasing prisoners to private industry as laborers.

Corporal punishment The infliction of physical pain, and sometimes mutilation.

Corporate model Managing prison industry according to successful businesslike principles.

Correctional accreditation A voluntary program to meet American Correctional Association standards.

Correctional standards Minimal conditions to be met to achieve accreditation.

Corrections The systematic and organized efforts directed by a society that attempt to punish offenders, protect the public from offenders, change offenders' behavior, and in some cases, compensate victims.

Court intervention The involvement by courts, especially federal district courts, into a variety of prison conditions marking a reversal of the hands-off doctrine.

Crime control philosophy A philosophy emphasizing punishment and incarceration rather than treatment for juveniles and adults.

Criminal history category Number and seriousness of prior offenses often considered as one axis on a sentencing grid.

Criminal justice The formal crime control apparatus comprising police, courts, and corrections that seeks to identify, convict, sentence, and supervise offenders.

Criminal law Defines conduct considered to be a serious public threat. A person violating the criminal law is subject to prosecution by the government.

Custodial model A corrections model emphasizing restraint and incarceration.

D

Deinstitutionalization The process whereby large traditional institutions have been closed in favor of smaller community-based facilities.

Deinstitutional model Utilizing smaller community-based facilities rather than large, isolated institutions.

Democratic government Power is separated and divided and resides with citizens through elected officials. Major concern for individual rights.

Deprivation model Considers the origin of a prison subculture to be from inside the prison due to negative living conditions.

Determinate sentence A sentence permitting limited discretion that includes a fixed range of prison time.

Determinism Views human behavior as the product of a multitude of environmental and cultural influences.

Deterrence Potential illegal behaviors prevented by a particular legal threat.

Detoxification Process by which an individual who is physiologically and/or psychologically dependent upon a drug is brought to a drug-free state.

Deviance To depart from the normal or acceptable standard.

Direct supervision Found in third-generation jails where officers spend time in the living area actively supervising and interacting with inmates.

Discretion Latitude of free choice within certain legal bounds; when decisions may be made that are not generally open to reexamination by others. Many decisions in corrections and criminal justice are discretionary in nature.

Discretionary release Release from prison determined by a parole authority.

Diversion A process whereby an alleged offender (usually a juvenile delinquent) is "turned away" from further movement into the justice system.

Dual system of courts One result of federalism. Dividing governmental power and authority led to the creation of separate systems of judicial authority.

Due process A fundamental idea wherein a person should not be deprived of life, liberty, or property without legal procedures that are fair and reasonable.

Due process philosophy A philosophy emphasizing legal safeguards, extended into juvenile matters beginning during the 1960s.

E

Electronic monitoring A newer technology to monitor and verify the whereabouts of offenders by use of electronic devices.

Equal protection clause An important portion of the Fourteenth Amendment providing equal protection of the laws to citizens.

Equitable treatment alternatives Providing equal access for women offenders to institutions, resources, and community corrections programs that display parity with programs for male offenders.

F

Federal Bail Reform Act Federal legislation that has provided a model for alternative release options.

Federal corrections A correctional system authorized in 1930 at the federal level of government.

Federalism Dividing rather than centralizing governmental power. In the United States, power is divided by federal, state, and local levels of government.

Federal Probation Act (1925) Authorized the hiring of probation officers at the federal level, thereby helping to establish the practice of granting probation.

Fee system Payment of a fee to a jailor, by a person in jail, to maintain the facility. This system was often corrupt and abusive.

Felony Considered to be a serious crime, such as armed robbery, assault, and murder, with punishment ranging from incarceration for one year or more through a death sentence.

Financial bond Payment to secure pretrial release from jail.

Formal reward system A system to encourage good behavior through increasing privileges, lowering security levels, and awarding good time.

Fourteenth Amendment Defines citizenship and provides equal protection of the law to *citizens* of the United States.

Fragmentation of corrections A feature of overall correctional organizations in the United States. There are many jurisdictions, types of facilities, and interactions with a variety of other systems, making coordination by management difficult.

Free will Central to the classical view that emphasizes freedom of choice and the pursuit of pleasure and the avoidance of pain.

Funnel effect The decrease in volume of cases remaining at successive stages in the criminal justice process.

Furlough programs Programs allowing trusted inmates visits to their home community.

G

Gaol Old English term meaning and pronounced the same as jail.

Good time A credit for appropriate behavior that reduces the length of a prison term.

Group counseling A planned activity in which three or more people are present for the purpose of solving personal and social problems.

Guided discretion Laws guiding the discretionary use of the death penalty, taking into consideration aggravating and mitigating circumstances.

H

Halfway house A variety of community-based programs designed for a variety of offenders, including probationers and parolees.

Hands-off doctrine Prior to the 1960s the refusal of federal courts to intervene on behalf of inmates in correctional operations. A marked decline in this policy began during the latter part of the 1960s.

Hawes-Cooper Act The 1934 federal legislation in which prison-made goods transported from one state to another were subject to the laws of the importing state.

Hawthorne effect Based on a set of findings from a series of experiments conducted in the Hawthorne studies in which production actually increased, even when advantages were eliminated, because the experimental situation created a positive atmosphere wherein close personal attention was being given to workers.

Hidden crime Criminal behavior that is unknown to authorities.

Hidden delinquency Delinquent acts that are committed but are unknown to police or juvenile officials that can be measured by the self-report method.

Hierarchy of needs A theory of human motivation developed by Abraham Maslow ranking human needs from basic to self-actualizing.

Hustling Illegitimate economic activities in prison.

I

Importation model Considers the origin of a prison subculture to come along with prisoners and their prior life experiences.

Incapacitation Any of society's attempts to render a criminal incapable of further illegal acts.

Incarceration Detention in a jail or prison.

Incarceration rate The ratio of people incarcerated to every 100,000 population (e.g., 426/100,000).

Indeterminate dispositions Juveniles remain under supervision of treatment authorities until deemed rehabilitated or legally an adult.

Indeterminate sentence A discretionary sentence permitting a wide range of sanctions by judges and parole authorities.

Index crimes The eight most serious offenses contained in the Uniform Crime Reports: homicide, rape, robbery, assault, burglary, larceny, motor vehicle theft, and arson.

Inmate code A code of conduct governing relationships in prison according to inmate-generated norms and values.

Inmate subculture An enduring complex of norms and values by which inmates seek to accommodate to life in prison.

Institutional model Utilizing larger, more secure, and more isolated institutions rather than smaller community-based facilities.

Intake The initial point of entry into the jail and the official entry of the accused offender into the adult criminal justice system.

Intake function Found at the juvenile level and usually staffed by a probation officer. This person screens each case and has wide discretionary powers in deciding whether to refer cases to court.

Intensive probation A form of supervision that features greater structure and more frequent and rigid monitoring than basic probation.

Intermediate sanctions A category containing a variety of sentencing options, ranging from conditions stricter than regular probation, to less than a longer prison term.

Interstate compact A cooperative arrangement whereby states may exchange the supervision of parolees and probationers.

Isolation process Historical stages in which women prisoners have received unequal treatment.

J

Jail A confinement facility authorized by local or state law to hold individuals for periods in excess of forty-eight hours.

Judicial reprieve An early English practice of suspending the imposition of a sentence. This practice has been considered an antecedent of probation.

Jurisdiction The authority to act in a criminal justice case.

Just deserts A rationale maintaining that punishment be administered in the amount deserved according to the seriousness of the offense.

Justice model A corrections model emphasizing that punishment should be fair and based on the seriousness of the crime.

Juvenile court An American legal institution designed to administer treatment and justice for juveniles.

Juvenile jurisdiction Based on the two factors of age and behavior.

Juvenile justice system Various laws and courts established having jurisdictional authority over younger persons, usually specified as under the age of eighteen. This system is concerned with violation of criminal law by youths, status offenses, and dependent and neglected children.

Juvenile law Those laws affecting the lives of children.

L

Least restrictive alternative model A corrections model advocating minimum intervention and fitting offenders into the community.

Lex talionis Latin phrase that embodies the concept of retaliation and revenge—an eye for an eye, a tooth for a tooth.

Lockup A temporary holding facility that by law or practice can incarcerate an individual for less than forty-eight hours.

M

Mala in se Conduct that may be considered wrong in itself (for example, murder, incest).

Mala prohibita Conduct considered wrong because of the law (for example, status offenses, traffic laws).

Management functions Constituted by basic functional activities of planning, organizing, staffing, leading, and controlling.

Managers Persons, such as warden, superintendent, or shift lieutenant, who perform management activities.

Mandatory release Release from prison based on earned good time credit rather than on parole authority.

Mandatory sentence A nondiscretionary sentence requiring a fixed term of incarceration that can only be reduced by earning good time.

Maximum (or close) custody prisons Facilities designed for inmates who require maximum control and continuous supervision (individuals who have demonstrated behavior that is assaultive, predaceous, riotous, or who pose serious escape risks).

Medium custody prison Facilities for inmates with a history of conduct showing some degree of trustworthiness.

Minimum security Nonsecure facilities for trustworthy inmates.

Misdemeanor Considered a less serious violation of the criminal law, commonly penalized by a fine and/or a short jail sentence.

Mitigating circumstances Factors that could lessen a sentence.

Mother/child programs Efforts made in a few institutions to retain ties between an incarcerated mother and her children.

Mushfake A prison-made copy of something that is available on the streets.

N

National Crime Victim Survey (NCVS) Provides a measurement of crime as based on victimization surveys of United States householders.

National Criminal Justice Reference Service (NCJRS) Provides updated corrections and criminal justice information as a free service for registered users.

Nonintervention philosophy A philosophy seeking to avoid or minimize stigma and labeling as a delinquent.

Nonresidential program Juvenile offenders remain living in their own homes while involved in treatment programs.

Norms of behavior Expectations regarding what behavior is considered socially acceptable; guidelines for behavior appropriate and applicable to particular social situations.

O

Offense level Seriousness of crime often considered as one axis on a sentencing grid.

Open market Prison-made goods in direct competition with private-sector products.

Open system Viewing the operation of prisons with numerous input and output exchanges with other governmental units such as central office, courts, and legislatures.

Ordered segmentation Small cliques and friendship groups often based on racial, ethnic, and gang membership, reflecting the diversified composition of contemporary prison populations.

Organizational chart Formal depictions of positions and relationship of persons in organizations showing such items as position titles, chain of command, and levels of authority. Such charts do not indicate informal organization and power.

Organizations Provide a structure that coordinates work activities of its people in pursuing goals, such as a state department of corrections or local probation department.

P

Pardon Granted to an individual by a chief executive removing blame and protecting from future prosecution for act in question.

Parens patriae A doctrine from English law in which the state assumed authority and responsibility to oversee neglected and abused children. In the United States, this doctrine was extended to also include delinquent children, with the establishment of juvenile courts beginning in 1899.

Parole The conditional release of an adult from incarceration in a correctional institution by a paroling authority, after having served less than a full sentence, and placed under the supervision of a correctional agency empowered with the authority to reincarcerate the releasee upon determining that conditions of parole have been violated.

Parole agency A corrections agency charged with the responsibility to supervise those placed on parole.

Parole authority Typically a parole board or other body that has the responsibility for deciding whether an inmate should be released from incarceration.

Participative management style Concentrates on the development of human resources and involves employees to have a say in running an organization.

Partnership model A joint undertaking between the public and private sector in the operation of prison industries.

Penitentiary Early prisons where offenders were considered to be placed in a state of penitence to regret their wrongdoing and become a contrite and penitent person.

Persistent felon An identifiable group of prisoners who have been convicted of repeated violations. In some states these offenders are subject to longer and mandatory prison terms.

Petition An official document that brings an alleged juvenile offender's case before the juvenile court.

Philosophies Conceptual frameworks designed to account for particular behaviors, actions, or events.

Play families Considered by Rose Giallombardo as the principal adaptation by women to life in prison.

Plea negotiation A procedure in which the defense and prosecution agree to a reduction of punishment. This is commonly accomplished by the accused pleading guilty to a lesser charge.

Positivist view Considers the multitude of factors that help determine criminal behavior.

Presentence investigation A written document, usually prepared by probation officers, used primarily to provide information to the court when deciding on a sentence.

Presumptive sentence A sentence providing limited discretion with only a narrow range permitted between minimum and maximum lengths of incarceration.

Pretrial release program A screening program to identify those suspects eligible for pretrial release.

Prison community The mix of inmates and staff living in prison who in many respects have the same daily needs and required services as found in the outside community.

Prisoners' rights Numerous rights, most of which have been won by inmate suits through case law established by court decisions.

Prison industry Utilizing prisoners to produce goods for some market to meet a variety of goals such as making a profit for the prison, reducing idle time, and enforcing discipline.

Private-sector model Seeks to maximize private enterprise in prison industries.

Privatization Correctional facilities financed and operated by private agencies aimed to make a profit.

Probation A sentencing alternative in which incarceration is avoided and the offender remains in the community under the supervision of a probation officer.

Probation officer Positions of employment at the federal, state, and local level. In working with clients, officers typically perform tasks in investigative, service, and surveillance roles.

Procedural law Outlines due process requirements to be followed in enforcing the law.

Program inequality Underpar resources for female prisoners due to sexist attitudes and relative smaller numbers.

Protective custody Designated living areas within correctional institutions for residents who choose not to live with the general inmate population, usually out of fear.

Public offense Behaviors engaged in by juveniles that violate the criminal law.

Punishment The infliction on a person by the state of consequences normally considered unpleasant in response to having been convicted of a crime.

R

Rate of imprisonment The number of persons incarcerated for each 100,000 persons in the population. This rate for prisons and jails in the United States is 426 per 100,000, which is the highest known rate in the world.

Recidivism A general label assigned to statistical data that reports rates of repeated criminal activity.

Recognizance (ROR) A personal pledge or promise by a defendant before a court. The person remained at liberty with certain conditions imposed on behavior. Some features of this earlier practice have been incorporated into present probation.

Reformatory Facilities for mainly younger first-time offenders that originally stressed education, programs, and release based on behavior, but whose real promise was never completely realized.

Regional jails Sharing a jail among two or more jurisdictions, usually to reduce costs.

Rehabilitation model A corrections model that concentrates on treatment and changing behavior.

Rehabilitation philosophy A philosophy on which treatment for juveniles was founded.

Reintegration A correctional model that places responsibility for change not only on offenders but also upon the community.

Release eligibility date The date usually established by statute when an inmate with a particular sentence will first become eligible for consideration for release on parole.

Reprieve Temporary stay in carrying out a sentence.

Residential programs Juvenile offenders relocated away from their homes for close supervision and treatment.

Resocialization Directed toward relearning processes that remove motivations for future crime.

Restitution The repayment by the offender to victims who have suffered financial losses as a result of the offender's crime.

Retribution Punishment to fit the crime as a payment of debt to society.

Revocation Withdrawing the status of probation or parole.

Right to treatment Courts have ruled that juveniles have a right to treatment but have not indicated what constitutes minimum treatment.

Risk assessment guidelines A statistical prediction approach based on risk factors, such as early crime and early incarceration, utilized by some paroling authorities.

Role conflict A situation in which incompatibility exists between two or more roles that an individual is expected to perform. For example, probation officers often experience role conflict between what is expected from them in the service and surveillance roles.

S

Scientific method An objective, logical, and systematic method of analysis of events devised to permit the accumulation of reliable knowledge.

Section 1983 A portion of the Civil Rights Act of 1871 granting protections that prisoners have used to challenge the constitutionality of various aspects of their imprisonment.

Self-report A method begun during the 1950s to measure delinquency. It has provided more complete information and understanding about juvenile offenders primarily by measuring the extent and distribution of hidden delinquency.

Sentencing The process of placing an authorized judicial penalty upon a person who pleads guilty or is convicted of a crime. Sentencing decisions have direct implications for corrections.

Sentencing disparity Differences in sentences given to offenders for no apparent reason.

Sentencing guidelines Designed to reduce disparity and provide more determinate sentences usually by considering offense level and criminal history.

Separate system Inmates both worked and lived in separate cells, each with its own exercise yard (Eastern Penitentiary in Philadelphia).

Separation from children A special situation for many incarcerated women presenting enormous pain, problems, and challenges.

Sexual deprivation An intended consequence of prison life in the United States.

Sheltered market Prison labor used only for the public good and not in direct competition with private industry.

Shock incarceration A short period of incarceration designed to frighten offenders into law-abiding behavior.

Situationally specific management style Recognizes that no one management style fits every condition and urges adaptation of specific approaches as situational needs of organizations and members warrant.

Slave of the state During the hands-off era, prisoners were considered to have forfeited most rights and were considered to be slaves of the state.

Social control The complex of formal and informal means to promote socially acceptable behavior.

Split sentence Short-term incarceration coupled with a following period of probation.

State and local corrections One consequence of federalism. Correctional authority and functions exist at the state and local levels, separate from federal jurisdiction.

State-use system A common form of sheltered market where inmate-produced products, such as desks or license plates, can only be used by other governmental units.

Status offense Noncriminal behavior, such as truancy and running away from home, that is in violation of law applicable only to juveniles (usually through age seventeen).

Substantive law Defines illegal behavior and prescribes penalties for violation.

Supervision A broad term used to describe various tasks that parole and probation officers perform when monitoring the parolees.

T

Technical violation A term used when a probationer violates a conditional rule of probation. This could result in the revocation of probation but is uncommon in many jurisdictions.

Three generations of jails The different architectural designs and philosophies of supervision that have evolved from intermittent supervision, to remote supervision, to direct supervision.

Ticket of leave Conditional pardons granted to imprisoned offenders.

Total institutions Isolated and dehumanizing facilities designed to fulfill only the official aims of the institution, described by Erving Goffman.

Totalitarian government Power is centralized and controlled by a small number of people. There is little concern for individual rights.

Training schools Traditional facilities for juveniles based on an institutional model. They continue to house large numbers of juvenile offenders nationwide.

Transitional adjustment Problems faced by a juvenile upon returning to home and community after release from a residential program.

Turnkey fee A per-resident fee paid by the local or state government to cover the cost of jail operation.

U

UNICOR The trade name for Federal Prison Industries, Inc., which employs about 30 percent of all federal inmates, manufacturing and selling a variety of products.

Uniform Crime Report (UCR) Provides a measurement of crime based on crimes known to the law enforcement agencies as voluntarily reported to the Federal Bureau of Investigation.

Utilitarian doctrine Associated with Jeremy Bentham. Actions are right insofar as they contribute to maximizing the happiness of people, wrong insofar as they decrease that happiness.

W

Waiver of jurisdiction Changing a juvenile case from juvenile court to adult criminal court.

Weekend jail Convicted persons live at home and work during the week and report to serve jail time on weekends.

Writ of certiorari A written order from a higher court to a lower court requiring that a case be brought forward for review.

Writ of habeas corpus A written document presented to the court to determine the legality of imprisonment.

Index of Cases Cited

Banks v. Havener 1964 234 F. Supp. 27 E.D.Va: 155
Banning v. Looney 1954 213 F. 2d 771 (10th Cir.) Cert. denied, 348 U.S. 859: 150
Barefield v. Leach 1974 10282 D.N.M.: 212
Baxter v. Palmigiano 1976 425 U.S. 308: 156
Bell v. Wolfish 1979 441 U.S. 520: 157, 158
Bounds v. Smith 1977 430 U.S. 817: 158
Brown v. Allen 1953 344 U.S. 443: 152
Brown v. Johnson 1985 743 F. 2d 408 (6th Cir.): 155
Canterino v. Wilson 1982 546 F. Supp 174 W.D.Ky: 212
Cooper v. Pate 1964 378 U.S. 546: 150, 152
Cordero v. Coughlin 1984 607 F. Supp. 9 S.D.N.Y.: 127
Crug v. Beto 1972 405 U.S. 319: 155
Crug v. Hauck 1971 404 U.S. 59: 153
Edwards v. Duncan 1966 355 F. 2d 993, 994 (4th Cir.): 154
Estelle v. Gamble 1976 429 U.S. 97: 90, 154, 158
Ex parte Hull 1941 312 U.S. 546: 152
Fulwood v. Clemmer 1962 206 F. Supp. 370 D.D.C.: 153
Furman v. Georgia 1972 408 U.S. 238: 346
Gagnon v. Scarpelli 1973 411 U.S. 788: 156, 260, 291
Glover v. Johnson 1979 478 F. Supp 1075 E.D. Mich: 212
Greenholtz v. Inmates of Nebraska Penal and Correctional Complex 1979 442 U.S. 1: 283
Gregg v. Georgia 1976 428 U.S. 153: 347
Grosso v. Lally 1977 4–74–447 D.Md: 212
Gutherie v. Evans 1973 Civil Action No. 73–3068, S.D. Ga: 156
Holt v. Sarver 1971 442 F. 2d 308 (8th Cir.): 154
Hudson v. McMillian 1992 60 U.S.L.W. 4151: 154
Hudson v. Palmer 1984 468 U.S. 517: 156
Hutto v. Finney 1978 437 U.S. 678: 153
Inmates of Attica Correctional Facility v. Rockefeller 1971 453 F. 2d 12, 23 (2d Cir.): 153
Inmates of the Boys' Training School v. Affleck 1972 346 F. Supp. 1354: 332
In re Gault 1967 376 U.S. 1: 311, 312
In re Green 1981 669 F. 2d 779: 152
In re Winship 1970 397 U.S. 358: 311
Jackson v. Bishop 1968 404 F. 2d 571 (8th Cir.): 153
Jackson v. Godwin 1968 400 F. 2d 529 (5th Cir.): 154
Johnson v. Avery 1969 393 U.S. 483, 485: 153
Jones v. North Carolina Prisoners' Labor Union 1977 433 U.S. 119: 156
Kent v. United States 1966 383 U.S. 541: 307, 311
King v. United States 1974 492 F. 2d 1337 (7th Cir.): 283
Knuckles v. Prasse 1969 302 F. Supp. 1036 E.D.Pa; affd. 435 F. 2d 1255 (CA3 1970): 155
Landman v. Royster 1972 346 F. Supp. 297, 312 E.D.Va: 153
LaRocca v. Dalshein 1983 467 N.Y.S. 2d 302 (App.Div.): 127
Lockett v. Ohio 1978 438 U.S. 536: 347
McClesky v. Kemp 1987 481 U.S. 279: 348
McCray v. Burrell 1975 516 F. 2d 357, 367 (4th Cir.): 153
McLaughlin v. Royster 1972 346 F. Supp. 297, 311 E.D.Va: 154
Meachum v. Fano 1976 427 U.S. 215: 156, 158
Medina v. O'Neill 1984 589 F. Supp. 1028: 344
Mempa v. Rhay 1967 389 U.S. 128: 260
Mills v. Oliver 1973 366 F. Supp. 77 E.D.Va: 90
Milonar v. Williams 1982 691 F. 2d 931: 344
Mistretta v. United States 1989 109 U.S. 647: 73
Morrissey v. Brewer 1972 408 U.S. 471: 156, 260, 290

Nelson v. Heyne 1974 491 F. Supp. 353: 332
Newman v. Alabama 1979 466 F. Supp. 628 M.D.Ala: 151
Padgett v. Stein 1976 406 F. Supp. 287 MDPO: 172
People ex rel. Forsyth v. Court of Sessions 1894 141 N.Y. 288, 36 N.E. 386: 250
People ex rel. Ward v. Moran 1973 54 Ill. 2d 552, 302 N.E. 2d 300: 257
Procunier v. Martinez 1974 416 U.S. 396: 155
Rhodes v. Chapman 1981 452 U.S. 337: 157, 158
Roberts v. Louisiana 1977 431 U.S. 633: 347
Ruffin v. Commonwealth 1871 62 Va. (21 Gratt) 790, 796: 150
Rufo v. Inmates of Suffolk County Jail 1992 60 U.S.L.W. 4100: 157, 158
Ruiz v. Estelle 1980 503 F. Supp. 1265 S.D. Texas Cert. denied, 1035, Ct. 1438: 154
Santos v. New York States Board of Parole 441 F. 2d 1216: 287
Stanford v. Kentucky 1989 109 S. Ct. 2969: 348
Stroud v. Swope 1951 187 F. 2d 850 (9th Cir.): 150
Sutton v. Settle 1962 302 F. 2d 286 (8th Cir.): 150
Thompson v. Oklahoma 1988 487 U.S. 815: 348
Turner v. Safley 1987 107 S. Ct. 2254: 157, 158
Washington v. Harper 1990 58 U.S.L.W. 4249: 156
Whitley v. Albers 1986 106 S. Ct. 1078: 154
Wildwording v. Swenson 1971 404 U.S. 249: 152
Wilson v. Seiter 1991 59 U.S.L.W. 4671: 157, 158
Wolf v. McDonnell 1974 418 U.S. 539: 155, 158
Wright v. McMann 1972 460 F. 2d 126, 129 (2d Cir.): 153
Younger v. Gilmore 1971 404 U.S. 15: 153

Index

A

Abadinsky, Howard, 276
Access to Court, 152
Administrative Segregation, 144
Aftercare, 334
Aggravating Circumstances, 65, 347
AIDS, 125
Alternative Release Options, 100
Alternatives to Incarceration, 234
American Bar Association, 251
American Correctional Association, 31, 32, 33, 38, 276
American Probation and Parole Association, 39
Amnesty, 297
Appeal, 15, 149
Arraignment, 14
Arrest, 12
Ashurst-Summers Act, 167
Auburn Prison, 113
Augustus, John, 249
Autocratic Management Style, 185

B

Bail, 98, 249
Banishment, 55
Basic Probation, 235
Baunach, Phyllis Jo, 210
Beccaria, Cesare B., 47, 57, 58
Benefit of Clergy, 245
Bentham, Jeremy, 47
Bill of Rights, 25
Blended Philosophies, 322
Bloomberg, Seth Allan, 184
Bondsperson, 98
Boot Camp, 224, 225
Bortner, M. A., 313
Branches of Government, 20
Brockway, Zebulon R., 116, 277
Brokerage Approach, 261
Burger, Warren E., 164

C

Capital Punishment, 43, 56, 345
Careers in Corrections, 34, 35, 36
Carlson, Eric, 223
Carter, Madeline M., 224
Carter, Robert M., 29
Caseflow into Corrections, 12, 13
Caseflow into Juvenile Justice, 309, 310
Casework Approach, 261
Centralized Management Structure, 179
Chemically Dependent, 91
Chronic Offenders, 87
Citizens' Advisory Board, 230
Civil Law, 7
Civil Rights, 154
Civil Rights Act of 1871, 97
Classical View, 47
Classification, 93, 124
Code of Ethics, 37, 38, 39
Coed Prisons, 213
Community Acceptance, 228
Community-based Corrections, 213, 220
Community-based Sentences, 66
Community Involvement, 220
Community Resource Management Team (CRMT), 262
Community Service, 66, 236
Commutation, 297
Conditional Release, 272
Conditions of Parole, 285
Conditions of Probation, 257
Congregate System, 112
Conjugal Visits, 142
Construction Costs, 120
Contempt Powers, 156
Contraband, 95, 141
Controlling Outcomes, 181
Convicted Status, 86
Convict Lease, 167
Cool, 206
Corporal Punishment, 43, 56
Corporate Model, 168
Correctional Accreditation, 32
Correctional Education Programs, 173
Correctional Officers, 182
Correctional Standards, 31
Corrections, 4, 44
Corrections at the State Level, 29
Cost-effectiveness, 223
Court Appearances, 94
Court Intervention, 152
Courts, 4
Crime Control Philosophy, 322
Criminal History Category, 71
Criminal Justice Procedures, 12, 13, 14, 15
Criminal Justice System, 4
Criminal Law, 7
Crofton, Sir Walter, 115, 276
Cruel and Unusual Punishment, 153
Custodial Model, 50, 51

D

Day Fines, 224
Deinstitutionalization, 222
Deinstitutionalized Model, 322, 323, 324
Democratic Government, 20
Deprivation Model, 137, 206
Detention Centers, 237
Determinate Sentences, 68, 294
Determinism, 48

Deterrence, 57
de Toqueville, Alexis, 112
Detoxification, 91
Development of Prisons in the United States, 111
Deviance, 45
Dilulio, John I., Jr., 184, 189, 190
Direct Supervision, 103
Disciplinary Actions, 144
Discretion, 6
Discretionary Release, 272
Disparity in Sentences, 68, 69, 70, 294
Diversion, 311
Diversion Centers, 237
Drug and Alcohol Testing, 230
Dual System of Courts, 23, 24
Due Process, 20
Due Process Philosophy, 321
Due Process Rights, 311
Duffee, David, 179

E

Early Parole, 296
Eclectic Approach, 176
Economic Punishment, 57
Eighth Amendment, 25, 151
Einstadter, Werner, 265
Elderly Inmates, 129
Electronic Monitoring, 224, 238
Elmira Reformatory, 116, 277
Emergency Release, 296
Empy, LaMar T., 304
Enforcement, 309
Enforcing, 289
Equal Protection Clause, 154
Equitable Treatment Alternatives, 212
Eskridge, Chris, 223
Essential Services, 94
Exclusionary Rule, 9
Executive Branch, 22
Executive Clemency, 296

F

Federal Bail Reform Act, 100
Federal Bureau of Prisons, 27, 28
Federal Coed Correctional Institution, Fort Worth, TX, 213, 214
Federal Coed Correctional Institution, Lexington, KY, 214
Federal Sentencing Guidelines, 71
Federalism, 23, 24
Fee System, 80
Felony, 8
Female Inmate Subculture, 202
Ferri, Enrico, 48
Fifth Amendment, 25
Filing of Cases, 249
Final Release, 287
Financial Bond, 98
Fines, 66
First Amendment, 25, 151
Fogel, David, 50
Formal Reward Systems, 149
Fourteenth Amendment, 25, 26, 151
Fourth Amendment, 25
Fragmentation of Corrections, 26
Free Will, 47
Fry, Elizabeth, 196
Funnel Effect, 15
Furlough Programs, 142

G

Gangs, 130
Garcitoral, Alicio, 20
Georgia's Alternatives to Incarceration, 234
Giallombardo, Rose, 202
Gilmore, Gary, 346
Glasser, William, 333
Glossary, 353
Goffman, Erving, 136, 200
Good Time, 149, 278
Group Counseling, 175, 333
Grusky, O., 188
Guided Discretion, 347

H

Hands-off Doctrine, 150, 151, 152, 153
Harer, Miles D., 137, 174, 189
Hart, Hasting H., 84
Hawes-Cooper Act, 167
Hawthorne Effect, 188
Hawthorne Studies, 187
Heffernan, Esther, 205
Helping, 289
Helping Role, 263
Hidden Crime, 11
Hidden Delinquency, 314
HIV, 125
Holmes, Oliver Wendell, 54
Holt, Norman, 188
Howard, John, 82
Hustling, 140
Hypothesis, 34

I

Importation Model, 138, 206
Incapacitation, 50, 51
Incarceration, 57, 68, 80, 110
Incarceration Rate, 123
Incarceration Sentences, 68
Increased Privileges, 149
Indeterminate Disposition, 325
Indeterminate Sentences, 68, 294
Index Crimes, 9
Indictment, 14
Individual Counseling, 175, 333
Initial Appearance, 12
Inmate Code, 139
Inmate Rights, 97
Inmate Subculture, 137, 138
Institutional Classification, 124
Institutional Model, 322
Institutional Parole Officer, 277
Intake, 92
Intake Function, 309
Intensive Probation, 224, 236
Intermediate Sanctions, 220, 224
Interstate Compact, 272
Interviewing, 287
Investigating, 288
Investigation, 12
Investigative Role, 263
Irwin, John, 88, 131
Isolation Process, 195

J

Jacobs, James B., 138
Jail, 80
Jail Census, 85
Jail Operational Procedures, 92
Jail Populations, 85
Jail Standards, 106
Judicial Branch, 22
Judicial Reprieve, 246
Jurisdiction, 24
Just Deserts, 50
Justice Model, 50
Juvenile Court, 308
Juvenile Jurisdiction, 305
Juvenile Justice System, 304
Juvenile Law, 305
Juvenile Legal Terms, 308

K

Kassebaum, Gene G., 202
Killits Case, 251

L

Leading, 181
Least Restrictive Alternative, 50, 52, 325
Legislative Branch, 22
Leisure Time Activities, 176
Levels of Government, 20, 23
Lex Talionis, 50
Life, 205
Likert, Rensis, 186
Local Control, 84
Lockups, 84
Lombardo, Lucien X., 182
Lombroso, Cesare, 34, 48
Lowered Security Levels, 149
Lynds, Elam, 113

M

Maconochie, Alexander, 115, 275
Mahan, Sue, 214
Mala in Se, 46
Mala Prohibita, 46
Management Functions, 181
Management of Prisons, 177
Managers, 178
Mandatory Release, 272
Mandatory Sentences, 68
Manhattan Bail Project, 99
Maslow, Abraham, 186, 187
Masters, 156
Mauer, Marc, 345, 346, 347
Maximum Custody, 117, 125
McGarry, Peggy, 224
McGee, Richard A., 29
McGregor, Douglas, 186
McKelvey, Blake, 29
Medium Custody, 118, 125

Mental Health Programs, 174
Meritorious Good Time, 280
Minimum Custody, 118, 125
Misdemeanor, 8
Mitigating Circumstances, 65, 347
Monitors, 156
Mother/Child Programs, 209
Mount Pleasant Female Prison, 196
Mushfake, 140

N

National Advisory Commission on Corrections, 222, 257, 267
National Crime Victim Survey, 11
National Criminal Justice Reference Service (NCJRS), 16
National Issues, 340
National Probation Act, 250
Nelson, E. Kim, 29
New Gate Prison, 112
Niches, 140
Nonincarceration Sentences, 66
Nonintervention Philosophy, 321
Nonresidential Programs, 330
Norfolk Penal Colony, 275
Norms of Behavior, 45

O

Offense Level, 71
Ombudsmen, 156
Open Markets, 166
Open System, 179
Operating Costs, 120
Orange County Corrections Division, 39
Ordered Segmentation, 139
Organ, James P., 276
Organizational Categories, 29
Organizational Chart, 178
Organizations, 178
Organizing, 181
Orientation, 93
Outcomes of Probation, 264

P

Pardon, 297
Parens Patriae, 308, 321
Parole, 271
Parole Agency, 272
Parole Authority, 271
Parole Board, 281
Parole Hearing, 281
Parole Officers, 271
Parole Outcomes, 291
Parole Plan, 280
Parole Success, 287
Participative Management Style, 186
Partnership Model, 168
Penal Servitude Act, 276
Penal Transportation, 275
Penitentiary, 29, 115
Penn, William, 82
Persistent Felons, 131
Petersilia, Joan, 265
Petition, 311
Planning, 181
Play Families, 202
Plea Negotiation, 14
Police, 4
Pooling Prison Resources, 213
Popper, Alice, 142
Positivist View, 47, 48
Preliminary Hearing, 14, 281
Presentence Investigation Report, 65, 251
Presidential Task Force on Corrections, 221
Presumptive Sentence, 68
Pretrial Release Program, 100
Prison Chaplains, 176
Prison Community, 177
Prison Costs, 120
Prison Design, 119
Prison Discipline, 143
Prison Gangs, 130
Prison Homosexuality, 142
Prison Industry, 164
Prison Population, 121
Prisoners' Rights, 150
Private Facilities, 326
Private-sector Model, 168
Privatization, 343
Probation, 66, 244, 326
Probation/Parole Officers, 262
Probation Population, 244
Procedural Law, 9
Professional Ethics, 37
Program Inequality, 216
Programs, 95
Psychological Service, 174
Public Facilities, 326
Public Offense, 306
Public Protection, 228
Punishment, 54

R

Race and Imprisonment, 341
Racial Discrimination, 347
Rafter, Nicole, 195, 214
Ragen, Joseph, 185
Rate of Imprisonment, 340
Receivers, 156
Recidivism, 11
Reclassification, 124
Recognizance (ROR), 100, 248
Reformatory, 115
Regional Jails, 104
Rehabilitation Model, 50, 53
Rehabilitation Philosophy, 321
Rehabilitation Programs, 172
Reintegration, 50, 52, 220
Reiss, Albert J., Jr., 6
Release, 94
Release Eligibility Date, 277
Reprieve, 297
Residential Programs, 331
Resocialization, 50, 53
Restitution, 66, 224
Retaining Mother/Child Ties, 209
Retribution, 50, 348
Revocation of Parole, 289
Revocation of Probation, 260
Right to Treatment, 172, 332
Risk Assessment Guidelines, 295
Role Conflict, 263
Rousseau, Jean Jacques, 47

S

Scientific Method, 33, 34
Section 1983, 151
Seiter, Richard, 223
Self-help Groups, 177
Self-reports, 314
Sentence Reductions, 296
Sentencing, 14, 65
Sentencing Disparity, 68
Sentencing Guidelines, 71
Separate System, 115
Separation of Children, 206
Sexual Deprivation, 141
Sheltered Markets, 167
Shock Incarceration, 68, 237
Shock Probation, 267
Shover, Neal, 265
Significant Others, 143
Situationally Specific Management Style, 190
Sixth Amendment, 25
Slave of the State, 150
Social Control, 45
Social Learning, 45
Social Science, 33
Socioeconomic Characteristics, 87
Solitary Confinement, 144
Solzhenitsyn, Alexander, 56
Special Needs, 90
Split Sentence, 266
Square, 205
Staffing, 181
Stanley, David T., 295
State Sentencing Laws, 73
State-use System, 167
Status Offense, 306
Statutory Good Time, 278
Stigma, 284
Substantive Law, 9
Supervised Release, 100
Supervision, 287
Surveillance and Control, 230
Surveillance/Enforcement Role, 263
Systemwide Classification, 124

T

Tappan, Paul W., 27
Technical Violation, 260, 289
Three Generations of Jails, 103
Ticket of Leave, 276
Total Institution, 136, 200
Totalitarian Government, 20
Training Schools, 323
Transitional Adjustments, 334
Treatment, 311
Treatment Outcomes, 334
Trial, 14
Turnkey Fee, 82

U

Unconditional Release, 273
Unconvicted Status, 86
Unequal Justice, 348
UNICOR, 168
Uniform Crime Report, 11
United States Constitution, 21
Utilitarian Doctrine, 47

V

Vengeance, 348
Vocational Education, 173

W

Waiver of Jurisdiction, 307
Ward, David A., 202
Warden, 178
Weekend Jail, 96
Weisheit, Ralph, 214
Wheeler, Stanton, 144
Wilson, T., 188
Women's Correctional Institution, Bedford Hills, NY, 209
Writ of Certiorari, 151
Writ of Habeas Corpus, 151